Applied Probability
Control
Economics
Information and Communication
Modeling and Identification
Numerical Techniques
Optimization

Applications of Mathematics

21

Edited by A.V. Balakrishnan

Applications of Mathematics

1 Fleming/Rishel, **Deterministic and Stochastic Optimal Control** (1975)
2 Marchuk, **Methods of Numerical Mathematics,** Second Ed. (1982)
3 Balakrishnan, **Applied Functional Analysis,** Second Ed. (1981)
4 Borovkov, **Stochastic Processes in Queueing Theory** (1976)
5 Liptser/Shiryayev, **Statistics of Random Processes I: General Theory** (1977)
6 Liptser/Shiryayev, **Statistics of Random Processes II: Applications** (1978)
7 Vorob'ev, **Game Theory: Lectures for Economists and Systems Scientists** (1977)
8 Shiryayev, **Optimal Stopping Rules** (1978)
9 Ibragimov/Rozanov, **Gaussian Random Processes** (1978)
10 Wonham, **Linear Multivariable Control: A Geometric Approach,** Third Edn. (1985)
11 Hida, **Brownian Motion** (1980)
12 Hestenes, **Conjugate Direction Methods in Optimization** (1980)
13 Kallianpur, **Stochastic Filtering Theory** (1980)
14 Krylov, **Controlled Diffusion Processes** (1980)
15 Prabhu, **Stochastic Storage Processes: Queues, Insurance Risk, and Dams** (1980)
16 Ibragimov/Has'minskii, **Statistical Estimation: Asymptotic Theory** (1981)
17 Cesari, **Optimization: Theory and Applications** (1982)
18 Elliott, **Stochastic Calculus and Applications** (1982)
19 Marchuk/Shaidourov, **Difference Methods and Their Extrapolations** (1983)
20 Hijab, **Stabilization of Control Systems** (1986)
21 Protter, **Stochastic Integration and Differential Equations** (1990)
22 Benveniste/Métivier/Priouret, **Adaptive Algorithms and Stochastic Approximations** (1990)
23 Kloeden/Platen, **Numerical Solution of Stochastic Differential Equations** (1992)
24 Kushner/Dupuis, **Numerical Methods for Stochastic Control Problems in Continuous Time** (1992)

Philip Protter

Stochastic Integration and Differential Equations

A New Approach

Springer-Verlag Berlin Heidelberg GmbH

Philip Protter
Mathematics and Statistics Departments
Purdue University
West Lafayette, IN 47907
USA

Managing Editor

A.V. Balakrishnan
Systems Science Department
University of California
Los Angeles, CA 90024
USA

Second, Corrected Printing 1992

Mathematics Subject Classification (1980): 60H05, 60H20, 60G44

DOI 10.1007/978-3-662-02619-9

Library of Congress Cataloging-in-Publication Data
Protter, Philip E.
Stochastic integration and differential equations: a new approach / Philip Protter. - 2nd corr. print.
p. cm. - (Applications of mathematics; 21)
Includes bibliographical references and indexes.

1. Stochastic integrals. 2. Martingales (Mathematics). 3. Stochastic differential equations. I. Title. II. Series. QA274.22.P76 1992. 519.2-dc20 92-26019 CIP

Originally published by Springer-Verlag Berlin Heidelberg New York in 1990.
MyCopy version of the original edition 1990
Data conversion: EDV-Beratung Mattes, Heidelberg
2141/3140-543210 - Printed on acid-free paper
www.springer.com/mycopy

To Diane

Preface

The idea of this book began with an invitation to give a course at the Third Chilean Winter School in Probability and Statistics, at Santiago de Chile, in July, 1984. Faced with the problem of teaching stochastic integration in only a few weeks, I realized that the work of C. Dellacherie [**2**] provided an outline for just such a pedagogic approach. I developed this into a series of lectures (Protter [**6**]), using the work of K. Bichteler [**2**], E. Lenglart [**3**] and P. Protter [**7**], as well as that of Dellacherie. I then taught from these lecture notes, expanding and improving them, in courses at Purdue University, the University of Wisconsin at Madison, and the University of Rouen in France. I take this opportunity to thank these institutions and Professor Rolando Rebolledo for my initial invitation to Chile.

This book assumes the reader has some knowledge of the theory of stochastic processes, including elementary martingale theory. While we have recalled the few necessary martingale theorems in Chap. I, we have not provided proofs, as there are already many excellent treatments of martingale theory readily available (e.g., Breiman [**1**], Dellacherie-Meyer [**1**, **2**], or Ethier-Kurtz [**1**]). There are several other texts on stochastic integration, all of which adopt to some extent the usual approach and thus require the general theory. The books of Elliott [**1**], Kopp [**1**], Métivier [**1**], Rogers-Williams [**1**] and to a much lesser extent Letta [**1**] are examples. The books of McKean [**1**], Chung-Williams [**1**], and Karatzas-Shreve [**1**] avoid the general theory by limiting their scope to Brownian motion (McKean) and to continuous semimartingales.

Our hope is that this book will allow a rapid introduction to some of the deepest theorems of the subject, without first having to be burdened with the beautiful but highly technical "general theory of processes."

Many people have aided in the writing of this book, either through discussions or by reading one of the versions of the manuscript. I would like to thank J. Azema, M. Barlow, A. Bose, M. Brown, C. Constantini, C. Dellacherie, D. Duffie, M. Emery, N. Falkner, E. Goggin, D. Gottlieb, A. Gut, S. He, J. Jacod, T. Kurtz, J. Sam Lazaro, R. Leandre, E. Lenglart, G. Letta, S. Levantal, P.A. Meyer, E. Pardoux, H. Rubin, T. Sellke, R. Stockbridge, C. Stricker, P. Sundar, and M. Yor. I would especially like to thank J. San Martin for his careful reading of the manuscript in several of its versions.

Svante Janson read the entire manuscript in several versions, giving me support, encouragement, and wonderful suggestions, all of which improved the book. He also found, and helped to correct, several errors. I am extremely grateful to him, especially for his enthusiasm and generosity.

The National Science Foundation provided me with partial support throughout the writing of this book.

I wish to thank Judy Snider for her cheerful and excellent typing of several versions of this book.

Philip Protter

Contents

Introduction

In this book we present a new approach to the theory of modern stochastic integration. The novelty is that we define a semimartingale as a stochastic process which is a "good integrator" on an elementary class of processes, rather than as a process that can be written as the sum of a local martingale and an adapted process with paths of finite variation on compacts: This approach has the advantage over the customary approach of not requiring a close analysis of the structure of martingales as a prerequisite. This is a significant advantage because such an analysis of martingales itself requires a highly technical body of knowledge known as "the general theory of processes". Our approach has a further advantage of giving traditionally difficult and non-intuitive theorems (such as Stricker's theorem) transparently simple proofs. We have tried to capitalize on the natural advantage of our approach by systematically choosing the simplest, least technical proofs and presentations. As an example we have used K.M. Rao's proofs of the Doob-Meyer decomposition theorems in Chap. III, rather than the more abstract but less intuitive Doléans-Dade measure approach.

In Chap. I we present preliminaries, including the Poisson process, Brownian motion, and Lévy processes. Naturally our treatment presents those properties of these processes that are germane to stochastic integration.

In Chap. II we define a semimartingale as a good integrator and establish many of its properties and give examples. By restricting the class of integrands to adapted processes having left continuous paths with right limits, we are able to give an intuitive Riemann-type definition of the stochastic integral as the limit of sums. This is sufficient to prove many theorems (and treat many applications) including a change of variables formula ("Itô's formula").

Chapter III is devoted to developing a minimal amount of "general theory" in order to prove the Bichteler-Dellacherie theorem, which shows that our "good integrator" definition of a semimartingale is equivalent to the usual one as a process X having a decomposition $X = M + A$, into the sum of a local martingale M and an adapted process A having paths of finite variation on compacts. We reintroduce Meyer's original notion of a process being natural, allowing for less abstract and more intuitive proofs. However in what is essentially an optional last section (Sect. 8) we give a simple proof that a process with paths of integrable variation is natural if and only if it is predictable, since natural processes are referred to as predictable in the literature.

Using the results of Chap. III we extend the stochastic integral by continuity to predictable integrands in Chap. IV, thus making the stochastic integral a Lebesgue-type integral. These more general integrands allow us to give a presentation of the theory of semimartingale local times.

Chapter V serves as an introduction to the enormous subject of stochastic differential equations. We present theorems on the existence and uniqueness of solutions as well as stability results. Fisk-Stratonovich equations are presented, as well as the Markov nature of the solutions when the differentials have Markov-type properties. The last part of the chapter is an introduction to the theory of flows. Throughout Chap. V we have tried to achieve a balance between maximum generality and the simplicity of the proofs.

CHAPTER I

Preliminaries

1. Basic Definitions and Notation

We assume as given a complete probability space $(\Omega, \mathcal{F}, P)$. In addition we are given a *filtration* $(\mathcal{F}_t)_{0 \leq t \leq \infty}$. By a filtration we mean a family of σ-algebras $(\mathcal{F}_t)_{0 \leq t \leq \infty}$ that is increasing: $\mathcal{F}_s \subset \mathcal{F}_t$ if $s \leq t$.

Definition. A filtered complete probability space $(\Omega, \mathcal{F}, P, (\mathcal{F}_t)_{0 \leq t \leq \infty})$ is said to satisfy the **usual hypotheses** if

(i) $\mathcal{F}_0$ contains all the P-null sets of $\mathcal{F}$;
(ii) $\mathcal{F}_t = \bigcap_{u>t} \mathcal{F}_u$, all t, $0 \leq t < \infty$; that is, the filtration $(\mathcal{F}_t)_{0 \leq t \leq \infty}$ is *right continuous*.

We always assume that the usual hypotheses hold.

Definition. A random variable $T : \Omega \to [0, \infty]$ is a **stopping time** if the event $\{T \leq t\} \in \mathcal{F}_t$, every t, $0 \leq t \leq \infty$.

One important consequence of the right continuity of the filtration is the following theorem:

Theorem 1. *The event* $\{T < t\} \in \mathcal{F}_t$, $0 \leq t \leq \infty$, *if and only if* T *is a stopping time.*

Proof. Since $\{T \leq t\} = \bigcap_{t+\epsilon>u>t} \{T < u\}$, any $\epsilon > 0$, we have $\{T \leq t\} \in \bigcap_{u>t} \mathcal{F}_u = \mathcal{F}_t$, so T is a stopping time. For the converse, $\{T < t\} = \bigcup_{t>\epsilon>0} \{T \leq t - \epsilon\}$, and $\{T \leq t - \epsilon\} \in \mathcal{F}_{t-\epsilon}$, hence also in $\mathcal{F}_t$. □

A **stochastic process** X on $(\Omega, \mathcal{F}, P)$ is a collection of random variables $(X_t)_{0 \leq t < \infty}$. The process X is said to be **adapted** if $X_t \in \mathcal{F}_t$ (that is, is $\mathcal{F}_t$-

measurable) for each t. We must take care to be precise about the concept of equality of two stochastic processes.

Definition. Two stochastic processes X and Y are **modifications** if $X_t = Y_t$ a.s., each t. Two processes X and Y are **indistinguishable** if a.s., for all t, $X_t = Y_t$.

If X and Y are *modifications* there exists a null set, N_t, such that if $\omega \notin N_t$, then $X_t(\omega) = Y_t(\omega)$. The null set N_t depends on t. Since the interval $[0,\infty)$ is uncountable the set $N = \bigcup_{0 \le t < \infty} N_t$ could have any probability between 0 and 1, and it could even be non-measurable. If X and Y are *indistinguishable*, however, then there exists one null set N such that if $\omega \notin N$, then $X_t(\omega) = Y_t(\omega)$, for all t. In other words, the functions $t \mapsto X_t(\omega)$ and $t \mapsto Y_t(\omega)$ are the same for all $\omega \notin N$, where $P(N) = 0$. The set N is in $\mathcal{F}_t$, all t, since $\mathcal{F}_0$ contains all the P-null sets of $\mathcal{F}$. The functions $t \mapsto X_t(\omega)$ mapping $[0,\infty)$ into $\mathbf{R}$ are called the **sample paths** of the stochastic process X.

Definition. A stochastic process X is said to be **càdlàg** if it a.s. has sample paths which are right continuous, with left limits. (The nonsensical word *càdlàg* is an acronym from the French "continu à droite, limites à gauche".)

Theorem 2. *Let X and Y be two stochastic processes, with X a modification of Y. If X and Y have right continuous paths a.s., then X and Y are indistinguishable.*

Proof. Let A be the null set where the paths of X are not right continuous, and let B be the analogous set for Y. Let $N_t = \{\omega : X_t(\omega) \neq Y_t(\omega)\}$, and let $N = \bigcup_{t \in \mathbf{Q}} N_t$, where $\mathbf{Q}$ denotes the rationals in $[0,\infty)$. Then $P(N) = 0$. Let $M = A \cup B \cup N$, and $P(M) = 0$. We have $X_t(\omega) = Y_t(\omega)$ for all $t \in \mathbf{Q}$, $\omega \notin M$. If t is not rational, let t_n decrease to t through $\mathbf{Q}$. For $\omega \notin M$, $X_{t_n}(\omega) = Y_{t_n}(\omega)$, each n, and $X_t(\omega) = \lim_{n\to\infty} X_{t_n}(\omega) = \lim_{n\to\infty} Y_{t_n}(\omega) = Y_t(\omega)$. Since $P(M) = 0$, X and Y are indistinguishable. □

Corollary. *Let X and Y be two stochastic processes which are càdlàg. If X is a modification of Y, then X and Y are indistinguishable.*

Càdlàg processes provide natural examples of stopping times.

Definition. Let X be a stochastic process and let Λ be a Borel set in $\mathbf{R}$. Define

$$T(\omega) = \inf\{t > 0 : X_t \in \Lambda\}.$$

Then T is called a **hitting time of Λ for X**.

Theorem 3. *Let X be an adapted càdlàg stochastic process; and let Λ be an open set. Then the hitting time of Λ is a stopping time.*

Proof. By Theorem 1 it suffices to show that $\{T < t\} \in \mathcal{F}_t$, $0 \leq t < \infty$. But

$$\{T < t\} = \bigcup_{s \in \mathbf{Q} \cap [0,t)} \{X_s \in \Lambda\},$$

since Λ is open and X has right continuous paths. Since $\{X_s \in \Lambda\} = X_s^{-1}(\Lambda) \in \mathcal{F}_s$, the result follows. □

Theorem 4. *Let X be an adapted càdlàg stochastic process, and let Λ be a closed set. Then the random variable*

$$T(\omega) = \inf\{t > 0 : X_t(\omega) \in \Lambda \text{ or } X_{t-}(\omega) \in \Lambda\}$$

is a stopping time.

Proof. By $X_{t-}(\omega)$ we mean $\lim_{s \to t,\, s<t} X_s(\omega)$. Let $A_n = \{x : d(x,\Lambda) < \frac{1}{n}\}$, where $d(x,\Lambda)$ denotes the distance from a point x to Λ. Then A_n is an open set and

$$\{T \leq t\} = \{X_t \in \Lambda \text{ or } X_{t-} \in \Lambda\} \cup \{\bigcap_n \bigcup_{s \in \mathbf{Q} \cap [0,t)} \{X_s \in A_n\}\}. \qquad \square$$

It is a very deep result that the hitting time of a *Borel set* is a stopping time. We do not have need of this result.

The next theorem collects elementary facts about stopping times; we leave the proof to the reader.

Theorem 5. *Let S, T be stopping times. Then the following are stopping times:*

(i) $S \wedge T = \min(S,T)$
(ii) $S \vee T = \max(S,T)$
(iii) $S + T$
(iv) αS, *where* $\alpha > 1$.

The σ-algebra $\mathcal{F}_t$ can be thought of as representing all (theoretically) observable events up to and including time t. We would like to have an analogous notion of events that are observable before a random time.

Definition. Let T be a stopping time. **The stopping time σ-algebra**, $\mathcal{F}_T$, is defined to be

$$\{\Lambda \in \mathcal{F} : \Lambda \cap \{T \leq t\} \in \mathcal{F}_t, \text{ all } t \geq 0\}.$$

The previous definition is not especially intuitive. However it does well represent "knowledge" up to time T, as the next theorem illustrates.

Theorem 6. *Let T be a finite stopping time. Then $\mathcal{F}_T$ is the smallest σ-algebra containing all càdlàg processes sampled at T. That is,*

$$\mathcal{F}_T = \sigma\{X_T; X \textit{ all adapted càdlàg processes}\}.$$

Proof. Let $\mathcal{G} = \sigma\{X_T; X$ all adapted càdlàg processes$\}$. Let $\Lambda \in \mathcal{F}_T$. Then $X_t = 1_\Lambda 1_{\{t \geq T\}}$[1] is a càdlàg process, and $X_T = 1_\Lambda$; hence $\Lambda \in \mathcal{G}$, and $\mathcal{F}_T \subset \mathcal{G}$.

Next let X be an adapted càdlàg process. We need to show X_T is $\mathcal{F}_T$-measurable. Consider $X(s, \omega)$ as a function from $[0, \infty) \times \Omega$ into $\mathbf{R}$. Construct $\varphi : \{T \leq t\} \mapsto [0, \infty) \times \Omega$ by $\varphi(\omega) = (T(\omega), \omega)$. Then since X is adapted and càdlàg, we have $X_T = X \circ \varphi$, is a measurable mapping from $(\{T \leq t\}, \mathcal{F}_t \cap \{T \leq t\})$ into $(\mathbf{R}, \mathcal{B})$, where $\mathcal{B}$ are the Borel sets of $\mathbf{R}$. Therefore

$$\{\omega : X(T(\omega), \omega) \in B\} \cap \{T \leq t\}$$

is in $\mathcal{F}_t$, and this implies $X_T \in \mathcal{F}_T$. Therefore $\mathcal{G} \subset \mathcal{F}_T$. □

We leave it to the reader to check that if $S \leq T$ a.s., then $\mathcal{F}_S \subset \mathcal{F}_T$, and the less obvious (and less important) fact that $\mathcal{F}_S \cap \mathcal{F}_T = \mathcal{F}_{S \wedge T}$.

If X and Y are càdlàg, then $X_t = Y_t$ a.s. each t implies that X and Y are indistinguishable, as we have already noted. Since fixed times are stopping times, obviously if $X_T = Y_T$ a.s. for each finite stopping time T, then X and Y are indistinguishable. If X is càdlàg, let ΔX denote the process $\Delta X_t = X_t - X_{t-}$. Then ΔX is not càdlàg, though it is adapted and for a.s. ω, $t \to \Delta X_t = 0$ except for at most countably many t. We record here a useful result.

Theorem 7. *Let X be adapted and càdlàg. If $\Delta X_T 1_{\{T<\infty\}} = 0$ a.s. for each stopping time T, then ΔX is indistinguishable from the zero process.*

Proof. It suffices to prove the result on $[0, t_0]$ for $0 < t_0 < \infty$. The set $\{t : |\Delta X_t| > 0\}$ is countable a.s. since X is càdlàg. Moreover

$$\{t : |\Delta X_t| > 0\} = \bigcup_{n=1}^{\infty} \{t : |\Delta X_t| > \frac{1}{n}\}$$

and the set $\{t : |\Delta X_t| > \frac{1}{n}\}$ must be finite for each n, since $t_0 < \infty$. Using Theorem 4 we define stopping times for each n inductively as follows:

$$T^{n,1} = \inf\{t > 0 : |\Delta X_t| > \frac{1}{n}\}$$

$$T^{n,k} = \inf\{t > T^{n,k-1} : |\Delta X_t| > \frac{1}{n}\}.$$

[1] 1_A is the indicator function of $A : 1_A(\omega) = \begin{cases} 1 & \omega \in A \\ 0 & \omega \notin A \end{cases}$.

Then $T^{n,k} > T^{n,k-1}$ a.s. on $\{T^{n,k-1} < \infty\}$. Moreover

$$\{|\Delta X_t| > 0\} = \bigcup_{n,k} \{|\Delta X_{T^{n,k}} 1_{\{T^{n,k}<\infty\}}| > 0\},$$

where the right side of the equality is a countable union. The result follows. □

Corollary. *Let X and Y be adapted and càdlàg. If $\Delta X_T 1_{\{T<\infty\}} = \Delta Y_T 1_{\{T<\infty\}}$ a.s. for each stopping time T, then ΔX and ΔY are indistinguishable.*

A much more general version of Theorem 7 is true, but it is a very deep result which uses Meyer's "section theorems", and we will nct have need of it. See, for example, Dellacherie [1] or Dellacherie-Meyer [**1**].

A fundamental theorem of measure theory that we will need from time to time is known as the Monotone Class Theorem. Actually there are several such theorems, but the one given here is sufficient for our needs.

Definition. A **monotone vector space** $\mathcal{H}$ on a space Ω is defined to be the collection of bounded, real-valued functions f on Ω satisfying the three conditions:

(i) $\mathcal{H}$ is a vector space over $\mathbf{R}$;
(ii) $1_\Omega \in \mathcal{H}$ (i.e., constant functions are in $\mathcal{H}$)
(iii) If $(f_n)_{n\geq 1} \subset \mathcal{H}$ and $0 \leq f_1 \leq f_2 \leq \cdots \leq f_n \leq \ldots$ and $\lim_{n\to\infty} f_n = f$ and f is bounded, then $f \in \mathcal{H}$.

Definition. A collection $\mathcal{M}$ of real functions defined on a space Ω is said to be **multiplicative** if $f, g \in \mathcal{M}$ implies that $fg \in \mathcal{M}$.

For a collection of real-valued functions $\mathcal{M}$ defined on Ω, we let $\sigma(\mathcal{M})$ denote the space of functions defined on Ω which are measurable with respect to the σ-algebra on Ω generated by $\{f^{-1}(\Lambda); \Lambda \in \mathcal{B}(\mathbf{R}), f \in \mathcal{M}\}$.

Theorem 8 (Monotone Class Theorem). *Let $\mathcal{M}$ be a multiplicative class of bounded real-valued functions defined on a space Ω, and let $\mathcal{A} = \sigma(\mathcal{M})$. If $\mathcal{H}$ is a monotone vector space containing $\mathcal{M}$, then $\mathcal{H}$ contains all bounded, $\mathcal{A}$-measurable functions.*

Theorem 8 is proved in Dellacherie-Meyer [**1**, p. 14] with the additional hypothesis that $\mathcal{H}$ is closed under uniform convergence. This extra hypothesis is unnecessary, however, since *every monotone vector space is closed under uniform convergence.* (See Sharpe [**1**, p. 365]).

2. Martingales

In this section we give, mostly without proofs, only the essential results from the theory of continuous time martingales. The reader can consult any of a large number of texts to find excellent proofs; for example Dellacherie-Meyer [**2**], or Ethier-Kurtz [**1**]. Also, recall that we will always assume as given a filtered, complete probability space $(\Omega, \mathcal{F}, (\mathcal{F}_t)_{0\le t\le\infty}, P)$, where the filtration $(\mathcal{F}_t)_{0\le t\le\infty}$ is assumed to be right continuous.

Definition. A real valued, adapted process $X = (X_t)_{0\le t<\infty}$ is called a **martingale** (resp. **supermartingale, submartingale**) with respect to the filtration $(\mathcal{F}_t)_{0\le t\le\infty}$ if

(i) $X_t \in L^1(dP)$; that is, $E\{|X_t|\} < \infty$;
(ii) if $s \le t$, then $E\{X_t|\mathcal{F}_s\} = X_s$, a.s. (resp. $E\{X_t|\mathcal{F}_s\} \le X_s$, resp. $\ge X_s$).

Note that martingales are only defined on $[0,\infty)$; that is, for finite t and not $t = \infty$. It is often possible to extend the definition to $t = \infty$.

Definition. A martingale X is said to be **closed** by a random variable Y if $E\{|Y|\} < \infty$ and $X_t = E\{Y|\mathcal{F}_t\}$, $0 \le t < \infty$.

A random variable Y closing a martingale is not necessarily unique. We give a sufficient condition for a martingale to be closed (as well as a construction for closing it) in Theorem 12.

Theorem 9. *Let X be a supermartingale. The function $t \to E\{X_t\}$ is right continuous if and only if there exists a unique modification, Y, of X, which is càdlàg. Such a modification is unique.*

By uniqueness we mean up to indistinguishability. Our standing assumption that the "usual hypotheses" are satisfied is used implicitly in the statement of Theorem 9. Also, note that the process Y is, of course, also a supermartingale. Theorem 9 is proved using Doob's Upcrossing Inequalities. If X is a martingale then $t \to E\{X_t\}$ is constant, and hence it has a right continuous modification.

Corollary 1. *If $X = (X_t)_{0\le t<\infty}$ is a martingale then there exists a unique modification Y of X which is càdlàg.*

Since all martingales have right continuous modifications, *we will always assume that we are taking the right continuous version*, without any special mention. Note that it follows from Corollary 1 that a right continuous martingale is càdlàg.

Theorem 10 (Martingale Convergence Theorem). *Let X be a right continuous supermartingale, $\sup_{0\leq t<\infty} E\{|X_t|\} < \infty$. Then the random variable $Y = \lim_{t\to\infty} X_t$ a.s. exists, and $E\{|Y|\} < \infty$. Moreover if X is a martingale closed by a random variable Z, then Y also closes X and $Y = E\{Z|\bigvee_{0\leq t<\infty} \mathcal{F}_t\}$.*[2]

A condition known as uniform integrability is sufficient for a martingale to be closed.

Definition. *A family of random variables $(U_\alpha)_{\alpha\in A}$ is* **uniformly integrable** *if*

$$\lim_{n\to\infty} \sup_\alpha \int_{\{|U_\alpha|\geq n\}} |U_\alpha| dP = 0.$$

Theorem 11. *Let $(U_\alpha)_{\alpha\in A}$ be a subset of L^1. The following are equivalent:*

(i) $(U_\alpha)_{\alpha\in A}$ *is uniformly integrable.*
(ii) $\sup_{\alpha\in A} E\{|U_\alpha|\} < \infty$, *and whatever $\epsilon > 0$ there exists $\delta > 0$ such that $\Lambda \in \mathcal{F}$, $P(\Lambda) \leq \delta$, imply $E\{|U_\alpha 1_\Lambda|\} < \epsilon$.*
(iii) *There exists a positive, increasing, convex function $G(x)$ defined on $[0,\infty)$ such that $\lim_{x\to\infty} \frac{G(x)}{x} = +\infty$ and $\sup_\alpha E\{G\circ|U_\alpha|\} < \infty$.*

The assumption that G is convex is not needed for the implications (iii)⇒(ii) and (iii)⇒(i).

Theorem 12. *Let X be a right continuous martingale which is uniformly integrable. Then $Y = \lim_{t\to\infty} X_t$ a.s. exists, $E\{|Y|\} < \infty$, and Y closes X as a martingale.*

Theorem 13. *Let X be a (right continuous) martingale. Then $(X_t)_{t\geq 0}$ is uniformly integrable if and only if $Y = \lim_{t\to\infty} X_t$ exists a.s., $E\{|Y|\} < \infty$, and $(X_t)_{0\leq t\leq\infty}$ is a martingale, where $X_\infty = Y$.*

If X is a uniformly integrable martingale, then X_t converges to $X_\infty = Y$ in L^1 as well as almost surely. The next theorem we use only once (in the proof of Theorem 28), but we give it here for completeness. The notation $(X_n)_{n\leq 0}$ refers to a process indexed by the nonpositive integers: $\cdots X_{-2}, X_{-1}, X_0$.

---

[2] $\bigvee_{0\leq t<\infty} \mathcal{F}_t$ denotes the smallest σ-algebra generated by $(\mathcal{F}_t)$, all t, $0 \leq t < \infty$.

Theorem 14 (Backwards Convergence). *Let $(X_n)_{n\leq 0}$ be a martingale. Then $\lim_{n\to-\infty} X_n = E\{X_0 | \bigcap_{n=-\infty}^{0} \mathcal{F}_n\}$ a.s. and in L^1.*

A less probabilistic interpretation of martingales uses Hilbert space theory. Let $Y \in L^2(\Omega, \mathcal{F}, P)$. Since $\mathcal{F}_t \subseteq \mathcal{F}$, the spaces $L^2(\Omega, \mathcal{F}_t, P)$ form a family of Hilbert subspaces of $L^2(\Omega, \mathcal{F}, P)$. Let ${}^{\pi_t}Y$ denote the Hilbert space projection of Y onto $L^2(\Omega, \mathcal{F}_t, P)$.

Theorem 15. *Let $Y \in L^2(\Omega, \mathcal{F}, P)$. The process $X_t = {}^{\pi_t}Y$ is a uniformly integrable martingale.*

Proof. It suffices to show $E\{Y|\mathcal{F}_t\} = {}^{\pi_t}Y$. The random variable $E\{Y|\mathcal{F}_t\}$ is the unique $\mathcal{F}_t$-measurable r.v. such that $\int_A Y dP = \int_A E\{Y|\mathcal{F}_t\} dP$, for any event $A \in \mathcal{F}_t$. We have $\int_A Y dP = \int_A {}^{\pi_t}Y dP + \int_A (Y - {}^{\pi_t}Y) dP$. But $\int_A (Y - {}^{\pi_t}Y) dP = \int 1_A (Y - {}^{\pi_t}Y) dP$. Since $1_A \in L^2(\Omega, \mathcal{F}_t, P)$, and $(Y - {}^{\pi_t}Y)$ is in the orthocomplement of $L^2(\Omega, \mathcal{F}_t, P)$, we have $\int 1_A(Y - {}^{\pi_t}Y) dP = 0$, and thus by uniqueness $E\{Y|\mathcal{F}_t\} = {}^{\pi_t}Y$. Since $\|{}^{\pi_t}Y\|_{L^2} \leq \|Y\|_{L^2}$, by part (iii) of Theorem 11 we have that X is uniformly integrable (take $G(x) = x^2$). □

The next theorem is one of the most useful martingale theorems for our purposes.

Theorem 16 (Doob's Optional Sampling Theorem). *Let X be a right continuous martingale, which is closed by a random variable X_∞. Let S and T be two stopping times such that $S \leq T$ a.s. Then X_S and X_T are integrable and*

$$X_S = E\{X_T | \mathcal{F}_S\} \qquad a.s.$$

Theorem 16 has a similar version for supermartingales.

Theorem 17. *Let X be a right continuous supermartingale (martingale), and let S and T be two bounded stopping times such that $S \leq T$ a.s. Then X_S and X_T are integrable and*

$$X_S \geq E\{X_T | \mathcal{F}_S\} \qquad a.s.\ (=).$$

If T is a stopping time, then so is $t \wedge T = \min(t, T)$, for each $t \geq 0$.

Definition. Let X be a stochastic process and let T be a random time. X^T is said to be **the process stopped at** T if $X_t^T = X_{t\wedge T}$.

Note that if X is adapted and càdlàg and if T is a stopping time, then

$$X_t^T = X_{t\wedge T} = X_t 1_{\{t<T\}} + X_T 1_{\{t\geq T\}}$$

is also adapted. A martingale stopped at a stopping time is still a martingale, as the next theorem shows.

Theorem 18. *Let X be a uniformly integrable right continuous martingale, and let T be a stopping time. Then $X^T = (X_{t\wedge T})_{0\le t\le\infty}$ is also a uniformly integrable right continuous martingale.*

Proof. X^T is clearly right continuous. By Theorem 16

$$\begin{aligned} X_{t\wedge T} &= E\{X_T|\mathcal{F}_{t\wedge T}\} \\ &= E\{X_T 1_{\{T<t\}} + X_T 1_{\{T\ge t\}}|\mathcal{F}_{t\wedge T}\} \\ &= X_T 1_{\{T<t\}} + E\{X_T 1_{\{T\ge t\}}|\mathcal{F}_{t\wedge T}\} \end{aligned}$$

However for $H \in \mathcal{F}_t$ we have $H1_{\{T\ge t\}} \in \mathcal{F}_T$. Thus:

$$= X_T 1_{\{T<t\}} + E\{X_T|\mathcal{F}_t\}1_{\{T\ge t\}}.$$

Therefore

$$\begin{aligned} X_{t\wedge T} &= X_T 1_{\{T<t\}} + E\{X_T|\mathcal{F}_t\}1_{\{T\ge t\}} \\ &= E\{X_T|\mathcal{F}_t\}, \end{aligned}$$

since $X_T 1_{\{T<t\}}$ is $\mathcal{F}_t$-measurable. Thus X^T is a uniformly integrable $(\mathcal{F}_t)$-martingale by Theorem 13. □

Observe that the difficulty in Theorem 18 is to show that X^T is a martingale for the filtration $(\mathcal{F}_t)_{0\le t\le\infty}$. It is a trivial consequence of Theorem 16 that $X^T = X_{t\wedge T}$ is a martingale for the filtration $(\mathcal{G}_t)_{0\le t\le\infty}$ given by $\mathcal{G}_t = \mathcal{F}_{t\wedge T}$.

Corollary. *Let Y be an integrable random variable and let S, T be stopping times. Then*

$$\begin{aligned} E\{E\{Y|\mathcal{F}_S\}|\mathcal{F}_T\} &= E\{E\{Y|\mathcal{F}_T\}|\mathcal{F}_S\} \\ &= E\{Y|\mathcal{F}_{S\wedge T}\}. \end{aligned}$$

Proof. Let $Y_t = E\{Y|\mathcal{F}_t\}$. Then Y^T is a uniformly integrable martingale and

$$\begin{aligned} Y_{S\wedge T} = Y_S^T &= E\{Y_T|\mathcal{F}_S\} \\ &= E\{E\{Y|\mathcal{F}_T\}|\mathcal{F}_S\}. \end{aligned}$$

Interchanging the roles of T and S yields

$$\begin{aligned} Y_{S\wedge T} = Y_T^S &= E\{Y_S|\mathcal{F}_T\} \\ &= E\{E\{Y|\mathcal{F}_S\}|\mathcal{F}_T\}. \end{aligned}$$

Finally, $E\{Y|\mathcal{F}_{S\wedge T}\} = Y_{S\wedge T}$. □

The next inequality is elementary, but indispensable.

Theorem 19 (Jensen's Inequality). *Let $\varphi : \mathbf{R} \to \mathbf{R}$ be convex, and let X and $\varphi(X)$ be integrable random variables. For any σ-algebra $\mathcal{G}$,*

$$\varphi \circ E\{X|\mathcal{G}\} \leq E\{\varphi(X)|\mathcal{G}\}.$$

Corollary 1. *Let X be a martingale, and let φ be convex such that $\varphi(X_t)$ is integrable, $0 \leq t < \infty$. Then $\varphi(X)$ is a submartingale. In particular, if M is a martingale, then $|M|$ is a submartingale.*

Corollary 2. *Let X be a submartingale and let φ be convex, nondecreasing, and such that $\varphi(X_t)_{0 \leq t < \infty}$ is integrable. Then $\varphi(X)$ is also a submartingale.*

We end our review of martingale theory with Doob's inequalities; the most important is when $p = 2$.

Theorem 20. *Let X be a positive submartingale. For all $p > 1$, with q conjugate to p (i.e., $\frac{1}{p} + \frac{1}{q} = 1$), we have*

$$\| \sup_t |X_t| \|_{L^p} \leq q \sup_t \|X_t\|_{L^p}.$$

We let X^* denote $\sup_s |X_s|$. Note that if M is a martingale with $M_\infty \in L^2$, then $|M|$ is a positive submartingale, and taking $p = 2$ we have

$$E\{(M^*)^2\} \leq 4E\{M_\infty^2\}.$$

This last inequality is called **Doob's maximal quadratic inequality**.

An elementary but useful result concerning martingales is the following.

Theorem 21. *Let $X = (X_t)_{0 \leq t \leq \infty}$ be an adapted process with càdlàg paths. Suppose $E\{|X_T|\} < \infty$ and $E\{X_T\} = 0$ for any stopping time T, finite or not. Then X is a uniformly integrable martingale.*

Proof. Let $0 \leq s < t < \infty$, and let $\Lambda \in \mathcal{F}_s$. Let

$$u_\Lambda = \begin{cases} u & \text{if } \omega \in \Lambda \\ \infty & \text{if } \omega \notin \Lambda. \end{cases}$$

Then u_Λ are stopping times for all $u \geq s$. Moreover

$$\begin{aligned} \int_\Lambda X_u dP &= \int X_{u_\Lambda} dP - \int_{\Omega \setminus \Lambda} X_\infty dP \\ &= - \int_{\Omega \setminus \Lambda} X_\infty dP \end{aligned}$$

since $E\{X_{u_\Lambda}\} = 0$ by hypothesis, for $u \geq s$. Thus for $\Lambda \in \mathcal{F}_s$ and $s < t$, $E\{X_t 1_\Lambda\} = E\{X_s 1_\Lambda\} = -E\{X_\infty 1_{\Omega\setminus\Lambda}\}$, which implies $E\{X_t|\mathcal{F}_s\} = X_s$, and X is a martingale, $0 \leq t \leq \infty$. □

3. The Poisson Process and Brownian Motion

The Poisson process and Brownian motion are the two fundamental examples in the theory of continuous time stochastic processes. The Poisson process is the simpler of the two, and we begin with it. We recall that we assume given a filtered probability space $(\Omega, \mathcal{F}, (\mathcal{F}_t)_{0\leq t\leq\infty}, P)$ satisfying the usual hypotheses.

Let $(T_n)_{n\geq 0}$ be a strictly increasing sequence of positive random variables. We always take $T_0 = 0$ a.s. Recall that the indicator function $1_{\{t\geq T_n\}}$ is defined as:

$$1_{\{t\geq T_n\}} = \begin{cases} 1 & \text{if } t \geq T_n(\omega) \\ 0 & \text{if } t < T_n(\omega) \end{cases}$$

Definition. The process $N = (N_t)_{0\leq t\leq\infty}$ defined by

$$N_t = \sum_{n\geq 1} 1_{\{t\geq T_n\}}$$

with values in $\mathbb{N} \cup \{\infty\}$ ($\mathbb{N} = \{0, 1, 2, \dots\}$) is called **the counting process associated to the sequence** $(T_n)_{n\geq 1}$.

If we set $T = \sup_n T_n$, then

$$[T_n, \infty) = \{N \geq n\} = \{(t, \omega) : N_t(\omega) \geq n\}$$

as well as

$$[T_n, T_{n+1}) = \{N = n\}, \quad \text{and} \quad [T, \infty) = \{N = \infty\}.$$

The random variable T is the **explosion time** of N. If $T = \infty$ a.s., then N is a counting process *without explosions*. For $T = \infty$, note that for $0 \leq s < t < \infty$ we have

$$N_t - N_s = \sum_{n\geq 1} 1_{\{s<T_n\leq t\}}.$$

The increment $N_t - N_s$ counts the number of random times T_n that occur between the fixed times s and t.

As we have defined a counting process it is not necessarily adapted to the filtration $(\mathcal{F}_t)_{t\geq 0}$. Indeed, we have the following:

Theorem 22. *A counting process N is adapted if and only if the associated random variables $(T_n)_{n\geq 1}$ are stopping times.*

Proof. If the $(T_n)_{n\geq 0}$ are stopping times (with $T_0 = 0$ a.s.), then the event

$$\{N_t = n\} = \{\omega : T_n(\omega) \leq t < T_{n+1}(\omega)\} \in \mathcal{F}_t,$$

for each n; thus $N_t \in \mathcal{F}_t$ and N is adapted. If N is adapted, then $\{T_n \leq t\} = \{N_t \geq n\} \in \mathcal{F}_t$, each t, and therefore T_n is a stopping time. □

Note that a counting process without explosions has right continuous paths with left limits; hence a counting process without explosions is càdlàg.

Definition. An adapted counting process N without explosions is a **Poisson process** if

(i) for any s, t, $0 \leq s < t < \infty$, $N_t - N_s$ is independent of $\mathcal{F}_s$;
(ii) for any s, t, u, v, $0 \leq s < t < \infty$, $0 \leq u < v < \infty$, $t - s = v - u$, then the distribution of $N_t - N_s$ is the same as that of $N_v - N_u$.

Properties (i) and (ii) are known respectively as *increments independent of the past*, and *stationary increments*.

Theorem 23. *Let N be a Poisson process. Then*

$$P(N_t = n) = \frac{e^{-\lambda t}(\lambda t)^n}{n!},$$

$n = 0, 1, 2, \ldots$, *for some $\lambda \geq 0$. That is, N_t has the Poisson distribution with parameter λt. Moreover, N is continuous in probability.*

Proof. The proof of Theorem 23 is standard and is often given in more elementary courses (cf. eg. Çinlar [1, pp. 71ff]). We sketch it here.

Step 1: For all $t \geq 0$, $P(N_t = 0) = e^{-\lambda t}$, for some constant $\lambda \geq 0$.

Since $\{N_t = 0\} = \{N_s = 0\} \cap \{N_t - N_s = 0\}$ for $0 \leq s < t < \infty$, by the independence of the increments

$$\begin{aligned} P(N_t = 0) &= P(N_s = 0)P(N_t - N_s = 0) \\ &= P(N_s = 0)P(N_{t-s} = 0), \end{aligned}$$

by the stationarity of the increments. Let $\alpha(t) = P(N_t = 0)$. We have $\alpha(t) = \alpha(s)\alpha(t-s)$, for all $0 \leq s < t < \infty$. Since $\alpha(t)$ can be easily seen to be right continuous in t, we deduce that either $\alpha(t) = 0$ for all $t \geq 0$ or

$$\alpha(t) = e^{-\lambda t} \text{ for some } \lambda \geq 0.$$

If $\alpha(t) = 0$ it would follow that $N_t(\omega) = \infty$ a.s. for *all* t which would contradict that N is a counting process. Note that $\lim_{u\to t} P(|N_u - N_t| > \varepsilon) = \lim_{u\to t} P(|N_{u-t}| > \varepsilon) = \lim_{v\to 0} P(N_v > \varepsilon) = \lim_{v\to 0} 1 - e^{-\lambda v} = 0$; hence N is continuous in probability.

Step 2: $P(N_t \geq 2)$ is $o(t)$. (That is, $\lim_{t\to 0}\frac{1}{t}P(N_t \geq 2) = 0$.)

Let $\beta(t) = P(N_t \geq 2)$. Since the paths of N are nondecreasing, β is also nondecreasing. One readily checks that showing $\lim_{t\to 0}\frac{1}{t}\beta(t) = 0$ is equivalent to showing that $\lim_{n\to\infty} n\beta(\frac{1}{n}) = 0$. Divide $[0,1]$ into n subintervals of equal length, and let S_n denote the number of subintervals containing at least two arrivals. By the independence and stationarity of the increments S_n is the sum of n i.i.d. zero-one valued random variables, and hence has a Binomial distribution (n,p), where $p = \beta(\frac{1}{n})$. Therefore $E\{S_n\} = np = n\beta(\frac{1}{n})$.

Since N is a counting process, we know the arrival times are *strictly* increasing; that is, $T_n < T_{n+1}$ a.s. Therefore for fixed ω, for n sufficiently large no subinterval has more than one arrival (otherwise there would be an explosion). This implies $\lim_{n\to\infty} S_n(\omega) = 0$ a.s. Since $S_n \leq N_1$, if $E\{N_1\} < \infty$ we can use the Dominated Convergence Theorem to conclude $\lim_{n\to\infty} n\beta(\frac{1}{n}) = \lim_{n\to\infty} E\{S_n\} = 0$. (That $E\{N_1\} < \infty$ is a consequence of Theorem 34, established in Sect. 4).

Step 3: $\lim_{t\to 0}\frac{1}{t}P\{N_t = 1\} = \lambda$.

Since $P\{N_t = 1\} = 1 - P\{N_t = 0\} - P\{N_t \geq 2\}$, $\lim_{t\to 0}\frac{1}{t}P\{N_t = 1\} = \lim_{t\to 0}\frac{1-e^{-\lambda t}+o(t)}{t} = \lambda$.

Step 4: Conclusion.

We write $\varphi(t) = E\{\alpha^{N_t}\}$, for $0 \leq \alpha \leq 1$. Then for $0 \leq s < t < \infty$, the independence and stationarity of the increments implies that $\varphi(t+s) = \varphi(t)\varphi(s)$ which in turn implies that $\varphi(t) = e^{t\psi(\alpha)}$. But

$$\begin{aligned}\varphi(t) &= \sum_{n=0}^{\infty} \alpha^n P(N_t = n) \\ &= P(N_t = 0) + \alpha P(N_t = 1) + \sum_{n=2}^{\infty} \alpha^n P(N_t = n),\end{aligned}$$

and $\psi(\alpha) = \varphi'(0)$, the derivative of φ at 0. Therefore

$$\begin{aligned}\psi(\alpha) &= \lim_{t\to 0}\frac{\varphi(t)-1}{t} = \lim_{t\to 0}\left\{\frac{P(N_t=0)-1}{t} + \frac{\alpha P(N_t = 1)}{t} + \frac{1}{t}o(t)\right\} \\ &= -\lambda + \lambda\alpha.\end{aligned}$$

Therefore $\varphi(t) = e^{-\lambda t + \lambda\alpha t}$, hence

$$\varphi(t) = \sum_{n=0}^{\infty} \alpha^n P(N_t = n) = e^{-\lambda t}\sum_{n=0}^{\infty}\frac{(\lambda t)^n \alpha^n}{n!}.$$

Equating coefficients of the two infinite series yields

$$P(N_t = n) = e^{-\lambda t}\frac{(\lambda t)^n}{n!},$$

for $n = 0, 1, 2, \ldots$. □

Definition. The parameter λ associated to a Poisson process by Theorem 23 is called the **intensity**, or **arrival rate**, of the process.

Corollary. *A Poisson process N with intensity λ satisfies*

$$E\{N_t\} = \lambda t$$
$$Variance(N_t) = Var(N_t) = \lambda t.$$

The proof is trivial and we omit it.

There are other, equivalent definitions of the Poisson process. For example, a counting process N without explosion can be seen to be a Poisson process if for all s, t, $0 \leq s < t < \infty$, $E\{N_t\} < \infty$ and

$$E\{N_t - N_s | \mathcal{F}_s\} = \lambda(t - s).$$

Theorem 24. *Let N be a Poisson process with intensity λ. Then $N_t - \lambda t$ and $(N_t - \lambda t)^2 - \lambda t$ are martingales.*

Proof. Since λt is non-random, the process $N_t - \lambda t$ has mean zero and independent increments. Therefore

$$E\{N_t - \lambda t - (N_s - \lambda s) | \mathcal{F}_s\} = E\{N_t - \lambda t - (N_s - \lambda s)\} = 0,$$

for $0 \leq s < t < \infty$. The analogous statement holds for $(N_t - \lambda t)^2 - \lambda t$. □

Definition. Let H be a stochastic process. The **natural filtration of H denoted** $(\mathcal{F}_t^0)_{0 \leq t < \infty}$, is defined to be $\mathcal{F}_t^0 = \sigma\{H_s; s \leq t\}$. That is, the smallest filtration that makes H adapted.

Note that natural filtrations are *not* assumed to contain all the P-null sets of $\mathcal{F}$.

Theorem 25. *Let N be a counting process. The natural filtration of N is right continuous.*

Proof. Let $E = [0, \infty]$ and $\mathcal{B}$ be the Borel sets of E, and let Γ be the path space given by

$$\Gamma = \Big(\prod_{s \in [0,\infty)} E_s, \bigotimes_{s \in [0,\infty)} \mathcal{B}_s \Big).$$

Define the maps $\pi_t : \Omega \to \Gamma$ by

$$\pi_t(\omega) = s \mapsto N_{s \wedge t}(\omega).$$

(Thus the range of π_t is contained in the set of functions constant after t). The σ-algebra $\mathcal{F}_t^0$ is also generated by the single function space – valued random variable π_t.

Let Λ be an event in $\bigcap_{n\geq 1}\mathcal{F}^0_{t+\frac{1}{n}}$. Then there exists a set $A_n \in \bigotimes_{s\in[0,\infty)}\mathcal{B}_s$ such that $\Lambda = \{\pi_{t+\frac{1}{n}} \in A_n\}$. Next set $W_n = \{\pi_t = \pi_{t+\frac{1}{n}}\}$. For each ω, there exists an n such that $s \to N_s(\omega)$ is constant on $[t, t+\frac{1}{n}]$; therefore $\Omega = \bigcup_{n\geq 1} W_n$, where W_n is an increasing sequence of events. Therefore

$$\begin{aligned}
\Lambda &= \lim_n (W_n \cap \Lambda)\\
&= \lim_n (W_n \cap \{\pi_{t+\frac{1}{n}} \in A_n\})\\
&= \lim_n (W_n \cap \{\pi_t \in A_n\})\\
&= \lim_n \{\pi_t \in A_n\},
\end{aligned}$$

which implies $\Lambda \in \mathcal{F}_t^0$. We conclude $\bigcap_{n\geq 1}\mathcal{F}^0_{t+\frac{1}{n}} \subset \mathcal{F}_t^0$, which implies they are equal. □

We next turn our attention to the Brownian motion process. Recall that we are assuming as given a filtered probability space $(\Omega, \mathcal{F}, (\mathcal{F}_t)_{0\leq t<\infty}, P)$ that satisfies the usual hypotheses.

Definition. An adapted process $B = (B_t)_{0\leq t<\infty}$ taking values in $\mathbf{R}^n$ is called an **n-dimensional Brownian motion** if:

(i) for $0 \leq s < t < \infty$, $B_t - B_s$ is independent of $\mathcal{F}_s$ (*increments are independent of the past*);
(ii) for $0 < s < t$, $B_t - B_s$ is a Gaussian random variable with mean zero and variance matrix $(t-s)C$, for a given, non random matrix C.

The Brownian motion *starts at* x if $P(B_0 = x) = 1$.

The existence of Brownian motion is proved using a path-space construction, together with Kolmogorov's extension theorem. It is simple to check that a Brownian motion is a martingale as long as $E\{|B_0|\} < \infty$. Therefore by Theorem 9 there exists a version which has right continuous paths, a.s. Actually, more is true:

Theorem 26. *Let B be a Brownian motion. Then there exists a modification of B which has continuous paths a.s.*

Theorem 26 is proved in textbooks on probability theory (eg. Breiman [**1**]). It can also be proved as an elementary consequence of Kolmogorov's Lemma (Theorem 53 of Chap. IV). *We will always assume that we are using* *the version of Brownian motion with continuous paths.* We will also assume, unless stated otherwise, that C is the identity matrix. We then say that a Brownian motion B with continuous paths, with $C = I$ the identity matrix, and with $B_0 = x$ for some $x \in \mathbf{R}^n$, is a **standard Brownian motion**. Note that for an $\mathbf{R}^n$ standard Brownian motion B, writing $B_t = (B_t^1, \ldots, B_t^n)$, $0 \le t < \infty$, then each B^i is an $\mathbf{R}^1$ Brownian motion with continuous paths, and the B^i's are independent.

We have already observed that a Brownian motion B with $E\{|B_0|\} < \infty$ is a martingale. Another important elementary observation is that:

Theorem 27. *Let $B = (B_t)_{0 \le t < \infty}$ be a one dimensional standard Brownian motion with $B_0 = 0$. Then $M_t = B_t^2 - t$ is a martingale.*

Proof. $E\{M_t\} = E\{B_t^2 - t\} = 0$. Also

$$E\{M_t - M_s | \mathcal{F}_s\} = E\{B_t^2 - B_s^2 - (t-s) | \mathcal{F}_s\},$$

and

$$E\{B_t B_s | \mathcal{F}_s\} = B_s E\{B_t | \mathcal{F}_s\} = B_s^2,$$

since B is a martingale with B_s, $B_t \in L^2$. Therefore

$$\begin{aligned} E\{M_t - M_s | \mathcal{F}_s\} &= E\{B_t^2 - 2B_t B_s + B_s^2 - (t-s) | \mathcal{F}_s\} \\ &= E\{(B_t - B_s)^2 - (t-s) | \mathcal{F}_s\} \\ &= E\{(B_t - B_s)^2\} - (t-s) \\ &= 0, \end{aligned}$$

due to the independence of the increments from the past. □

Theorem 28. *Let π_n be a sequence of partitions of $[a, a+t]$. Suppose $\pi_m \subset \pi_n$ if $m > n$ (that is, the sequence is a refining sequence). Suppose moreover that $\lim_{n\to\infty} mesh(\pi_n) = 0$. Let $\pi_n B = \sum_{t_i \in \pi_n} (B_{t_{i+1}} - B_{t_i})^2$. Then $\lim_{n\to\infty} \pi_n B = t$ a.s., for a standard Brownian motion B.*

Proof. We first show convergence in mean square. We have

$$\begin{aligned} \pi_n B - t &= \sum_{t_i \in \pi_n} \{(B_{t_{i+1}} - B_{t_i})^2 - (t_{i+1} - t_i)\} \\ &= \sum_i Y_i, \end{aligned}$$

where Y_i are independent random variables with zero means. Therefore

$$E\{(\pi_n B - t)^2\} = E\{(\sum_i Y_i)^2\} = \sum_i E\{Y_i^2\}.$$

Next observe that $(B_{t_{i+1}} - B_{t_i})^2/(t_{i+1} - t_i)$ has the distribution of Z^2, where

Z is Gaussian with mean 0 and variance 1. Therefore

$$E\{(\pi_n B - t)^2\} = E\{(Z^2-1)^2\} \sum_{t_i \in \pi_n} (t_{i+1}-t_i)^2$$
$$\leq E\{(Z^2-1)^2\}\text{mesh}(\pi_n)t,$$

which tends to 0 as n tends to ∞. This establishes L^2 convergence (and hence convergence in probability as well).

To obtain the a.s. convergence we use the Backwards Martingale Convergence Theorem (Theorem 14). Define

$$N_n(\omega) = \pi_{-n}B = \sum_{t_i \in \pi_{-n}} (B_{t_{i+1}}(\omega) - B_{t_i}(\omega))^2,$$

for $n = -1, -2, -3, \ldots$. Then it is straightforward (though notationally messy) to show that

$$E\{N_n | N_{n-1}, N_{n-2}, \ldots\} = N_{n-1}.$$

Therefore N_n is a martingale relative to $\mathcal{G}_n = \sigma\{N_k, k \leq n\}$, $n = -1, -2, -3, \ldots$. By Theorem 14 we deduce $\lim_{n\to-\infty} N_n = \lim_{n\to\infty}\pi_n B$ exists a.s., and since $\pi_n B$ converges to t in L^2, we must have $\lim_{n\to\infty}\pi_n B = t$ a.s. as well. □

Comments. As noted in the proofs, the proof is simple (and half as long) if we conclude only L^2 convergence (and hence convergence in probability), instead of a.s. convergence. Also, we can avoid the use of the backwards martingale convergence theorem (Theorem 14) in the second half of the proof if we add the hypothesis that $\sum_n \text{mesh}(\pi_n) < \infty$. The result then follows, after having proved the L^2 convergence, by using the Borel-Cantelli lemma and Chebsyshev's inequality. Furthermore to conclude only L^2 convergence we do not need the hypothesis that the sequence of partitions be refining. □

Theorem 28 can be used to prove that the paths of Brownian motion are of unbounded variation on compacts. It is this fact that is central to the difficulties in defining an integral with respect to Brownian motion (and martingales in general).

Theorem 29. *For almost all ω, the sample paths $t \to B_t(\omega)$ of a standard Brownian motion B are of unbounded variation on any interval.*

Proof. Let $A = [a,b]$ be an interval. The variation of paths of B is defined to be:

$$V_A(\omega) = \sup_{\pi \in \mathcal{P}} \sum_{t_i \in \pi} |B_{t_{i+1}} - B_{t_i}|$$

where $\mathcal{P}$ are all finite partitions of $[a,b]$. Suppose $P(V_A < \infty) > 0$. Let π_n be a sequence of refining partitions of $[a,b]$ with $\lim_n \text{mesh}(\pi_n) = 0$. Then by Theorem 28 on $\{V_A < \infty\}$,

$$\begin{aligned} b-a &= \lim_{n\to\infty} \sum_{t_i\in\pi_n} (B_{t_{i+1}} - B_{t_i})^2 \\ &\le \lim_{n\to\infty} \sup_{t_i\in\pi_n} |B_{t_{i+1}} - B_{t_i}| \sum_{t_i\in\pi_n} |B_{t_{i+1}} - B_{t_i}| \\ &\le \lim_{n\to\infty} \sup_{t_i\in\pi_n} |B_{t_{i+1}} - B_{t_i}| V_A \\ &= 0, \end{aligned}$$

since $\sup_{t_i\in\pi_n} |B_{t_{i+1}} - B_{t_i}|$ tends to 0 a.s. as mesh(π_n) tends to 0 by the a.s. uniform continuity of the paths on A. Since $b-a\le 0$ is absurd, by Theorem 27 we conclude $V_A = \infty$ a.s. Since the null set can depend on the interval $[a,b]$, we only consider intervals with rational endpoints a, b with $a<b$. Such a collection is countable, and since any interval $(a,b) = \bigcup_{n=1}^{\infty}[a_n, b_n]$ with a_n, b_n rational, we can omit the dependence of the null set on the interval. □

We conclude this section by observing that not only are the increments of standard Brownian motion independent, they are also stationary. Thus Brownian motion is a Lévy process (as is the Poisson process), and the theorems of Sect. 4 apply to it. In particular, we can conclude by Theorem 31 of Sect. 4 that the *completed natural filtration of standard Brownian motion is right continuous.*

4. Lévy Processes

The Lévy processes, which include the Poisson process and Brownian motion as special cases, were the first class of stochastic processes to be studied in the modern spirit (by the French mathematician Paul Lévy). They still provide prototypic examples for Markov processes as well as for semimartingales. Most of the results of this section hold for $\mathbf{R}^n$-valued processes; for notational simplicity, however, we will consider only $\mathbf{R}^1$-valued processes.[3] Once again we recall that we are assuming given a filtered probability space $(\Omega, \mathcal{F}, (\mathcal{F}_t)_{0\le t\le\infty}, P)$ satisfying the usual hypotheses.

Definition. An adapted process $X = (X_t)_{0\le t<\infty}$ with $X_0 = 0$ a.s. is a **Lévy process** if

[3] $\mathbf{R}^n$ denotes n-dimensional Euclidean space. $\mathbf{R}_+$ denotes the positive real numbers; that is, $[0,\infty)$.

(i) X has *increments independent of the past*: that is, $X_t - X_s$ is independent of $\mathcal{F}_s$, $0 \leq s < t < \infty$;
(ii) X has *stationary increments*: that is, $X_t - X_s$ has the same distribution as X_{t-s}, $0 \leq s < t < \infty$;
(iii) X_t is *continuous in probability*: that is, $\lim_{t \to s} X_t = X_s$, where the limit is taken in probability.

If we take the Fourier transform of each X_t we get a function $f(t,u) = f_t(u)$:

$$f_t(u) = E\{e^{iuX_t}\},$$

where $f_0(u) = 1$, and $f_{t+s}(u) = f_t(u)f_s(u)$, and $f_t(u) \neq 0$ for every (t,u). Using the (right) continuity in probability we conclude $f_t(u) = \exp(-t\psi(u))$, for some continuous function $\psi(u)$ with $\psi(0) = 0$. (Bochner's Theorem can be used to show the converse: if ψ is continuous, $\psi(0) = 0$, and if for all $t \geq 0$, $f_t(u) = e^{-t\psi(u)}$ satisfies $\sum_{i,j} \alpha_i \overline{\alpha}_j f_t(u_i - u_j) \geq 0$, for all finite $(u_1, \dots, u_n; \alpha_1, \dots, \alpha_n)$, then there exists a Lévy process corresponding to f.)

In particular it follows that if X is a Lévy process then for each $t > 0$, X_t has an infinitely divisible distribution. Inversely it can be shown that for each infinitely divisible distribution μ there exists a Lévy process X such that μ is the distribution of X_1.

Theorem 30. *Let X be a Lévy process. There exists a unique modification Y of X which is càdlàg and which is also a Lévy process.*

Proof. Let $M_t^u = \frac{e^{iuX_t}}{f_t(u)}$. For each fixed u in $\mathbf{Q}$, the rationals in $\mathbf{R}$, the process $(M_t^u)_{0 \leq t < \infty}$ is a complex-valued martingale (relative to $(\mathcal{F}_t)$).

We first show that the paths of X cannot explode a.s. For any real u, $(M_t^u)_{t \geq 0}$ is a (complex-valued) martingale and thus for a.a. ω the functions $t \to M_t^u(\omega)$ and $t \to e^{iuX_t(\omega)}$, with $t \in \mathbf{Q}_+$, are the restrictions to $\mathbf{Q}_+$ of càdlàg functions. Let

$$\Lambda = \{(\omega, u) \in \Omega \times \mathbf{R} : e^{iuX_t(\omega)}, t \in \mathbf{Q}_+, \text{ is not the restriction of a càdlàg function}\}.$$

One can check that Λ is a measurable set and we have seen that $\int 1_\Lambda(\omega,u)P(d\omega) = 0$, each $u \in \mathbf{R}$. By Fubini's theorem

$$\int \int_{-\infty}^{\infty} 1_\Lambda(\omega,u)du P(d\omega) = \int_{-\infty}^{\infty} \int 1_\Lambda(\omega,u)P(d\omega)du = 0,$$

hence we conclude that for a.a. ω the function $t \to e^{iuX_t(\omega)}$, $t \in \mathbf{Q}_+$ is the restriction of a càdlàg function for almost all $u \in \mathbf{R}$. We can now conclude that the function $t \to X_t(\omega)$, $t \in \mathbf{Q}_+$, is the restriction of a càdlàg function for every such ω, with the help of the lemma that follows the proof of this theorem.

Next set $Y_t = \lim_{s \in \mathbf{Q}_+, s \downarrow t} X_s$ on $\Omega \setminus \Lambda$ and $Y_t = 0$ on Λ, all t. Since $\mathcal{F}_t$ contains all the P-null sets of $\mathcal{F}$ and $(\mathcal{F}_t)_{0 \le t < \infty}$ is right continuous, $Y_t \in \mathcal{F}_t$. Since X is continuous in probability, $P\{Y_t \neq X_t\} = 0$, hence Y is a modification of X. It is clear that Y is a Lévy process as well. □

The next lemma was used in the proof of Theorem 30. Although it is a pure analysis lemma, we give a proof using probability theory.

Lemma. *Let x_n be a sequence of real numbers such that e^{iux_n} converges as n tends to ∞ for almost all $u \in \mathbf{R}$. Then x_n converges to a finite limit.*

Proof. We will verify the following Cauchy criterion: x_n converges if for any increasing sequences n_k and m_k, then $\lim_{k\to\infty} x_{n_k} - x_{m_k} = 0$. Let U be a random variable which has the uniform distribution on $[0,1]$. For any real t, by hypothesis a.s. $e^{itUx_{n_k}}$ and $e^{itUx_{m_k}}$ converge to the same limit. Therefore

$$\lim_{k\to\infty} e^{itU(x_{n_k} - x_{m_k})} = 1 \quad \text{a.s.}$$

Therefore the characteristic functions converge:

$$\lim_{k\to\infty} E\{e^{it(x_{n_k} - x_{m_k})U}\} = 1,$$

for all $t \in \mathbf{R}$. Consequently $(x_{n_k} - x_{m_k})U$ converges to zero in probability, whence $\lim_{k\to\infty} x_{n_k} - x_{m_k} = 0$, as claimed. □

We will henceforth **always assume** that we are using the (unique) càdlàg version of any given Lévy process. Lévy processes provide us with examples of filtrations that satisfy the "usual hypotheses", as the next theorem shows.

Theorem 31. *Let X be a Lévy process and let $\mathcal{G}_t = \mathcal{F}_t^0 \vee \mathcal{N}$, where $(\mathcal{F}_t^0)_{0 \le t < \infty}$ is the natural filtration of X, and $\mathcal{N}$ are the P-null sets of $\mathcal{F}$. Then $(\mathcal{G}_t)_{0 \le t < \infty}$ is right continuous.*

Proof. We must show $\mathcal{G}_{t+} = \mathcal{G}_t$, where $\mathcal{G}_{t+} = \bigcap_{u>t} \mathcal{G}_u$. Note that since the filtration $\mathcal{G}$ is increasing, it suffices to show that $\mathcal{G}_t = \bigcap_{n \ge 1} \mathcal{G}_{t+\frac{1}{n}}$. Thus we can take countable limits and it follows that if $s_1, \dots, s_n \le t$, for $(u_1, \dots, u_n)$:

$$\begin{aligned} E\{e^{i(u_1 X_{s_1} + \cdots + u_n X_{s_n})} | \mathcal{G}_t\} &= E\{e^{i(u_1 X_{s_1} + \cdots + u_n X_{s_n})} | \mathcal{G}_{t+}\} \\ &= e^{i(u_1 X_{s_1} + \cdots + u_n X_{s_n})}. \end{aligned}$$

For $v_1, \dots, v_n > t$ and $(u_1, \dots, u_n)$, we give the proof for $n = 2$ for notational convenience. Therefore let $z > v > t$, and suppose given u_1 and u_2. We have

$$\begin{aligned} E\{e^{i(u_1X_v+u_2X_z)}|\mathcal{G}_{t+}\} &= \lim_{w\downarrow t} E\{e^{i(u_1X_v+u_2X_z)}|\mathcal{G}_w\} \\ &= \lim_{w\downarrow t} E\{e^{iu_1X_v}\frac{e^{iu_2X_z}}{f_z(u_2)}f_z(u_2)|\mathcal{G}_w\} \\ &= \lim_{w\downarrow t} E\{e^{iu_1X_v}\frac{e^{iu_2X_v}}{f_v(u_2)}f_z(u_2)|\mathcal{G}_w\}, \end{aligned}$$

using that $M_v^{u_2} = \frac{e^{iu_2X_v}}{f_v(u_2)}$ is a martingale. Combining terms the above becomes

$$= \lim_{w\downarrow t} E\{e^{i(u_1+u_2)X_v}f_{z-v}(u_2)|\mathcal{G}_w\}$$

and the same martingale argument yields

$$\begin{aligned} &= \lim_{w\downarrow t} e^{i(u_1+u_2)X_w}f_{v-w}(u_1+u_2)f_{z-v}(u_2) \\ &= e^{i(u_1+u_2)X_t}f_{v-t}(u_1+u_2)f_{z-v}(u_2) \\ &\ \ \vdots \\ &= E\{e^{i(u_1X_v+u_2X_z)}|\mathcal{G}_t\}. \end{aligned}$$

It follows that $E\{e^{i\Sigma u_jX_{s_j}}|\mathcal{G}_{t+}\} = E\{e^{i\Sigma u_jX_{s_j}}|\mathcal{G}_t\}$ for all $(s_1,\ldots,s_n)$ and all $(u_1,\ldots,u_n)$, whence $E\{Z|\mathcal{G}_{t+}\} = E\{Z|\mathcal{G}_t\}$ for every bounded $Z \in \bigvee_{0\le s<\infty}\mathcal{F}_s^0$. Hence $\mathcal{G}_{t+} = \mathcal{G}_t$. □

The next theorem shows that a Lévy process "renews itself" at stopping times.

Theorem 32. *Let X be a Lévy process and let T be a stopping time. On the set $\{T<\infty\}$ the process $Y=(Y_t)_{0\le t<\infty}$ defined by $Y_t = X_{T+t}-X_T$ is a Lévy process adapted to $\mathcal{H}_t=\mathcal{F}_{T+t}$, Y is independent of $\mathcal{F}_T$ and Y has the same distribution as X.*

Proof. First assume T is bounded. Let $A \in \mathcal{F}_T$ and let $(u_1,\ldots,u_n;t_0,t_1,\ldots,t_n)$ be given with $u_j \in \mathbb{Q}$ and $t_j \in \mathbb{R}_+$, t_j increasing with j.

Recall that $M_t^{u_j} = \frac{e^{iu_jX_t}}{f_t(u_j)}$ is a martingale, where $f_t(u_j) = E\{e^{iu_jX_t}\}$. Then

$$\begin{aligned} &E\{1_A\exp(i\sum_j u_j(X_{T+t_j}-X_{T+t_{j-1}}))\} \\ &\quad = E\left\{1_A\prod_j\frac{M^{u_j}_{T+t_j}}{M^{u_j}_{T+t_{j-1}}}\frac{f_{T+t_j}(u_j)}{f_{T+t_{j-1}}(u_j)}\right\} \\ &\quad = P(A)\prod_j f_{t_j-t_{j-1}}(u_j) \end{aligned}$$

by applying the Optional Sampling Theorem (Theorem 17) n times. Note that this shows the independence of $Y_t = X_{T+t} - X_T$ from $\mathcal{F}_T$ as well as showing that Y has independent and stationary increments and *that the distribution of* Y *is the same as that of* X.

If T is not bounded, we let $T^n = min(T, n) = T \wedge n$. The formula is valid for $\Lambda_n = A \cap \{T \leq n\}$ when $A \in \mathcal{F}_T$, since then $\Lambda_n \in \mathcal{F}_{T \wedge n}$. Taking limits and using the dominated convergence theorem we see that our formula holds for unbounded T as well, for events $\Lambda = A \cap \{T < \infty\}$, $A \in \mathcal{F}_T$. This gives the result. □

Since a standard Brownian motion is a Lévy process, Theorem 32 gives us *a fortiori* the strong Markov property for Brownian motion. This allows us to establish a pretty result for Brownian motion, known as the *reflection principle.* Let $B = (B_t)_{t \geq 0}$ denote a standard Brownian motion, $B_0 = 0$ a.s., and let $S_t = \sup_{0 \leq s \leq t} B_s$, the *maximum process* of Brownian motion. Since B is continuous, $S_t = \sup_{0 \leq u \leq t_{u \in \mathbb{Q}}} B_u$, where $\mathbb{Q}$ denotes the rationals; hence S_t is an adapted process with non-decreasing paths.

Theorem 33 (Reflection Principle for Brownian Motion). *Let* $B = (B_t)_{t \geq 0}$ *be standard Brownian motion* ($B_0 = 0$ *a.s.*) *and* $S_t = \sup_{0 \leq s \leq t} B_s$, *the Brownian maximum process. For* $y \geq 0$, $z > 0$,

$$P(B_t < z - y; S_t \geq z) = P(B_t > y + z).$$

Proof. Let $T = \inf\{t > 0 : B_t = z\}$. Then T is a stopping time by Theorem 4, and $P(T < \infty) = 1$. We next define a new process X by:

$$X_t = B_t 1_{\{t < T\}} + (2B_T - B_t) 1_{\{t \geq T\}}.$$

The process X is the Brownian motion B up to time T, and after time T it is the Brownian motion B "reflected" about the constant level z. Since B and $-B$ have the same distribution, it follows from Theorem 32 that X is also a standard Brownian motion.

Next we define

$$R = \inf\{t > 0 : X_t = z\}.$$

Then clearly

$$P(R \leq t; X_t < z - y) = P(T \leq t; B_t < z - y);$$

since (R, X) and (T, B) have the same distribution. However we also have that $R = T$ identically, whence

$$\{R \leq t; X_t < z - y\} = \{T \leq t; B_t > z + y\}$$

by the construction of X. Therefore

$$(*) \qquad P(T \le t; B_t > z + y) = P(T \le t; B_t < z - y).$$

The left side of (*) equals

$$P(S_t \ge z; B_t > z + y) = P(B_t > z + y),$$

where the last equality is a consequence of the containment $\{S_t \ge z\} \supset \{B_t > z + y\}$. Also the right side of (*) equals $P(S_t \ge z; B_t < z - y)$. Combining these yields

$$P(S_t \ge z; B_t < z - y) = P(B_t > z + y),$$

which is what was to be proved. □

Corollary. *Let $B = (B_t)_{t \ge 0}$ be standard Brownian motion ($B_0 = 0$ a.s.) and $S_t = \sup_{0 < s \le t} B_s$. For $z > 0$,*

$$P(S_t > z) = 2P(B_t > z).$$

Proof. Take $y = 0$ in Theorem 33. Then

$$P(B_t < z; S_t \ge z) = P(B_t > z).$$

Adding $P(B_t > z)$ to both sides and noting that $\{B_t > z\} = \{B_t > z\} \cap \{S_t \ge z\}$ yields the result since $P(S_t = z) = 0$. □

A Lévy process is càdlàg, and hence the only type of discontinuities it can have is jump discontinuities. Letting $X_{t-} = \lim_{s \uparrow t} X_s$, the left limit at t, we define

$$\Delta X_t = X_t - X_{t-},$$

the jump at t. If $\sup_t |\Delta X_t| \le C < \infty$ a.s., where C is a non-random constant, then **we say that X has bounded jumps.**

Our next result states that a Lévy process with bounded jumps has finite moments of all orders. This fact was used in Sect. 3 (Step 2 of the proof of Theorem 23) to show that $E\{N_1\} < \infty$ for a Poisson process N.

Theorem 34. *Let X be a Lévy process with bounded jumps. Then $E\{|X_t|^n\} < \infty$ for all $n = 1, 2, 3, \ldots$.*

Proof. Let C be a (non-random) bound for the jumps of X. Define the stopping times

$$T_1 = \inf\{t : |X_t| \ge C\}$$

$$\vdots$$

$$T_{n+1} = \inf\{t > T_n : |X_t - X_{T_n}| \ge C\}.$$

Since the paths are right continuous, the stopping times $(T_n)_{n\geq 1}$ form a strictly increasing sequence. Moreover $|\Delta X_T| \leq C$ by hypothesis for *any* stopping time T. Therefore $\sup_s |X_s^{T_n}| \leq 2nC$ by recursion. Theorem 32 implies that $T_n - T_{n-1}$ is independent of $\mathcal{F}_{T_{n-1}}$ and also that the distribution of $T_n - T_{n-1}$ is the same as that of T_1.

The above implies that

$$E\{e^{-T_n}\} = (E\{e^{-T_1}\})^n = \alpha^n,$$

for some α, $0 \leq \alpha < 1$. But also

$$P\{|X_t| > 2nC\} \leq P\{T_n < t\} \leq \frac{E(e^{-T_n})}{e^{-t}} \leq e^t \alpha^n,$$

which implies that X_t has an exponential moment and hence moments of all orders. □

We next turn our attention to an analysis of the *jumps of a Lévy process.* Let Λ be a Borel set in $\mathbf{R}$ bounded away from 0 (that is, $0 \notin \overline{\Lambda}$, where $\overline{\Lambda}$ is the closure of Λ). For a Lévy process X we define the random variables:

$$T_\Lambda^1 = \inf\{t > 0 : \Delta X_t \in \Lambda\}$$

$$\vdots$$

$$T_\Lambda^{n+1} = \inf\{t > T_\Lambda^n : \Delta X_t \in \Lambda\}$$

Since X has càdlàg paths and $0 \notin \overline{\Lambda}$, the reader can readily check that $\{T_\Lambda^n \geq t\} \in \mathcal{F}_{t+} = \mathcal{F}_t$ and therefore each T_Λ^n is a stopping time. Moreover $0 \notin \overline{\Lambda}$ and càdlàg paths further imply $T_\Lambda^1 > 0$ a.s. and that $\lim_{n\to\infty} T_\Lambda^n = \infty$ a.s. We define

$$N_t^\Lambda = \sum_{0<s\leq t} 1_\Lambda(\Delta X_s) = \sum_{n=1}^{\infty} 1_{\{T_\Lambda^n \leq t\}}$$

and observe that N^Λ is a counting process without an explosion. It is straightforward to check that for $0 \leq s < t < \infty$,

$$N_t^\Lambda - N_s^\Lambda \in \sigma\{X_u - X_v; s \leq v < u \leq t\},$$

and therefore $N_t^\Lambda - N_s^\Lambda$ is independent of $\mathcal{F}_s$; that is, N^Λ has independent increments. Note further that $N_t^\Lambda - N_s^\Lambda$ is the number of jumps that $Z_u = X_{s+u} - X_s$ has in Λ, $0 \leq u \leq t - s$. By the stationarity of the distributions of X, we conclude $N_t^\Lambda - N_s^\Lambda$ has the same distribution as N_{t-s}^Λ; therefore N^Λ is a counting process with stationary and independent increments. *We conclude that N^Λ is a Poisson process.* Let $\nu(\Lambda) = E\{N_1^\Lambda\}$ be the parameter of the Poisson process N^Λ ($\nu(\Lambda) < \infty$ by the proof of Theorem 34).

Theorem 35. *The set function $\Lambda \to N_t^\Lambda(\omega)$ defines a σ-finite measure on $\mathbf{R} \setminus \{0\}$ for each fixed (t, ω). The set function $\nu(\Lambda) = E\{N_1^\Lambda\}$ also defines a σ-finite measure on $\mathbf{R} \setminus \{0\}$.*

Proof. The set function $\Lambda \to N_t^\Lambda(\omega)$ is simply a counting measure: $\mu(\Lambda) = \{\text{number of } s \leq t : \ \Delta X_s(\omega) \in \Lambda\}$. It is then clear that ν is also a measure. □

Definition. The measure ν defined by

$$\nu(\Lambda) = E\{N_1^\Lambda\} = E\{\sum_{0<s\leq 1} 1_\Lambda(\Delta X_s)\}$$

is called the **Lévy measure** of the Lévy process X.

We wish to investigate further the role the Lévy measure plays in governing the jumps of X. To this end we establish a preliminary result. We let $N_t(\omega, dx)$ denote the random measure of Theorem 35. Since $N_t(\omega, dx)$ is a counting measure, the next result is obvious.

Theorem 36. *Let Λ be a Borel set of $\mathbf{R}$, $0 \notin \overline{\Lambda}$, f Borel and finite on Λ. Then*

$$\int_\Lambda f(x) N_t(\omega, dx) = \sum_{0<s\leq t} f(\Delta X_s) 1_\Lambda(\Delta X_s).$$

Just as we showed that N_t^Λ has independent and stationary increments, we have the following consequence.

Corollary. *Let Λ be a Borel set of $\mathbf{R}$ with $0 \notin \overline{\Lambda}$, and let f be Borel and finite on Λ. Then*

$$\int_\Lambda f(x) N_t(\cdot, dx)$$

is a Lévy process.

For a given set Λ (as always, $0 \notin \overline{\Lambda}$), we defined **the associated jump process**:

$$J_t^\Lambda = \sum_{0<s\leq t} \Delta X_s 1_\Lambda(\Delta X_s).$$

By Theorem 36 and its Corollary we conclude

$$J_t^\Lambda = \int_\Lambda x N_t(\cdot, dx),$$

and J_t^Λ is a Lévy process itself, and is defined, and $J_t^\Lambda < \infty$ a.s., each $t \geq 0$.

Theorem 37. *Given Λ, $0 \notin \overline{\Lambda}$, the process $X_t - J_t^\Lambda$ is a Lévy process.*

Proof. It is clear that we need only check the independence and stationarity of the increments. But

$$X_t - J_t^\Lambda - (X_s - J_s^\Lambda) = X_t - X_s - \sum_{s<u\leq t} \Delta X_u 1_\Lambda(\Delta X_u)$$

which is clearly $\sigma\{X_v - X_u; s \leq u < v \leq t\}$-measurable, and due to the stationarity of the increments of X it has the same law as $X_{t-s} - J_{t-s}^\Lambda$. □

We are now in a position to consider

$$Y_t^a = X_t - \sum_{0<s\leq t} \Delta X_s 1_{\{|\Delta X_s|\geq a\}},$$

for some constant $a > 0$. The advantage of doing this is that Y^a then has jumps bounded by a, and hence has finite moments of all orders (Theorem 34). We can choose any $a > 0$; we arbitrarily choose $a = 1$. Note that

$$\begin{aligned} Y_t^1 &= X_t - J_t^{(-\infty,-1]\cup[1,\infty)} \\ &= X_t - \int_{|x|\geq 1} x N_t(\cdot, dx). \end{aligned}$$

The next theorem gives an interpretation of the Lévy measure as the expected rate at which the jumps of the Lévy process fall in a given set.

Theorem 38. *Let Λ be Borel with $0 \notin \overline{\Lambda}$. Let ν be the Lévy measure of X, and let $f1_\Lambda \in L^2(d\nu)$. Then*

$$E\{\int_\Lambda f(x) N_t(\cdot, dx)\} = t\int_\Lambda f(x)\nu(dx)$$

and also

$$E\{(\int_\Lambda f(x)N_t(\cdot,dx) - t\int_\Lambda f(x)\nu(dx))^2\} = t\int_\Lambda f(x)^2\nu(dx).$$

Proof. First let $f = \sum_j a_j 1_{\Lambda_j}$, a simple function. Then

$$\begin{aligned} E\{\sum_j a_j N_t^{\Lambda_j}\} &= \sum a_j E\{N_t^{\Lambda_j}\} \\ &= t\sum_j a_j \nu(\Lambda_j), \end{aligned}$$

since $N_t^{\Lambda_j}$ is a Poisson process with parameter $\nu(\Lambda_j)$. The first equality follows easily.

For the second equality, let $M_t^i = N_t^{\Lambda_i} - t\nu(\Lambda_i)$. The M_t^i are L^p martingales, all $p \geq 1$, by the proof of Theorem 34. Moreover, $E(M_t^i) = 0$. Suppose Λ_i, Λ_j are disjoint. We have:

$$E\{M_t^i M_t^j\} = E\{\sum_k (M_{t_{k+1}}^i - M_{t_k}^i) \sum_\ell (M_{t_{\ell+1}}^j - M_{t_\ell}^j)\}$$

for any partition $0 = t_0 < t_1 < \cdots < t_n = t$. Using the martingale property we have

$$E\{M_t^i M_t^j\} = E\{\sum_k (M_{t_{k+1}}^i - M_{t_k}^i)(M_{t_{k+1}}^j - M_{t_k}^j)\}.$$

Using the inequality $|ab| \leq (a^2 + b^2)$, we have

$$\sum_k (M_{t_{k+1}}^i - M_{t_k}^i)(M_{t_{k+1}}^j - M_{t_k}^j) \leq (\sum_k (M_{t_{k+1}}^i - M_{t_k}^i)^2 + \sum_k (M_{t_{k+1}}^j - M_{t_k}^j)^2).$$

However $\sum_k (M_{t_{k+1}}^i - M_{t_k}^i)^2 \leq ((N_t^{\Lambda_i})^2 + \nu(\Lambda_i)^2 t^2)$; therefore the sums are dominated by an integrable random variable. Since M_t^i and M_t^j have paths of finite variation on $[0,t]$ it is easy to deduce that if we take a sequence $(\pi_n)_{n\geq 1}$ of partitions where the mesh tends to 0 we have

$$\lim_{n\to\infty} \sum_{t_k, t_{k+1} \in \pi_n} (M_{t_{k+1}}^i - M_{t_k}^i)(M_{t_{k+1}}^j - M_{t_k}^j) = \sum_{0<s\leq t} \Delta M_s^i \Delta M_s^j.$$

Using Lebesgue's dominated convergence theorem we conclude

$$E\{M_t^i M_t^j\} = E\{\sum_{0<s\leq t} \Delta M_s^i \Delta M_s^j\} = 0;$$

the expectation above is 0 because Λ_i and Λ_j are disjoint, implying that M^i and M^j jump at different times. The second equality is now easy to verify for simple functions. For general f, let f_n be a sequence of simple functions such that $f_n 1_\Lambda$ converges to f in $L^2(d\nu)$, and the result follows. □

Corollary. *Let $f : \mathbf{R} \to \mathbf{R}$ be bounded and vanish in a neighborhood of* 0. *Then*

$$E\{\sum_{0<s\leq t} f(\Delta X_s)\} = t\int_{-\infty}^{\infty} f(x)\nu(dx).$$

Proof. We need only combine Theorem 38 with Theorem 36. □

We now come to one of the fundamental results about Lévy processes.

Theorem 39. *Let Λ_1, Λ_2 be two disjoint Borel sets with $0 \notin \overline{\Lambda}_1$, $0 \notin \overline{\Lambda}_2$. Then the two processes*

$$J_t^1 = \sum_{0<s\leq t} \Delta X_s 1_{\Lambda_1}(\Delta X_s)$$

$$J_t^2 = \sum_{0<s\leq t} \Delta X_s 1_{\Lambda_2}(\Delta X_s)$$

are independent Lévy processes.

Proof. By Theorem 36 and its Corollary we have that J^1 and J^2 are Lévy processes. To show they are independent, we begin by forming for u, v in $\mathbb{R}$,

$$C_t^u = \frac{e^{iuJ_t^1}}{E\{e^{iuJ_t^1}\}} - 1$$

$$D_t^v = \frac{e^{ivJ_t^2}}{E\{e^{ivJ_t^2}\}} - 1.$$

Then C^u and D^v are both martingales, with $E(C_t^u) = E(D_t^v) = 0$. As in the proof of Theorem 38, let π_n: $0 = t_0 < t_1 < \cdots < t_n = t$ be a sequence of partitions of $[0,t]$ with $\lim_n \text{mesh}(\pi_n) = 0$. Then

$$E\{C_t^u D_t^v\} = E\{\sum_k (C_{t_{k+1}}^u - C_{t_k}^u) \sum_\ell (D_{t_{\ell+1}}^v - D_{t_\ell}^v)\}$$

$$= E\{\sum_k (C_{t_{k+1}}^u - C_{t_k}^u)(D_{t_{k+1}}^v - D_{t_k}^v)\}.$$

Since C^u and D^v have paths of finite variation on compacts, it follows by letting $\text{mesh}(\pi_n)$ tend to 0, that:

$$E\{C_t^u D_t^v\} = E\{\sum_{0<s\leq t} \Delta C_s^u \Delta D_s^v\}.$$

The expectation above equals zero because C^u and D^v jump at different times, due to the void intersection of Λ_1 and Λ_2.

We conclude that $E\{C_t^u D_t^v\} = 0$, and thus

$$E\{e^{iuJ_t^1} e^{ivJ_t^2}\} = E\{e^{iuJ_t^1}\} E\{e^{ivJ_t^2}\},$$

which in turn implies, because of the independence and stationarity of the increments:

$$E\{e^{i(u_1 J_{t_1}^1 + u_2(J_{t_2}^1 - J_{t_1}^1) + \cdots + u_n(J_{t_n}^1 - J_{t_{n-1}}^1))} e^{i(v_1 J_{t_1}^2 + \cdots + v_n(J_{t_n}^2 - J_{t_{n-1}}^2))}\}$$

$$= E\{e^{iu_1 J_{t_1}^1 + i\sum_{j=2}^n u_j(J_{t_j}^1 - J_{t_{j-1}}^1)}\} E\{e^{iv_1 J_{t_1}^2 + i\sum_{j=2}^n v_j(J_{t_j}^2 - J_{t_{j-1}}^2)}\}.$$

This is enough to give independence. □

The preceding results combine to yield the following useful result.

Theorem 40. *Let X be a Lévy process. Then $X_t = Y_t + Z_t$, where Y, Z are Lévy processes, Y is a martingale with bounded jumps, $Y_t \in L^p$ for all $p \geq 1$ and Z has paths of finite variation on compacts.*

Proof. Let $J_t = \sum_{0<s\leq t} \Delta X_s 1_{\{|\Delta X_s| \geq 1\}}$. Since X has càdlàg paths, for each fixed ω the function $s \to X_s(\omega)$ has only finitely many jumps bigger than one on $[0,t]$. Therefore J has paths of finite variation on compacts. J is also a Lévy process by Theorem 36 and its Corollary. The process $W = X - J$ is also a Lévy process (Theorem 37), and W has jumps bounded by one. We know therefore that, $n \geq 1$, $E\{|W_t|^n\}$ exists (Theorem 34), and the stationary increments of W implies $E\{W_t\} = \alpha t$, for $\alpha = E\{W_1\}$. (Recall $E\{W_0\} = 0$). We set $Y_t = W_t - E\{W_t\}$. Then Y has independent increments and mean 0; it is a martingale. Setting $Z_t = J_t + \alpha t$ completes the proof. □

While Theorem 40 is the most important result about Lévy processes from the standpoint of stochastic integration, the next two theorems provide a better understanding of Lévy processes themselves.

Theorem 41. *Let X be a Lévy process with jumps bounded by a : $\sup_s |\Delta X_s| \leq a$ a.s. Let $Z_t = X_t - E\{X_t\}$. Then Z is a martingale and $Z_t = Z_t^c + Z_t^d$ where Z^c is a martingale with continuous paths, Z^d is a martingale,*

$$Z_t^d = \int_{\{|x|\leq a\}} x(N_t(\cdot, dx) - t\nu(dx)),$$

and Z^c and Z^d are independent Lévy processes.

Proof. Z has mean zero and independent increments so it is a martingale, as well as a Lévy process. For a given set Λ we define

$$\begin{aligned} M_t^\Lambda &= \int_\Lambda x N_t(\cdot, dx) - t \int_\Lambda x\nu(dx) \\ &= \sum_{0<s\leq t} \Delta X_s 1_\Lambda(\Delta X_s) - t \int_\Lambda x\nu(dx). \end{aligned}$$

For this proof we take $a = 1$. Let $\Lambda_k = \{\frac{1}{k+1} < |x| \leq \frac{1}{k}\}$. Then M^{Λ_k} are pairwise independent Lévy processes and martingales (Theorem 39). Set $M^n = \sum_{k=1}^n M^{\Lambda_k}$. Then the martingales $Z - M^n$ and M^n are independent by an argument similar to the one in the proof of Theorem 39. Moreover $Var(Z_t) = Var(Z_t - M_t^n) + Var(M_t^n)$ where $Var(X)$ denotes the variance of a random variable X. Therefore $Var(M_t^n) \leq Var(Z_t) < \infty$ for all n. We deduce that M_t^n converges in L^2 as n tends to ∞ to a martingale Z_t^d, and $Z - M^n$ also converges to a martingale Z^c. Using Doob's maximal quadratic inequality (Theorem 20), we can find a subsequence converging a.s., uniformly in t on compacts, which permits the conclusion that Z^c has continuous paths. The independence of Z^d and Z^c follows from the independence of M^n and $Z - M^n$, for every n. □

Note that a consequence of the convergence of M_t^n to Z_t^d in L^2 in the proof of Theorem 41 is that the integral $\int_{[-1,0)\cup(0,1]} x^2\nu(dx)$ is finite. Note that this improves a bit on the conclusion in Theorem 38.

We recall that for a set Λ, $0 \notin \overline{\Lambda}$, the process $N_t^\Lambda = \int_\Lambda N_t(\cdot, dx)$ is a Poisson process with parameter $\nu(\Lambda)$, and thus $N_t^\Lambda - t\nu(\Lambda)$ is a martingale.

Definition. Let N be a Poisson process with parameter λ. Then $N_t - \lambda t$ is called a **compensated Poisson process.**

Theorem 41 can be interpreted as saying that a Lévy process with bounded jumps decomposes into the sum of a continuous martingale Lévy process and a martingale which is a mixture of compensated Poisson processes. It is not hard to show that $E\{e^{iuZ_t^c}\} = e^{-t\sigma^2u^2/2}$, which implies that Z^c must be a Brownian motion. The full decomposition theorem then follows easily. We state it here without proof (consult Bretagnolle [**1**] or Jacod-Shiryayev [**1**] for a proof).

Theorem 42 (Lévy Decomposition). *Let X be a Lévy process. Then X has a decomposition*

$$\begin{aligned} X_t &= B_t + \int_{\{|x|<1\}} x(N_t(\cdot, dx) - t\nu(dx)) \\ &\quad + tE\{X_1 - \int_{\{|x|\geq 1\}} xN_1(\cdot, dx)\} + \int_{\{|x|\geq 1\}} xN_t(\cdot, dx) \\ &= B_t + \int_{\{|x|<1\}} x(N_t(\cdot, dx) - t\nu(dx)) + \alpha t + \sum_{0<s\leq t} \Delta X_s 1_{\{|\Delta X_s|\geq 1\}} \end{aligned}$$

where B is a Brownian motion; for any set Λ, $0 \notin \overline{\Lambda}$, $N_t^\Lambda = \int_\Lambda N_t(\cdot, dx)$ is a Poisson process independent of B; N_t^Λ is independent of N_t^Γ if Λ and Γ are disjoint; N_t^Λ has parameter $\nu(\Lambda)$; and $\nu(dx)$ is a measure on $\mathbb{R} \setminus \{0\}$ such that $\int \min(1, x^2)\nu(dx) < \infty$.

Finally we observe that Theorem 42 gives us a formula (known as the **Lévy-Khintchine formula**) for the Fourier transform of a Lévy process.

Theorem 43. *Let X be a Lévy process with Lévy measure ν. Then*

$$E\{e^{iuX_t}\} = e^{-t\psi(u)},$$

where

$$\psi(u) = \frac{\sigma^2}{2}u^2 - i\alpha u + \int_{\{|x|\geq 1\}} (1 - e^{iux})\nu(dx) + \int_{\{|x|<1\}} (1 - e^{iux} + iux)\nu(dx).$$

Moreover given ν, σ^2, α, the corresponding Lévy process is unique in distribution.

5. Local Martingales

We recall that we assume given a filtered probability space $(\Omega, \mathcal{F}, (\mathcal{F}_t)_{0\le t<\infty}, P)$ *satisfying the usual hypotheses.* For a process X and a stopping time T we further recall that X^T denotes the *stopped process*

$$X_t^T = X_{t\wedge T} = X_t 1_{\{t<T\}} + X_T 1_{\{t\ge T\}}.$$

Definition. An adapted, càdlàg process X is a **local martingale** if there exists a sequence of increasing stopping times, T_n, with $\lim_{n\to\infty} T_n = \infty$ a.s. such that $X_{t\wedge T_n} 1_{\{T_n>0\}}$ is a uniformly integrable martingale for each n. Such a sequence (T_n) of stopping times is called a **fundamental sequence**.

Examples. Clearly any càdlàg martingale is a local martingale (take $T_n \equiv n$).

We give an example of a local martingale which is not a martingale. Let $(B_t)_{0\le t<\infty}$ be a Brownian motion in $\mathbf{R}^3$ with $B_0 = x$, where $x \neq 0$. Let $u(y) = \frac{1}{\|y\|}$, a superharmonic function on $\mathbf{R}^3$. As a consequence of Itô's Formula (cf. Theorems 32 and 40 of Chap. II) one can show that $X_t = u(B_t)$ is a positive supermartingale (indeed, it is even a uniformly integrable supermartingale). Next let $T_n = \inf\{t > 0 : \|B_t\| \le \frac{1}{n}\}$. Outside of the ball of radius $\frac{1}{n}$ centered at the origin the function u is harmonic. Again by Itô's Formula one can show that this implies $u(B_{t\wedge T_n})$ is a martingale. Since $u(B_{t\wedge T_n})$ is bounded by n it is uniformly integrable. On the other hand, if $B_0 = x \neq 0$ then $\lim_{t\to\infty} E\{u(B_t)\} = 0$ as is easily seen by a calculation, while $E\{u(B_0)\} = \frac{1}{\|x\|}$. Since the expectations of a martingale are constant, $u(B_t)$ is not a martingale. (For more on this example, see Sect. 6, Chap. II, following Corollary 4 of Theorem 27.)

The reason we multiply $X_{t\wedge T_n}$ by $1_{\{T_n>0\}}$ is to relax the integrability condition on X_0. This is useful, for example, in the consideration of stochastic integral equations with a non-integrable initial condition.

Definition. A stopping time T **reduces** a process M if M^T is a uniformly integrable martingale.

Theorem 44. *Let M, N be local martingales and let S and T be stopping times*

a) If T reduces M and $S \le T$ a.s., then S reduces M;
b) The sum $M + N$ is also a local martingale;
c) If S, T both reduce M, then $S \vee T$ also reduces M;
d) The processes M^T, $M^T 1_{\{T>0\}}$ are local martingales;

e) Let X be a càdlàg process and let T_n be a sequence of stopping times increasing to ∞ a.s. such that $X^{T_n}1_{\{T_n>0\}}$ is a local martingale for each n. Then X is a local martingale.

Proof. (a) follows from the Optional Sampling Theorem (Theorem 16). For (b), if $(S_n)_{n\geq 1}$, $(T_n)_{n\geq 1}$ are fundamental sequences for M and N, respectively, then $S_n \wedge T_n$ is a fundamental sequence for $M+N$. For (c), let $X_t = M_t - M_0$. Then $X^{S\vee T} = X^S + X^T - X^{S\wedge T}$ is a uniformly integrable martingale. But

$$|M_0|1_{\{S\vee T>0\}} \leq |M_0|1_{\{S>0\}} + |M_0|1_{\{T>0\}},$$

which is in L^1, therefore $X^{S\vee T} + M_0 1_{\{S\vee T>0\}} = M^{S\vee T}1_{\{S\vee T>0\}}$ is a uniformly integrable martingale. The proof of (d) consists of the observation that if $(T_n)_{n\geq 1}$ is a fundamental sequence for M, then it is also one for M^T and $M^T 1_{\{T>0\}}$ by the Optional Sampling Theorem (Theorem 16). For (e), we have $X^{T_n}1_{\{T_n>0\}} = M^n$ is a local martingale for each n. For fixed n we know there exists a fundamental sequence $U^{n,k}$ increasing to ∞ a.s. as k tends to ∞. For each n, choose $k = k(n)$ such that $P(U^{n,k(n)} < T_n \wedge n) < \frac{1}{2^n}$. Then $\lim_n U^{n,k(n)} = \infty$ a.s., and $U^{n,k(n)} \wedge T_n$ reduces X for each n. We take $R^m = \max(U^{1,k(1)} \wedge T_1, \ldots, U^{m,k(m)} \wedge T_m)$, and each R^m reduces X by part (c), the R^m are increasing, and $\lim R^m = \infty$ a.s. Therefore X is a local martingale. □

Corollary. *Local martingales form a vector space.*

We will often need to know that a reduced local martingale, M^T, is in L^p and not simply unformly integrable.

Definition. Let X be a stochastic process. **A property $\mathcal{P}$ is said to hold locally** if there exists a sequence of stopping times $(T_n)_{n\geq 1}$ increasing to ∞ a.s. such that $X^{T_n}1_{\{T_n>0\}}$ has property $\mathcal{P}$, each $n \geq 1$.

We see from Theorem 44(e) that a process which is locally a local martingale is also a local martingale. Other examples of local properties that arise frequently are *locally bounded* and *locally square integrable*. Theorems 45 and 46 are consequences of Theorem 44(e).

Theorem 45. *Let X be a process which is locally a square integrable martingale. Then X is a local martingale.*

The next theorem shows that the traditional "uniform integrability" assumption in the definition of local martingale is not really necessary.

Theorem 46. *Let M be adapted, càdlàg and let $(T_n)_{n\geq 1}$ be a sequence of stopping times increasing to ∞ a.s. If $M^{T_n} 1_{\{T_n>0\}}$ is a martingale for each n, then M is a local martingale.*

It is often of interest to determine when a local martingale is actually a martingale. A simple condition involves the maximal function. Recall that $X_t^* = \sup_{s\leq t} |X_s|$ and $X^* = \sup_s |X_s|$.

Theorem 47. *Let X be a local martingale such that $E\{X_t^*\} < \infty$ for every $t \geq 0$. Then X is a martingale. If $E\{X^*\} < \infty$, then X is a uniformly integrable martingale.*

Proof. Let $(T_n)_{n\geq 1}$ be a fundamental sequence of stopping times for X. If $s \leq t$, then $E\{X_{t\wedge T_n} | \mathcal{F}_s\} = X_{s\wedge T_n}$. The dominated convergence theorem yields $E\{X_t | \mathcal{F}_s\} = X_s$. If $E\{X^*\} < \infty$, since each $|X_t| \leq X^*$, it follows that $\{X_t; t \geq 0\}$ is uniformly integrable. □

Note that in particular a bounded local martingale is a uniformly integrable martingale. Other sufficient conditions for a local martingale to be a martingale are given in Corollaries 3 and 4 to Theorem 27 in Chap. II.

6. Stieltjes Integration and Change of Variables

Stochastic integration with respect to semimartingales can be thought of as an extension of path-by-path Stieltjes integration. We present here the essential elementary ideas of Stieltjes integration appropriate to our interests. We assume the reader is familiar with the Lebesgue theory of measure and integration on $\mathbb{R}_+$.

Definition. Let $A = (A_t)_{t\geq 0}$ be a càdlàg process. A is an **increasing process** if the paths of $A : t \to A_t(\omega)$ are non-decreasing for almost all ω. A is called a **finite variation process** (FV) if almost all of the paths of A are of finite variation on each compact interval of $\mathbb{R}_+$.

Let A be an increasing process. Fix an ω such that $t \to A_t(\omega)$ is right continuous and non-decreasing. This function induces a measure $\mu_A(\omega, ds)$ on $\mathbb{R}_+$. If f is a bounded, Borel function on $\mathbb{R}_+$, then $\int_0^t f(s)\mu_A(\omega, ds)$ is well defined for each $t > 0$. We denote this integral by $\int_0^t f(s) dA_s(\omega)$. If $F_s = F(s, \omega)$ is bounded and jointly measurable, we can define, ω-by-ω, the integral $I(t,\omega) = \int_0^t F(s,\omega)dA_s(\omega)$. I is right continuous in t and jointly measurable.

Proceeding analogously for A an FV process (except that the induced measure $\mu_A(\omega, ds)$ can have negative measure; that is, it is a *signed measure*), we can define a jointly measurable integral

$$I(t,\omega) = \int_0^t F(s,\omega)dA_s(\omega)$$

for F bounded and jointly measurable.

Let A be an FV process. We define

$$(*) \qquad |A|_t = \sup_{n\geq 1} \sum_{k=1}^{2^n} |A_{\frac{tk}{2^n}} - A_{\frac{t(k-1)}{2^n}}|.$$

Then $|A|_t < \infty$ a.s., and it is an increasing process.

Definition. For A an FV process, **the total variation process**, $|A| = (|A|_t)_{t\geq 0}$, is the increasing process defined in (*) above.

Notation. Let A be an FV process and let F be a jointly measurable process such that $\int_0^t F(s,\omega)dA_s(\omega)$ exists and is finite for all $t > 0$, a.s. We let

$$(F \cdot A)_t(\omega) = \int_0^t F(s,\omega)dA_s(\omega)$$

and we write $F \cdot A$ to denote the process $F \cdot A = (F \cdot A_t)_{t\geq 0}$. We will also write $\int_0^t F_s|dA_s|$ for $(F \cdot |A|)_t$. The next result is an absolute continuity result for Stieltjes integrals.

Theorem 48. *Let A, C be adapted, increasing processes such that $C - A$ is also an increasing process. Then there exists a jointly measurable, adapted process H (defined on $(0,\infty)$) such that $0 \leq H \leq 1$ and*

$$A = H \cdot C$$

or equivalently

$$A_t = \int_0^t H_s dC_s.$$

Proof. If μ and ν are two Borel measures on $\mathbf{R}_+$ with $\mu \ll \nu$, then we can set

$$\alpha(s,t) = \begin{cases} \frac{\mu((s,t])}{\nu((s,t])} & \text{if } \nu((s,t]) > 0 \\ 0 & \text{otherwise.} \end{cases}$$

Defining h and k by $h(t) = \lim_{s\uparrow t} \inf \alpha(s,t)$, $k(t) = \lim_{t\downarrow s} \sup\ \alpha(s,t)$, then h and k are both Borel measurable, and moreover they are each versions of the Radon-Nikodym derivative; that is:

$$d\mu = h\,d\nu; \qquad d\mu = k\,d\nu.$$

To complete the proof it suffices to show that we can follow the above procedure in a (t,ω)-measurable fashion. With the convention $\frac{0}{0} = 0$, it suffices to define:

$$H(t,\omega) = \lim\ \inf_{r\uparrow 1, r\in \mathbf{Q}_+} \frac{(A(t,\omega) - A(rt,\omega))}{(C(t,\omega) - C(rt,\omega))};$$

such an H is clearly adapted since both A and C are. □

Corollary. *Let A be an FV process. There exists a jointly measurable, adapted process H, $-1 \leq H \leq 1$, such that*

$$|A| = H \cdot A \qquad \textit{and} \qquad A = H \cdot |A|$$

or equivalently

$$|A|_t = \int_0^t H_s dA_s \qquad \textit{and} \qquad A_t = \int_0^t H_s |dA_s|.$$

Proof. We define $A_t^+ = \frac{1}{2}(|A|_t + A_t)$ and $A_t^- = \frac{1}{2}(|A|_t - A_t)$. Then A^+ and A^- are both increasing processes, and $|A| - A^+$ and $|A| - A^-$ are also increasing processes. By Theorem 48 there exist processes H^+ and H^- such that $A_t^+ = \int_0^t H_s^+ |dA_s|$, $A_t^- = \int_0^t H_s^- |dA_s|$. It then follows that $A_t = A_t^+ - A_t^- = \int_0^t (H_s^+ - H_s^-)|dA_s|$. Let $H_t \equiv H_t^+ - H_t^-$ and suppose H^+ and H^- are defined as in the proof of Theorem 48. Except for a P-null set, for a given ω it is clear that $|H_s(\omega)| = 1$ $dA_s(\omega)$ almost all s. We then consider $H \cdot A$:

$$\begin{aligned}\int_0^t H_s dA_s &= \int_0^t H_s d(\int_0^s H_u |dA_u|) \\ &= \int_0^t H_s H_s |dA_s| = \int_0^t 1|dA_s| = |A|_t.\end{aligned}$$

This completes the proof. □

When the integrand process H has *continuous paths*, the Stieltjes integral $\int_0^t H_s dA_s$ is also known as the **Riemann-Stieltjes integral** (for fixed ω). In this case we can define the integral as the limit of approximating sums. Such a result is proved in elementary textbooks on real analysis (e.g., Protter-Morrey [**1**, pp. 316–317]).

Theorem 49. *Let A be an FV process and let H be a jointly measurable process such that a.s. $s \to H(s,\omega)$ is continuous. Let π_n be a sequence of finite random partitions of $[0,t]$ with $\lim_{n\to\infty} mesh(\pi_n) = 0$. Then for $T_k \le S_k \le T_{k+1}$,*

$$\lim_{n\to\infty} \sum_{T_k, T_{k+1}\in\pi_n} H_{S_k}(A_{T_{k+1}} - A_{T_k}) = \int_0^t H_s dA_s \ a.s.$$

We next prove a change of variables formula when the FV process is *continuous.* Itô's Formula (Theorem 32 of Chap. II) is a generalization of this result.

Theorem 50 (Change of Variables). *Let A be an FV process with continuous paths, and let f be such that its derivative f' exists and is continuous. Then $(f(A_t))_{t\ge 0}$ is an FV process and*

$$f(A_t) - f(A_0) = \int_0^t f'(A_s)dA_s.$$

Proof. For fixed ω, the function $s \to f'(A_s(\omega))$ is continuous on $[0,t]$ and hence bounded. Therefore the integral $\int_0^t f'(A_s)dA_s$ exists. Fix t and let π_n be a sequence of partitions of $[0,t]$ with $\lim_{n\to\infty} \mathrm{mesh}(\pi_n) = 0$. Then

$$\begin{aligned} f(A_t) - f(A_0) &= \sum_{t_k, t_{k+1}\in\pi_n} \{f(A_{t_{k+1}}) - f(A_{t_k})\} \\ &= \sum_k f'(A_{S_k})(A_{t_{k+1}} - A_{t_k}), \end{aligned}$$

by the Mean Value Theorem, for some S_k, $t_k \le S_k \le t_{k+1}$. The result now follows by taking limits and Theorem 49. □

Comment. We will see in Chap. II that the sums

$$\sum_{t_k\in\pi_n[0,t]} f(A_{t_k})(A_{t_{k+1}} - A_{t_k})$$

converge in probability to

$$\int_0^t f(A_{s-})dA_s$$

for a continuous function f and an FV process A. This leads to the more general change of variables formula, valid for any FV process A, and $f \in C^1$:

$$f(A_t) - f(A_0) = \int_0^t f'(A_{s-})dA_s + \sum_{0<s\le t} \{f(A_s) - f(A_{s-}) - f'(A_{s-})\Delta A_s\}.$$

The next Corollary explains why Theorem 50 is known as a Change of Variables formula. The proof is an immediate application of Theorem 50.

Corollary. *Let g be continuous, and let $h(t) = \int_0^t g(u)du$. Let A be an FV process with continuous paths. Then*

$$\begin{aligned} h(A_t) - h(A_0) &= \int_{A_0}^{A_t} g(u)du \\ &= \int_0^t g(A_s)dA_s. \end{aligned}$$

We conclude this section with an example. Let N be a Poisson process with parameter λ. Then $M_t = N_t - \lambda t$, the compensated Poisson process, is a martingale, as well as an FV process. For a bounded (say), jointly measurable process H, we have

$$\begin{aligned} I_t = \int_0^t H_s dM_s &= \int_0^t H_s d(N_s - \lambda s) \\ &= \int_0^t H_s dN_s - \lambda \int_0^t H_s ds \\ &= \sum_{n=1}^{\infty} H_{T_n} 1_{\{t \geq T_n\}} - \lambda \int_0^t H_s ds \end{aligned}$$

where $(T_n)_{n\geq 1}$ are the arrival times of the Poisson process N. Now suppose the process H is bounded, adapted, and has continuous sample paths. We then have, for $0 \leq s < t < \infty$:

$$\begin{aligned} E\{I_t - I_s|\mathcal{F}_s\} &= E\{\int_s^t H_u dM_u|\mathcal{F}_s\} \\ &= E\{\lim_{n\to\infty} \sum_{t_k, t_{k+1}\in\pi_n} H_{t_k}(M_{t_{k+1}} - M_{t_k})|\mathcal{F}_s\} \\ &= \lim_{n\to\infty} \sum_{t_k, t_{k+1}\in\pi_n} E\{E\{H_{t_k}(M_{t_{k+1}} - M_{t_k})|\mathcal{F}_{t_k}\}|\mathcal{F}_s\} \\ &= 0. \end{aligned}$$

The interchange of limits can be justified by the dominated convergence theorem. We conclude that the integral process I is a martingale. This fact, that the stochastic Stieltjes integral of an adapted, bounded, continuous process with respect to a martingale is again a martingale, is true in much greater generality. We shall treat this systematically in Chap. II.

7. Naive Stochastic Integration Is Impossible

In Sect. 6 we saw that for an FV process A, and a continuous integrand process H, we could express the integral $\int_0^t H_s dA_s$ as the limit of sums (Theorem 49). The Brownian motion process, B, however, has paths of infinite variation on compacts. In this section we demonstrate, with the aid of the Banach-Steinhaus Theorem[4], some of the difficulties that are inherent in trying to extend the notion of Stieltjes integration to processes that have paths of unbounded variation, such as Brownian motion. For the reader's convenience we recall the Banach-Steinhaus Theorem:

Theorem 51. *Let X be a Banach space and let Y be a normed linear space. Let $\{T_\alpha\}$ be a family of bounded linear operators from X into Y. If for each $x \in X$ the set $\{T_\alpha x\}$ is bounded, then the set $\{\|T_\alpha\|\}$ is bounded.*

Let us put aside stochastic processes for the moment. Let $x(t)$ be a right continuous function on $[0,1]$, and let π_n be a refining sequence of dyadic rational partitions of $[0,1]$ with $\lim_{n\to\infty} \operatorname{mesh}(\pi_n) = 0$. We ask the question: what conditions on x are needed so that the sums

$$(*) \qquad S_n = \sum_{t_k, t_{k+1} \in \pi_n} h(t_k)(x(t_{k+1}) - x(t_k))$$

converge to a finite limit as $n \to \infty$ for all continuous functions h? From Theorem 49 of Sect. 6 we know that x of finite variation is sufficient. However it is also necessary:

Theorem 52. *If the sums S_n of $(*)$ converge to a limit for every continuous function h then x is of finite variation.*

Proof. Let X be the Banach space of continuous functions equipped with the supremum norm. Let Y be $\mathbb{R}$, equipped with absolute value as the norm. For $h \in X$, let

$$T_n(h) = \sum_{t_k, t_{k+1} \in \pi_n} h(t_k)(x(t_{k+1}) - x(t_k)).$$

For each fixed n it is simple to construct an h in X such that $h(t_k) = \operatorname{sign}\{x(t_{k+1}) - x(t_k)\}$, and $\|h\| = 1$. For such an h we have

$$T_n(h) = \sum_{t_k, t_{k+1} \in \pi_n} |x(t_{k+1}) - x(t_k)|.$$

[4] The Banach-Steinhaus Theorem is also known as the Principle of Uniform Boundedness.

Therefore

$$\|T_n\| \geq \sum_{t_k, t_{k+1} \in \pi_n} |x(t_{k+1}) - x(t_k)|,$$

each n, and $\sup_n \|T_n\| \geq$ Total Variation of x. On the other hand for each $h \in X$ we have $\lim_{n\to\infty} T_n(h)$ exists and therefore $\sup_n \|T_n(h)\| < \infty$. The Banach-Steinhaus theorem then implies that $\sup_n \|T_n\| < \infty$, hence the total variation of x is finite. □

Returning to stochastic processes, we might hope to circumvent the limitations imposed by Theorem 52 by appealing to convergence in probability: that is, if $X(s,\omega)$ is right continuous (or even continuous), can we have the sums

$$(**) \qquad \sum_{t_k, t_{k+1} \in \pi_n[0,1]} H_{t_k}(X_{t_{k+1}} - X_{t_k})$$

converging to a limit *in probability* for every continuous process H? Unfortunately the answer is that X must still have paths of finite variation, a.s. The reason is that one can make the procedure used in the proof of Theorem 52 measurable in ω, and hence a subsequence of the sums in (**) can be made to converge a.s. to $+\infty$ on the set where X is not of finite variation. If this set has positive probability, the sums cannot converge in probability.

The preceding discussion makes it appear impossible to develop a coherent notion of a stochastic integral $\int_0^t H_s X_s$ when X is a process with paths of infinite variation on compacts; for example a Brownian motion. Nevertheless this is precisely what we will do in Chap. II.

Bibliographic Notes

The basic definitions and notation presented here have become fundamental to the modern study of stochastic processes, and they can be found many places, such as Dellacherie-Meyer [**1**], Doob [**1**], and Jacod-Shiryaev [**1**]. Theorem 3 is true in much greater generality: For example the hitting time of a Borel set is a stopping time. This result is very difficult, and proofs can be found in Dellacherie [**1**] or [**2**].

The resume of martingale theory consists of standard theorems. The reader does not need results from martingale theory beyond what is presented here. Those proofs not given can be found in many places, for example Breiman [**1**], Dellacherie-Meyer [**2**], or Ethier-Kurtz [**1**].

The Poisson process and Brownian motion are the two most important stochastic processes for the theory of stochastic integration. Our treatment of the Poisson process follows Çinlar [**1**]. Theorem 25 is in Bremaud [**1**], and is due to Sam Lazaro. The facts about Brownian motion needed for the theory of stochastic integration are the only ones presented here. A good source for more detailed information on Brownian motion is Hida [**1**].

Lévy processes (processes with stationary and independent increments) are a crucial source of examples for the theory of semimartingales and stochastic integrals. Indeed in large part the theory is abstracted from the properties of these processes. There do not seem to be many presentations of Lévy processes concerned with their properties which are relevant to stochastic integration beyond that of Jacod-Shiryaev [**1**]. Our approach is inspired by Bretagnolle [**1**]. Local martingales were first proposed by K. Itô and S. Watanabe [**1**] in order to generalize the Doob-Meyer decomposition. Standard Stieltjes integration applied to finite variation stochastic processes was not well known before the fundamental work of Meyer [**8**]. Finally, the idea of using the Banach-Steinhaus theorem to show that naive stochastic integration is impossible is due to Meyer [**15**].

CHAPTER II

Semimartingales and Stochastic Integrals

1. Introduction to Semimartingales

The purpose of the theory of *stochastic integration* is to give a reasonable meaning to the idea of a differential to as wide a class of stochastic processes as possible. We saw in Sect. 7 of Chap. I that using Stieltjes integration on a path-by-path basis excludes such fundamental processes as Brownian motion, and martingales in general. Markov processes also in general have paths of unbounded variation and are similarly excluded. Therefore we must find an approach more general than Stieltjes integration.

We will define stochastic integrals first as a limit of sums. *A priori* this seems hopeless, since even by restricting our integrands to *continuous* processes we saw as a consequence of Theorem 52 of Chap. I that the differential must be of finite variation on compacts. However an analysis of the proof of Theorem 52 offers some hope: in order to construct a function h such that $h(t_k) = \text{sign}(x(t_{k+1}) - x(t_k))$, we need to be able to "see" the trajectory of x on $(t_k, t_{k+1}]$. The idea of K. Itô was to restrict the integrands to those that could not see into the future increments: that is, adapted processes.

The foregoing considerations lead us to define the stochastic processes that will serve as differentials as those that are "good integrators" on an appropriate class of adapted processes. We will, as discussed in Chap. I, assume that we are given a filtered, complete probability space $(\Omega, \mathcal{F}, (\mathcal{F}_t)_{t\geq 0}, P)$ satisfying the *usual hypotheses*.

Definition. A process H is said to be **simple predictable** if H has a representation

$$H_t = H_0 1_{\{0\}}(t) + \sum_{i=1}^{n} H_i 1_{(T_i, T_{i+1}]}(t)$$

where $0 = T_1 \leq \cdots \leq T_{n+1} < \infty$ is a finite sequence of stopping times, $H_i \in \mathcal{F}_{T_i}$ with $|H_i| < \infty$ a.s., $0 \leq i \leq n$. The collection of simple predictable processes is denoted **S**.

Note that we can take $T_1 = T_0 = 0$ in the above definition, so there is no "gap" between T_0 and T_1. We can topologize $\mathbf{S}$ by uniform convergence in (t, ω), and we denote $\mathbf{S}$ endowed with this topology by $\mathbf{S_u}$. We also write $\mathbf{L}^0$ for the space of finite-valued random variables *topologized by convergence in probability*.

Let X be a stochastic process. An operator, I_X, induced by X should have two fundamental properties to earn the name "integral". The operator I_X should be linear, and it should satisfy some version of the bounded convergence theorem. A particularly weak form of the bounded convergence theorem is that the uniform convergence of processes H^n to H implies only the convergence in probability of $I_X(H^n)$ to $I_X(H)$.

Inspired by the above considerations, for a given process X we define a linear mapping $I_X : \mathbf{S} \to \mathbf{L}^0$ as follows:

$$I_X(H) = H_0 X_0 + \sum_{i=1}^{n} H_i (X_{T_{i+1}} - X_{T_i}),$$

where $H \in \mathbf{S}$ has the representation

$$H_t = H_0 1_{\{0\}} + \sum_{i=1}^{n} H_i 1_{(T_i, T_{i+1}]}.$$

Since this definition is a path by path definition for the step functions $H_{.}(\omega)$, it does not depend on the choice of the representation of H in $\mathbf{S}$.

Definition. A process X is a **total semimartingale** if X is càdlàg, adapted, and $I_X : \mathbf{S}_u \to \mathbf{L}^0$ is continuous.

Recall that for a process X and a stopping time T, the notation X^T denotes the process $(X_{t \wedge T})_{t \geq 0}$.

Definition. A process X is called a **semimartingale** if, for each $t \in [0, \infty)$, X^t is a total semimartingale.

We postpone consideration of examples of semimartingales to Sect. 3.

2. Stability Properties of Semimartingales

We state a sequence of theorems giving some of the stability results which are particularly simple.

Theorem 1. *The set of (total) semimartingales is a vector space.*

Proof. This is immediate from the definition. □

Theorem 2. *If Q is a probability and absolutely continuous with respect to P, then every (total) P-semimartingale X is a (total) Q-semimartingale.*

Proof. Convergence in P-probability implies convergence in Q-probability. Thus the theorem follows from the definition of X. □

Theorem 3. *Let $(P_k)_{k\geq 1}$ be a sequence of probabilities such that X is a P_k-semimartingale for each k. Let $R = \sum_{k=1}^{\infty} \lambda_k P_k$, where $\lambda_k \geq 0$, each k, and $\sum_{k=1}^{\infty} \lambda_k = 1$. Then X is a semimartingale under R as well.*

Proof. Suppose $H^n \in \mathbf{S}$ converges uniformly to $H \in \mathbf{S}$. Since X is a P_k-semimartingale for all P_k, $I_X(H^n)$ converges to $I_X(H)$ in probability for every P_k. This then implies $I_X(H^n)$ converges to $I_X(H)$ under R. □

Theorem 4 (Stricker's Theorem). *Let X be a semimartingale for the filtration $(\mathcal{F}_t)_{t\geq 0}$. Let $(\mathcal{G}_t)_{t\geq 0}$ be a subfiltration of $(\mathcal{F}_t)_{t\geq 0}$, such that X is adapted to the $\mathcal{G}$-filtration. Then X is a $\mathcal{G}$-semimartingale.*

Proof. For a filtration $\mathcal{H}$, let $\mathbf{S}(\mathcal{H})$ denote the simple predictable processes for the filtration $\mathcal{H} = (\mathcal{H}_t)_{t\geq 0}$. In this case we have $\mathbf{S}(\mathcal{G})$ is contained in $\mathbf{S}(\mathcal{F})$. The theorem is then an immediate consequence of the definition. □

Theorem 4 shows that we can always shrink a filtration and preserve the property of being a semimartingale (as long as the process X is still adapted), since we are shrinking as well the possible integrands; this, in effect, makes it "easier" for the process X to be a semimartingale. Expanding the filtration, therefore, should be – and is – a much more delicate issue. We present here an elementary but useful result. Recall that we are given a filtered space $(\Omega, \mathcal{F}, (\mathcal{F}_t)_{t\geq 0}, P)$ satisfying the usual hypotheses.

Theorem 5. *Let $\mathcal{A}$ be a collection of events in $\mathcal{F}$ such that if $A_\alpha, A_\beta \in \mathcal{A}$ then $A_\alpha \cap A_\beta = \phi$ ($\alpha \neq \beta$). Let $\mathcal{H}_t$ be the filtration generated by $\mathcal{F}_t$ and $\mathcal{A}$. Then every $((\mathcal{F}_t)_{t\geq 0}, P)$-semimartingale is an $((\mathcal{H}_t)_{t\geq 0}, P)$-semimartingale also.*

Proof. Let $A_n \in \mathcal{A}$. If $P(A_n) = 0$, then A_n and A_n^c are in $\mathcal{F}_0$ by hypothesis. We assume, therefore, that $P(A_n) > 0$. Note that there can be at most a countable number of $A_n \in \mathcal{A}$ such that $P(A_n) > 0$. If $\Lambda = \bigcup_{n\geq 1} A_n$ is the union of all $A_n \in \mathcal{A}$ with $P(A_n) > 0$, we can also add Λ^c to $\mathcal{A}$ without loss of generality. Thus we can assume that $\mathcal{A}$ is a countable partition of Ω with $P(A_n) > 0$ for every $A_n \in \mathcal{A}$. Define a new probability Q_n by $Q_n(\cdot) = P(\cdot|A_n)$, for A_n fixed. Then $Q_n \ll P$, and X is a $((\mathcal{F}_t)_{t\geq 0}, Q_n)$-semimartingale by Theorem 2. If we enlarge the filtration $(\mathcal{F}_t)_{t\geq 0}$ by all the $\mathcal{F}$-measurable events that have Q_n-probability 0 or 1, we get a larger filtration $(\mathcal{J}_t^n)_{t\geq 0}$, and X is a $(\mathcal{J}_t^n, Q_n)$-semimartingale. Since $Q_n(A_m) = 0$

or 1 for $m \neq n$, we have $\mathcal{F}_t \subset \mathcal{H}_t \subset \mathcal{J}_t^n$, for $t \geq 0$, and for all n. By Stricker's theorem (Theorem 4) we conclude that X is an $((\mathcal{H}_t)_{t\geq 0}, Q_n)$-semimartingale. Finally, we have $dP = \sum_{n\geq 1} P(A_n)dQ_n$, where X is an $((\mathcal{H}_t)_{t\geq 0}, Q_n)$-semimartingale for each n. Therefore by Theorem 3 we conclude X is an $((\mathcal{H}_t)_{t\geq 0}, P)$-semimartingale. □

Corollary. *Let $\mathcal{A}$ be a finite collection of events in $\mathcal{F}$, and let $(\mathcal{H}_t)_{t\geq 0}$ be the filtration generated by $\mathcal{F}_t$ and $\mathcal{A}$. Then every $((\mathcal{F}_t)_{t\geq 0}, P)$-semimartingale is an $((\mathcal{H}_t)_{t\geq 0}, P)$-semimartingale also.*

Proof. Since $\mathcal{A}$ is finite, one can always find a (finite) partition $\mathcal{B}$ of Ω such that $\mathcal{H}_t = \mathcal{F}_t \vee \mathcal{B}$. The Corollary then follows by Theorem 5. □

Note that if $B = (B_t)_{t\geq 0}$ is a Brownian motion for a filtration $(\mathcal{F}_t)_{t\geq 0}$, by Theorem 5 we are able to add, in a certain manner, an infinite number of "future" events to the filtration and B will no longer be a martingale, but it will stay a semimartingale. This has interesting implications in Finance Theory (the theory of continuous trading). See for example Duffie-Huang [1].

The Corollary of the next theorem states that being a semimartingale is a "local" property; that is, a local semimartingale is a semimartingale. We get a stronger result by stopping at T_n- rather than at T_n in the next theorem. A process X is stopped at $T-$ if $X_t^{T-} = X_t 1_{(0\leq t<T)} + X_{T-}1_{(t\geq T)}$, where $X_{0-} = 0$.

Theorem 6. *Let X be a càdlàg, adapted process; let (T_n) be a sequence of positive r.v. increasing to ∞ a.s.; and let (X^n) be a sequence of semimartingales such that, for each n, $X^{T_n-} = (X^n)^{T_n-}$. Then X is a semimartingale.*

Proof. We wish to show X^t is a total semimartingale, each $t > 0$. Define $R_n = T_n 1_{(T_n\leq t)} + \infty 1_{(T_n>t)}$. Then

$$P\{|I_{X^t}(H)| \geq c\} \leq P\{|I_{(X^n)^t}(H)| \geq c\} + P(R_n < \infty).$$

But $P(R_n < \infty) = P(T_n \leq t)$, and since T_n increases to ∞ a.s., $P(T_n \leq t) \to 0$ as $n \to \infty$. Thus if H^k tends to 0 in $\mathbf{S}_u$, given $\epsilon > 0$, we choose n so that $P(R_n < \infty) < \epsilon/2$, and then choose k so large that $P\{|I_{(X^n)^t}(H^k)| \geq c\} < \epsilon/2$. Thus, for k large enough, $P\{|I_{X^t}(H^k)| \geq c\} < \epsilon$. □

Corollary. *Let X be a process. If there exists a sequence (T_n) of stopping times increasing to ∞ a.s., such that X^{T_n} (or $X^{T_n}1_{\{T_n>0\}}$) is a semimartingale, each n, then X is also a semimartingale.*

3. Elementary Examples of Semimartingales

The elementary properties of semimartingales established in Sect. 2 will allow us to see that many common processes are semimartingales. For example, the Poisson process, Brownian motion, and more generally all Lévy processes are semimartingales.

Theorem 7. *Each adapted process with càdlàg paths of finite variation on compacts (of finite total variation) is a semimartingale (a total semimartingale).*

Proof. It suffices to observe that $|I_X(H)| \leq \|H\|_u \int_0^\infty |dX_s|$, where $\int_0^\infty |dX_s|$ denotes the Lebesgue-Stieltjes total variation and where $\|H\|_u = \sup_{(t,w)} |H(t,w)|$. □

Theorem 8. *Each square integrable martingale with càdlàg paths is a semimartingale.*

Proof. Let X be a square integrable martingale with $X_0 = 0$, and let $H \in \mathbf{S}$. It suffices to observe that

$$\begin{aligned} E\{(I_X(H))^2\} &= E\{(\sum_{i=0}^{n} H_i(X_{T_{i+1}} - X_{T_i}))^2\} \\ &= E\{\sum_{i=0}^{n} H_i^2(X_{T_{i+1}} - X_{T_i})^2\} \\ &\leq \|H\|_u^2 E\{\sum_{i=0}^{n}(X_{T_{i+1}} - X_{T_i})^2\} \\ &= \|H\|_u^2 E\{\sum_{i=0}^{n}(X_{T_{i+1}}^2 - X_{T_i}^2)\} \\ &= \|H\|_u^2 E\{X_{T_{n+1}}^2\} \\ &\leq \|H\|_u^2 E\{X_\infty^2\}. \end{aligned}$$

□

Corollary 1. *Each càdlàg, locally square integrable local martingale is a semimartingale.*

Proof. Apply Theorem 8 together with the Corollary to Theorem 6. □

Corollary 2. *A local martingale with continuous paths is a semimartingale.*

Proof. Apply Corollary 1 together with Theorem 47 in Chap. I. □

Corollary 3. *The Wiener process (that is, Brownian motion) is a semimartingale.*

Proof. The Wiener process is a martingale with continuous paths if W_0 is integrable. It is always a continuous local martingale. □

Definition. We will say an adapted process X with càdlàg paths is **decomposable** if it can be decomposed $X_t = X_0 + M_t + A_t$, where $M_0 = A_0 = 0$, M is a locally square integrable martingale, and A is càdlàg, adapted, with paths of finite variation on compacts.

Theorem 9. *A decomposable process is a semimartingale.*

Proof. Let $X_t = X_0 + M_t + A_t$ be a decomposition of X. Then M is a semimartingale by Corollary 1 of Theorem 8, and A is a semimartingale by Theorem 7. Since semimartingales form a vector space (Theorem 1) we have the result. □

Corollary. *A Lévy process is a semimartingale.*

Proof. By Theorem 40 of Chap. I we know that a Lévy process is decomposable. Theorem 9 then gives the result. □

Since Lévy processes are prototypic strong Markov processes, one may well wonder if all $\mathbb{R}^n$-valued strong Markov processes are semimartingales. Simple examples, such as $X_t = B_t^{1/3}$, where B is standard Brownian motion, show this is not the case (while this example is simple, the proof that X is not a semimartingale is not elementary). However if one is willing to "regularize" the Markov process by a transformation of the space (in the case of this example using the "scale function" $S(x) = x^3$), "most reasonable" strong Markov processes are semimartingales. Indeed, Dynkin's formula, which states that if f is in the domain of the infinitesimal generator G of the strong Markov process Z, then the process

$$M_t = f(Z_t) - f(Z_0) - \int_0^t Gf(Z_s)ds$$

is well defined and is a local martingale, hints strongly that if the domain of G is rich enough, the process Z is a semimartingale. In this regard see section seven of Çinlar, Jacod, Protter, and Sharpe [1].

4. Stochastic Integrals

In Sect. 1 we defined semimartingales as adapted, càdlàg processes that acted as "good integrators" on the simple predictable processes. We now wish to enlarge the space of processes we can consider as integrands. In Chap. IV we will consider a large class of processes, namely those that are "predictably measurable" and have appropriate finiteness properties. Here, however, by keeping our space of integrands small – yet large enough to be interesting – we can keep the theory free of technical problems, as well as intuitive.

A particularly nice class of processes for our purposes is the class of adapted processes with left continuous paths that have right limits (the French acronym would be càglàd).

Definitions. We let: $\mathbb{D}$ denote the space of adapted processes with càdlàg paths; $\mathbb{L}$ denote the space of adapted processes with càglàd paths (left continuous with right limits); $b\mathbb{L}$ denote processes in $\mathbb{L}$ with bounded paths.

We have previously considered $\mathbf{S}_u$, the space of simple, predictable processes endowed with the topology of uniform convergence; and $\mathbf{L}^0$, the space of finite-valued random variables topologized by convergence in probability. We need to consider a third type of convergence:

Definition. A sequence of processes $(H^n)_{n\geq 1}$ converges to a process H **uniformly on compacts in probability** (abbreviated **ucp**) if, for each $t > 0$, $\sup_{0\leq s\leq t}|H^n_s - H_s|$ converges to 0 in probability.

We write $H^*_t = \sup_{0\leq s\leq t}|H_s|$. Then if $Y^n \in \mathbb{D}$ we have $Y^n \to Y$ in ucp if $(Y^n - Y)^*_t$ converges to 0 in probability for each $t > 0$. We write $\mathbb{D}_{ucp}$, $\mathbb{L}_{ucp}$, $\mathbf{S}_{ucp}$ to denote the respective spaces endowed with the *ucp* topology. We observe that $\mathbb{D}_{ucp}$ is a metrizable space; indeed, a compatible metric is given by $(X, Y \in \mathbb{D})$:

$$d(X,Y) = \sum_{n=1}^{\infty} \frac{1}{2^n} E\{\min(1, (X-Y)^*_n)\}.$$

The above metric space $\mathbb{D}_{ucp}$ is complete. For a semimartingale X and a process $H \in \mathbf{S}$, we have defined the appropriate notion $I_X(H)$ of a stochastic integral. The next result is key to extending this definition.

Theorem 10. *The space* $\mathbf{S}$ *is dense in* $\mathbb{L}$ *under the ucp topology.*

Proof. Let $Y \in \mathbb{L}$. Let $R_n = \inf\{t : |Y_t| > n\}$. Then R_n is a stopping time and $Y^n = Y^{R_n}1_{\{R_n>0\}}$ are in $b\mathbb{L}$ and converge to Y in ucp. Thus $b\mathbb{L}$ is dense in $\mathbb{L}$. Without loss we now assume $Y \in b\mathbb{L}$. Define Z by $Z_t = \lim_{\substack{u\to t\\ u>t}} Y_u$. Then $Z \in \mathbb{D}$. For $\epsilon > 0$, define

$$T^\epsilon_0 = 0$$
$$T^\epsilon_{n+1} = \inf\{t : t > T^\epsilon_n \text{ and } |Z_t - Z_{T^\epsilon_n}| > \epsilon\}.$$

Since Z is càdlàg, the T^ϵ_n are stopping times increasing to ∞ a.s. as n increases. Let $Z^\epsilon = \sum_n Z_{T^\epsilon_n} 1_{[T^\epsilon_n, T^\epsilon_{n+1})}$, for each $\epsilon > 0$. Then Z^ϵ are bounded and converge uniformly to Z as ϵ tends to 0. Let $U^\epsilon = Y_0 1_{\{0\}} + \sum_n Z_{T_n^\epsilon} 1_{(T_n^\epsilon, T_{n+1}^\epsilon]}$. Then $U^\epsilon \in \mathbf{bL}$ and the preceding implies U^ϵ converges uniformly on compacts to $Y_0 1_{\{0\}} + Z_- = Y$.

Finally, define

$$Y^{n,\epsilon} = Y_0 1_{\{0\}} + \sum_{i=1}^{n} Z_{T_i^\epsilon} 1_{(T_i^\epsilon \wedge n, T_{i+1}^\epsilon \wedge n]}$$

and this can be made arbitrarily close to $Y \in \mathbf{bL}$ by taking ϵ small enough and n large enough. □

We defined a semimartingale X as a process that induced a continuous operator, I_X, from $\mathbf{S}_u$ into $\mathbf{L}^0$. This I_X maps *processes* into *random variables.* We next define an operator (the stochastic integral operator), induced by X, that will map *processes* into *processes.*

Definition. For $H \in \mathbf{S}$ and X a càdlàg process, *define the (linear) mapping* $J_X : \mathbf{S} \to \mathbf{D}$ by:

$$J_X(H) = H_0 X_0 + \sum_{i=1}^{n} H_i (X^{T_{i+1}} - X^{T_i})$$

for H in $\mathbf{S}$ with the representation

$$H = H_0 1_{\{0\}} + \sum_{i=1}^{n} H_i 1_{(T_i, T_{i+1}]},$$

$H_i \in \mathcal{F}_{T_i}$ and $0 = T_0 \le T_1 \le \cdots \le T_{n+1} < \infty$ stopping times.

Definition. For $H \in \mathbf{S}$ and X an adapted càdlàg process, we call $J_X(H)$ **the stochastic integral of H with respect to X.**

We use interchangeably three notations for the stochastic integral:

$$J_X(H) = \int H_s dX_s = H \cdot X.$$

Observe that $J_X(H)_t = I_{X^t}(H)$. Indeed, I_X plays the role of a definite integral: For $H \in \mathbf{S}$, $I_X(H) = \int_0^\infty H_s dX_s$.

Theorem 11. *Let X be a semimartingale. Then the mapping $J_X : \mathbf{S}_{ucp} \to \mathbf{D}_{ucp}$ is continuous.*

Proof. Since we are only dealing with convergence on compact sets, without loss of generality we take X to be a total semimartingale. First suppose H^k

in $\mathbf{S}$ tends to 0 uniformly and is uniformly bounded. We will show $J_X(H^k)$ tends to 0 ucp. Let $\delta > 0$ be given and define stopping times T^k by

$$T^k = \inf\{t : |(H^k \cdot X)_t| \geq \delta\}.$$

Then $H^k 1_{[0,T^k]} \in \mathbf{S}$ and tends to 0 uniformly as k tends to ∞. Thus for every t,

$$\begin{aligned} P\{(H^k \cdot X)_t^* > \delta\} &\leq P\{|H^k \cdot X_{T^k \wedge t}| \geq \delta\} \\ &= P\{|(H^k 1_{[0,T^k]} \cdot X)_t| \geq \delta\} \\ &= P\{|I_X(H^k 1_{[0,T^k \wedge t]})| \geq \delta\} \end{aligned}$$

which tends to 0 by the definition of total semimartingale.

We have just shown that $J_X : \mathbf{S_u} \to \mathbf{D}_{ucp}$ is continuous. We now use this to show $J_X : \mathbf{S}_{ucp} \to \mathbf{D}_{ucp}$ is continuous. Suppose H^k goes to 0 ucp. Let $\delta > 0$, $\epsilon > 0$, $t > 0$. We have seen that there exists η such that $\|H\|_u \leq \eta$ implies $P(J_X(H)_t^* > \delta) < \frac{\epsilon}{2}$. Let $R_k = \inf\{s : |H_s^k| > \eta\}$, and set $\widetilde{H}^k = H^k 1_{[0,R_k]} 1_{\{R_k > 0\}}$. Then $\widetilde{H}^k \in \mathbf{S}$ and $\|\widetilde{H}^k\|_u \leq \eta$ by left continuity. Since $R^k \geq t$ implies $(\widetilde{H}^k \cdot X)_t^* = (H^k \cdot X)_t^*$, we have

$$\begin{aligned} P((H^k \cdot X)_t^* > \delta) &\leq P((\widetilde{H}^k \cdot X)_t^* > \delta) + P(R^k < t) \\ &\leq \frac{\epsilon}{2} + P((H^k)_t^* > \eta) \\ &< \epsilon \end{aligned}$$

if k is large enough, since $\lim_{k\to\infty} P((H^k)_t^* > \eta) = 0$. □

We have seen that when X is a semimartingale, the integration operator J_X is continuous on $\mathbf{S}_{ucp}$, and also that $\mathbf{S}_{ucp}$ is dense in $\mathbf{L}_{ucp}$. Hence we are able to extend the linear integration operator J_X from $\mathbf{S}$ to $\mathbf{L}$ by continuity, since $\mathbf{D}_{ucp}$ is a complete metric space.

Definition. Let X be a semimartingale. The continuous linear mapping $J_X : \mathbf{L}_{ucp} \to \mathbf{D}_{ucp}$ obtained as the extension of $J_X : \mathbf{S} \to \mathbf{D}$ is called the **stochastic integral**.

The preceding definition is rich enough for us to give an *immediate example* of a surprising stochastic integral. First recall that if a process $(A_t)_{t\geq 0}$ has continuous paths of finite variation with $A_0 = 0$, then the Riemann-Stieltjes integral of $\int_0^t A_s dA_s$ yields the formula (see Theorem 50 of Chap. I):

$$(*) \qquad \int_0^t A_s dA_s = \frac{1}{2} A_t^2.$$

Let us now consider a standard Brownian motion $B = (B_t)_{t\geq 0}$ with $B_0 = 0$. The process B does not have paths of finite variation on compacts, but it is a semimartingale. Let (π_n) be a refining sequence of partitions of $[0,\infty)$ with $\lim_{n\to\infty}\text{mesh}(\pi_n) = 0$. Let $B_t^n = \sum_{t_k\in\pi_n} B_{t_k} 1_{(t_k,t_{k+1}]}$. Then $B^n \in \mathbf{L}$ for each n. Moreover, B^n converges to B in ucp. Fix $t \geq 0$ and assume that t is a partition point of each π_n. Then

$$J_B(B^n)_t = \sum_{\substack{t_k\in\pi_n\\ t_k<t}} B_{t_k}(B^{t_{k+1}} - B^{t_k})$$

and

$$\begin{aligned} J_B(B)_t &= \lim_{n\to\infty} J_B(B^n)_t \\ &= \lim_n \sum_{t_k\in\pi_n, t_k<t} B_{t_k}(B_{t_{k+1}} - B_{t_k}) \\ &= \lim_n \sum_{t_k\in\pi_n, t_k<t} \Big\{\frac{1}{2}(B_{t_{k+1}} + B_{t_k})(B_{t_{k+1}} - B_{t_k}) \\ &\quad - \frac{1}{2}(B_{t_{k+1}} - B_{t_k})(B_{t_{k+1}} - B_{t_k})\Big\} \\ &= \frac{1}{2}B_t^2 - \frac{1}{2}\lim_n \sum_{t_k\in\pi_n} (B_{t_{k+1}} - B_{t_k})^2 \end{aligned}$$

where we have used telescoping sums. The last term on the right, however, converges a.s. to t, by Theorem 28 of Chap. I. We therefore conclude that

$$(**) \qquad \int_0^t B_s dB_s = \frac{1}{2}B_t^2 - \frac{1}{2}t,$$

a formula distinctly different from the Riemann-Stieltjes formula (*). The change of variables formula (Theorem 32) presented in Sect. 7, will give a deeper understanding of the difference in formulas (*) and (**).

5. Properties of Stochastic Integrals

Throughout this paragraph X will denote a semimartingale and H will denote an element of $\mathbf{L}$. Recall that the stochastic integral defined in Sect. 4 will be denoted by the three notations $J_X(H) = H \cdot X = \int H_s dX_s$. Evaluating these processes at t, we have

$$H \cdot X_t = \int_0^t H_s dX_s = \int_{[0,t]} H_s dX_s.$$

To exclude 0 in the integral we write

$$\int_{0+}^t H_s dX_s = \int_{(0,t]} H_s dX_s.$$

The integral $\int_0^\infty H_s dX_s$ is defined to be $\lim_{t\to\infty} \int_0^t H_s dX_s$ when the limit exists. Note that $\int_0^t H_s dX_s = H_0 X_0 + \int_{0+}^t H_s dX_s$. For a process $Y \in \mathsf{D}$, we recall that $\Delta Y_t = Y_t - Y_{t-}$, the jump at t. Also since $Y_{0-} = 0$ we have $\Delta Y_0 = Y_0$. Recall further that for a process Z and stopping time T, we let Z^T denote the stopped process $(Z_t^T) = Z_{t\wedge T}$. We will establish in this section several elementary properties of the stochastic integral. These properties will help us to understand the integral and to examine examples at the end of the section. Two processes Y and Z are **indistinguishable** if $P\{\omega : t \to X_t(\omega) \text{ and } t \to Y_t(\omega) \text{ are the same functions}\} = 1$.

Theorem 12. *Let T be a stopping time. Then $(H \cdot X)^T = H1_{[0,T]} \cdot X = H \cdot (X^T)$.*

Theorem 13. *The jump process $\Delta(H{\cdot}X)_s$ is indistinguishable from $H_s(\Delta X_s)$.*

Proofs. Both properties are clear when $H \in \mathsf{S}$, and they follow when $H \in \mathsf{L}$ by passing to the limit (convergence in ucp). □

Let Q denote another probability law, and let $H_Q{\cdot}X$ denote the stochastic integral of H with respect to X computed under the law Q.

Theorem 14. *Let $Q \ll P$. Then $H_Q{\cdot}X$ is Q-indistinguishable from $H_P{\cdot}X$.*

Proof. Note that by Theorem 2, X is a Q-semimartingale. The theorem is clear if $H \in \mathsf{S}$, and it follows for $H \in \mathsf{L}$ by passage to the limit in the ucp topology, since convergence in P-probability implies convergence in Q-probability. □

Theorem 15. *Let P_k be a sequence of probabilities such that X is a P_k-semimartingale for each k. Let $R = \sum_{k=1}^\infty \lambda_k P_k$ where $\lambda_k \geq 0$, each k, and $\sum_{k=1}^\infty \lambda_k = 1$. Then $H_R \cdot X = H_{P_k} \cdot X$, P_k a.s., for all k such that $\lambda_k > 0$.*

Proof. If $\lambda_k > 0$ then $P_k \ll R$, and the result follows by Theorem 14. Note that by Theorem 3 we know that X is an R-semimartingale. □

Corollary. *Let P and Q be any probabilities and suppose X is a semimartingale relative to both P and Q. Then there exists a process $H \cdot X$ which is a version of both $H_P \cdot X$ and $H_Q \cdot X$.*

Proof. Let $R = \frac{P+Q}{2}$. Then $H_R \cdot X$ is such a process by Theorem 15. □

Theorem 16. *Let $(\mathcal{G}_t)_{t\geq 0}$ be another filtration such that H is in both $\mathsf{L}(\mathcal{G})$ and $\mathsf{L}(\mathcal{F})$, and such that X is also a $\mathcal{G}$-semimartingale. Then $H_\mathcal{G} \cdot X = H_\mathcal{F} \cdot X$.*

Proof. $\mathsf{L}(\mathcal{G})$ denotes left continuous processes adapted to the filtration $(\mathcal{G}_t)_{t\geq 0}$. As in the proof of Theorem 10, we can construct a sequence of processes H^n converging to H where the construction of the H^n depends only on H. Thus $H^n \in \mathbf{S}(\mathcal{G}) \cap \mathbf{S}(\mathcal{F})$ and converges to H in ucp. Since the result is clear for $\mathbf{S}$, the full result follows by passing to the limit. □

Remark. While Theorem 16 is a simple result in this context, it is far from simple if the integrands are predictably measurable processes, rather than processes in $\mathbb{L}$. See the comment following Theorem 33 of Chap. IV.

The next two theorems are especially interesting because they show – at least for integrands in $\mathbb{L}$ – that the stochastic integral agrees with the path-by-path Lebesgue-Stieltjes integral, whenever it is possible to do so.

Theorem 17. *If the semimartingale X has paths of finite variation on compacts, then $H \cdot X$ is indistinguishable from the Lebesgue-Stieltjes integral, computed path by path.*

Proof. The result is evident for $H \in \mathbf{S}$. Let $H^n \in \mathbf{S}$ converge to H in ucp. Then there exists a subsequence n_k such that $\lim_{n_k \to \infty}(H^{n_k} - H)^*_t = 0$ a.s., and the result follows by interchanging limits, justified by the uniform a.s. convergence. □

Theorem 18. *Let $X, \overline{X}$ be two semimartingales, and let $H, \overline{H} \in \mathbb{L}$. Let $A = \{\omega : H_\cdot(\omega) = \overline{H}_\cdot(\omega)$ and $X_\cdot(\omega) = \overline{X}_\cdot(\omega)\}$, and let $B = \{\omega : t \to X_t(\omega)$ is of finite variation on compacts$\}$. Then $H \cdot X = \overline{H} \cdot \overline{X}$ on A, and $H \cdot X$ is equal to a path by path Lebesgue-Stieltjes integral on B.*

Proof. Without loss of generality we assume $P(A) > 0$. Define a new probability law Q by $Q(\Lambda) = P(\Lambda | A)$. Then under Q we have that H and $\overline{H}$ as well as X and $\overline{X}$ are indistinguishable. Thus $H_Q \cdot X = \overline{H}_Q \cdot \overline{X}$, and hence $H \cdot X = \overline{H} \cdot \overline{X}$ P – a.s. on A by Theorem 14, since $Q \ll P$.

As for the second assertion, if $B = \Omega$ the result is merely Theorem 17. Define R by $R(\Lambda) = P(\Lambda | B)$, assuming without loss that $P(B) > 0$. Then $R \ll P$ and $B = \Omega$, R- a.s. Hence $H_R \cdot X$ equals the Lebesgue-Stieljes integral R- a.s. by Theorem 17, and the result follows by Theorem 14. □

The preceding theorem and following corollary are known as **the local behavior of the integral.**

Corollary. *With the notation of Theorem 18, let S, T be two stopping times with $S < T$. Define*

$$C = \{\omega : H_t(\omega) = \overline{H}_t(\omega); X_t(\omega) = \overline{X}_t(\omega); S(\omega) < t \leq T(\omega)\}$$

$$D = \{\omega : t \to X_t(\omega) \text{ is of finite variation on } S(\omega) < t < T(\omega)\}.$$

Then $H \cdot X^T - H \cdot X^S = \overline{H} \cdot \overline{X}^T - \overline{H} \cdot \overline{X}^S$ on C and $H \cdot X^T - H \cdot X^S$ equals a path by path Lebesgue-Stieltjes integral on D.

Proof. Let $Y_t = X_t - X_{t\wedge S}$. Then $H \cdot Y = H \cdot X - H \cdot X^S$, and Y does not charge the set $[0, S]$, which is evident, or which – alternatively – can be viewed as an easy consequence of Theorem 18. One now applies Theorem 18 to Y^T to obtain the result. □

Theorem 19 (Associativity). *The stochastic integral process $Y = H \cdot X$ is itself a semimartingale, and for $G \in \mathbb{L}$ we have*

$$G \cdot Y = G \cdot (H \cdot X) = (GH) \cdot X.$$

Proof. Suppose we know $Y = H \cdot X$ is a semimartingale. Then $G \cdot Y = J_Y(G)$. If G, H are in $\mathbf{S}$, then it is clear that $J_Y(G) = J_X(GH)$. The associativity then extends to $\mathbb{L}$ by continuity.

It remains to show that $Y = H \cdot X$ is a semimartingale. Let (H^n) be in $\mathbf{S}$ converging in ucp to H. Then $H^n \cdot X$ converges to $H \cdot X$ in ucp. Thus there exists a subsequence (n_k) such that $H^{n_k} \cdot X$ converges a.s. to $H \cdot X$.

Let $G \in \mathbf{S}$ and let $Y^{n_k} = H^{n_k} \cdot X$, $Y = H \cdot X$. The Y^{n_k} are semimartingales converging pointwise to the process Y. For $G \in \mathbf{S}$, $J_Y(G)$ is defined for any process Y; so we have

$$\begin{aligned} J_Y(G) = G \cdot Y &= \lim_{n_k \to \infty} G \cdot Y^{n_k} \\ &= \lim_{n_k \to \infty} G \cdot (H^{n_k} \cdot X) \\ &= \lim_{n_k \to \infty} (GH^{n_k}) \cdot X \end{aligned}$$

which equals $\lim_{n_k \to \infty} J_X(GH^{n_k}) = J_X(GH)$, since X is a semimartingale. Therefore $J_Y(G) = J_X(GH)$ for $G \in \mathbf{S}$.

Let G^n converge to G in $\mathbf{S}_u$. Then $G^n H$ converges to GH in $\mathbb{L}_{ucp}$, and since X is a semimartingale, $\lim_{n \to \infty} J_Y(G^n) = \lim_{n \to \infty} J_X(G^n H) = J_X(GH) = J_Y(G)$. This implies Y^t is a total semimartingale, and so $Y = H \cdot X$ is a semimartingale. □

Theorem 19 shows that *the property of being a semimartingale is preserved by stochastic integration.* Also by Theorem 17 if the semimartingale X is an FV process, then the stochastic integral agrees with the Lebesgue-Stieltjes integral, and by the theory of Lebesgue-Stieltjes integration we are able to conclude the stochastic integral is an FV process also; that is, *the property of being an FV process is preserved by stochastic integration.*

One may well ask if other properties are preserved by stochastic integration; in particular, are the stochastic integrals of martingales and local martingales still martingales and local martingales? Local martingales are indeed preserved by stochastic integration, but we are not yet able easily to prove it. Instead we show that *locally square integrable local martingales are preserved by stochastic integration for integrands in* $\mathbf{L}$.

Theorem 20. *Let X be a locally square integrable local martingale, and let $H \in \mathbf{L}$. Then the stochastic integral $H \cdot X$ is also a locally square integrable local martingale.*

Proof. We have seen that a locally square integrable local martingale is a semimartingale (Corollary 1 of Theorem 8), so we can formulate $H \cdot X$. Without loss of generality, assume $X_0 = 0$. Also, if T^k increases to ∞ a.s. and $(H \cdot X)^{T_k}$ is a locally square integrable local martingale for each k, it is simple to check that $H \cdot X$ itself is one. Thus without loss we assume X is a square integrable martingale. By stopping H, we may further assume H is bounded, by ℓ. Let $H^n \in \mathbf{S}$ be such that H^n converges to H in ucp. We can then modify H^n, call it $\widetilde{H}^n$, such that $\widetilde{H}^n$ is bounded by ℓ, $\widetilde{H}^n \in \mathbf{S}$, and $\widetilde{H}^n$ converges uniformly to H in probability on $[0,t]$. Since $\widetilde{H}^n \in b\mathbf{S}$, one can check that $\widetilde{H}^n \cdot X$ is a martingale. Moreover

$$
\begin{aligned}
E\{(\widetilde{H}^n \cdot X)_t^2\} &= E\{\sum_{i=1}^{k_n}(\widetilde{H}_i^n(X_t^{T_{i+1}} - X_t^{T_i}))^2\} \\
&\leq \ell^2 E\{\sum_{i=1}^{k_n}(X_{T_{i+1}}^2 - X_{T_i}^2)\} \\
&\leq \ell^2 E\{X_\infty^2\},
\end{aligned}
$$

and hence $(\widetilde{H}^n \cdot X)_t$ are uniformly bounded in L^2 and thus uniformly integrable. Passing to the limit then shows both that $H \cdot X$ is a martingale and that it is square integrable. □

In Theorem 17 of Chap. III we show the more general result that if M is a local martingale and $H \in \mathbf{L}$, then $H \cdot M$ is again a local martingale.

A classical result from the theory of Lebesgue measure and integration (on $\mathbb{R}$) is that a bounded, measurable function f mapping an interval $[a,b]$ to $\mathbb{R}$ is Riemann integrable if and only if the set of discontinuities of f has Lebesgue measure zero (eg. Kingman and Taylor [1, p. 129]). Therefore we cannot hope to express the stochastic integral as a limit of sums unless the integrands have reasonably smooth sample paths. The spaces $\mathbf{D}$ and $\mathbf{L}$ consist of processes which jump at most countably often. As we will see in Theorem 21, this is smooth enough.

Definition. Let σ denote a finite sequence of finite stopping times:

$$0 = T_0 \leq T_1 \leq \cdots \leq T_k < \infty.$$

The sequence σ is called a **random partition**. A sequence of random partitions σ_n

$$\sigma_n : T_0^n \leq T_1^n \leq \cdots \leq T_{k_n}^n$$

is said to **tend to the identity** if

(i) $\lim_n \sup_k T_k^n = \infty$ a.s.
(ii) $\|\sigma_n\| = \sup_k |T_{k+1}^n - T_k^n|$ converges to 0 a.s.

Let Y be a process and let σ be a random partition. We define the process Y sampled at σ to be:

$$Y^\sigma \equiv Y_0 1_{\{0\}} + \sum_k Y_{T_k} 1_{(T_k, T_{k+1}]}.$$

It is easy to check that

$$\int Y_s^\sigma dX_s = Y_0 X_0 + \sum_i Y_{T_i}(X^{T_{i+1}} - X^{T_i}),$$

for any semimartingale X, any process Y in $\mathbf{S}$, $\mathbf{D}$, or $\mathbf{L}$.

Theorem 21. *Let X be a semimartingale, and let Y be a process in $\mathbf{D}$ or in $\mathbf{L}$. Let (σ_n) be a sequence of random partitions tending to the identity. Then the processes $\int_{0+} Y_s^{\sigma_n} dX_s = \sum_i Y_{T_i^n}(X^{T_{i+1}^n} - X^{T_i^n})$ tend to the stochastic integral $(Y_-) \cdot X$ in ucp.*

Proof. (The notation Y_- means the process whose value at s is given by $(Y_-)_s = \lim_{u \to s,\, u<s} Y_u$; also, $(Y_-)_0 = 0$, by convention). We prove the theorem for the case where Y is càdlàg, the other case being analogous. Also without loss of generality we can take $X_0 = 0$. Y càdlàg implies $Y_- \in \mathbf{L}$. Let $Y^k \in \mathbf{S}$ such that Y^k converges to Y_- (ucp). We have:

$$\int (Y_- - Y^{\sigma_n})_s dX_s = \int (Y_- - Y^k)_s dX_s + \int (Y^k - (Y_+^k)^{\sigma_n})_s dX_s$$
$$+ \int ((Y_+^k)^{\sigma_n} - Y^{\sigma_n})_s dX_s.$$

where Y_+^k denotes the càdlàg version of Y^k. The first term on the right side equals $J_X(Y_- - Y^k)$, and since J_X is continuous in $\mathbf{L}_{ucp}$ and since $Y_- - Y^k \to 0$, we have $\int (Y_- - Y^k)_s dX_s \to 0 (ucp)$. The same reasoning applies to the third term, for fixed n, as $k \to \infty$. Indeed, the convergence to 0 of $((Y_+^k)^{\sigma_n} - Y^{\sigma_n})$ as $k \to \infty$ is uniform in n.

It remains to consider the middle term on the right side above. Since the Y^k are simple predictable we can write the stochastic integrals in closed form, and since X is right continuous the integrals (for fixed (k, ω)) $\int (Y^k - (Y_+^k)^{\sigma_n})_s dX_s$ tend to 0 as $n \to \infty$. Thus one merely chooses k so large that the first and third terms are small, and then for fixed k, the middle term can be made small for large enough n. □

We consider here another example. We have already seen at the end of Sect. 4 that if B is a standard Wiener process, then

$$\int_0^t B_s dB_s = \frac{1}{2} B_t^2 - \frac{1}{2} t,$$

showing that the semimartingale calculus does not formally generalize the Riemann-Stieltjes calculus.

We have seen as well that the stochastic integral agrees with the Lebesgue-Stieltjes integral when possible (Theorems 17 and 18) and that the stochastic integral also preserves the martingale property (Theorem 20; at least for locally square integrable local martingales). The following example shows that the restriction of integrands to $\mathbb{L}$ is not as innocent as it may seem if we want to have both of these properties.

Let $M_t = N_t - \lambda t$, a compensated Poisson process (and hence a martingale with $M_t \in L^p$ for all $t \geq 0$ and all $p \geq 1$). Let $(T_i)_{i \geq 1}$ be the jump times of M. Let $H_t = 1_{[0,T_1)}(t)$. Then $H \in \mathbb{D}$. The Lebesgue-Stieltjes integral is:

$$\begin{aligned}\int_0^t H_s dM_s &= \int_0^t H_s dN_s - \lambda \int_0^t H_s ds \\ &= \sum_{i=1}^{\infty} H_{T_i} 1_{\{t \geq T_i\}} - \lambda \int_0^t H_s ds \\ &= -\lambda(t \wedge T_1).\end{aligned}$$

This process is not a martingale. We conclude that the space of integrands cannot be expanded even to $\mathbb{D}$, in general, and preserve the structure of the theory already established.

6. The Quadratic Variation of a Semimartingale

The quadratic variation process of a semimartingale, also known as the bracket process, is a simple object that nevertheless plays a fundamental role.

Definition. Let X, Y be semimartingales. The **quadratic variation process** of X, denoted $[X,X] = ([X,X]_t)_{t \geq 0}$, is defined by:

$$[X,X] = X^2 - 2\int X_- dX$$

(recall: $X_{0-} = 0$); The **quadratic covariation** of X, Y, also called the **bracket process** of X, Y, is defined by:

$$[X,Y] = XY - \int X_- dY - \int Y_- dX.$$

It is clear that the operation $(X,Y) \to [X,Y]$ is bilinear and symmetric. We therefore have a **polarization identity**:

$$[X,Y] = \frac{1}{2}([X+Y, X+Y] - [X,X] - [Y,Y]).$$

The next theorem gives some elementary properties of $[X,X]$. (*X is assumed to be a given semimartingale throughout this section*).

Theorem 22. *The quadratic variation process of X is a càdlàg, increasing, adapted process. Moreover it satisfies:*

(i) $[X,X]_0 = X_0^2$ and $\Delta[X,X] = (\Delta X)^2$
(ii) If σ_n is a sequence of random partitions tending to the identity, then

$$X_0^2 + \sum_i (X^{T_{i+1}^n} - X^{T_i^n})^2 \to [X,X]$$

with convergence in ucp, where σ_n is the sequence $0 = T_0^n \le T_1^n \le \dots \le T_i^n \le \dots, \le T_{k_n}^n$ and where T_i^n are stopping times.
(iii) If T is any stopping time, then $[X^T, X] = [X, X^T] = [X^T, X^T] = [X,X]^T$.

Proof. X is càdlàg, adapted, and so also is $\int X_- dX$ by its definition; thus $[X,X]$ is càdlàg, adapted as well. Recall the property of the stochastic integral: $\Delta(X_- \cdot X) = X_- \Delta X$. Then

$$\begin{aligned}(\Delta X)_s^2 &= (X_s - X_{s-})^2 = X_s^2 - 2X_s X_{s-} + X_{s-}^2 \\ &= X_s^2 - X_{s-}^2 + 2X_{s-}(X_{s-} - X_s) \\ &= \Delta(X^2)_s - 2X_{s-}(\Delta X_s),\end{aligned}$$

from which part (i) follows.

For part (ii), by replacing X with $\widetilde{X} = X - X_0$, we may assume $X_0 = 0$. Let $R_n = \sup_i T_i^n$. Then $R_n < \infty$ a.s. and $\lim_n R_n = \infty$ a.s, and thus by telescoping series:

$$(X^2)^{R_n} = \sum_i \{(X^2)^{T_{i+1}^n} - (X^2)^{T_i^n}\}$$

converges ucp to X^2. Moreover, the series $\sum_i X_{T_i^n}(X^{T_{i+1}^n} - X^{T_i^n})$ converges in ucp to $\int X_- dX$ by Theorem 21, since X is càdlàg. Since $b^2 - a^2 - 2a(b-a) = (b-a)^2$, and since $X_{T_i^n}(X^{T_{i+1}^n} - X^{T_i^n}) = X^{T_i^n}(X^{T_{i+1}^n} - X^{T_i^n})$, we can combine the two series convergences above to obtain the result. Finally, note that if $s < t$, then the approximating sums in part (ii) include more terms (all nonnegative), so it is clear that $[X,X]$ is nondecreasing. (Note that, *a priori*, one only has $[X,X]_s \le [X,X]_t$ a.s., with the null set depending on s and t; it is the property that $[X,X]$ has càdlàg paths that allows one to eliminate the dependence of the null set on s and t.) Part (iii) is a simple consequence of part (ii). □

An immediate consequence of Theorem 22 is the observation that if B is a Brownian motion, then $[B,B]_t = t$, since in Theorem 28 of Chap. I we showed the a.s. convergence of sums of the form in part (ii) of Theorem 22 when the partitions are refining.

Another consequence of Theorem 22 is that if X is a semimartingale with continuous paths of finite variation, then $[X,X]$ is the constant process equal to X_0^2. To see this one need only observe that

$$\begin{aligned}\sum(X^{T^n_{i+1}} - X^{T^n_i})^2 &\leq \sup_i |X^{T^n_{i+1}} - X^{T^n_i}| \sum_i |X^{T^n_{i+1}} - X^{T^n_i}| \\ &\leq \sup_i |X^{T^n_{i+1}} - X^{T^n_i}| V,\end{aligned}$$

where V is the total variation. Therefore the sums tend to 0 as $\|\sigma_n\| \to 0$. Theorem 22 has several more consequences which we state as Corollaries.

Corollary 1. *The bracket process $[X,Y]$ of two semimartingales has paths of finite variation on compacts, and it is also a semimartingale.*

Proof. By the polarization identity $[X,Y]$ is the difference of two increasing processes, hence its paths are of finite variation. Moreover, the paths are clearly càdlàg, and the process is adapted. Hence by Theorem 7 it is a semimartingale. □

Corollary 2 (Integration by Parts). *Let X, Y be two semimartingales. Then XY is a semimartingale and*

$$XY = \int X_- dY + \int Y_- dX + [X,Y].$$

Proof. The formula follows trivially from the definition of $[X,Y]$. That XY is a semimartingale follows from the formula, Theorem 19, and Corollary 1 above. □

In the integration by parts formula above, we have $(X_-)_0 = (Y_-)_0 = 0$, hence evaluating at 0 yields

$$X_0 Y_0 = (X_-)_0 Y_0 + (Y_-)_0 X_0 + [X,Y]_0.$$

Since $[X,Y]_0 = \Delta X_0 \Delta Y_0 = X_0 Y_0$, the formula is valid. Without the convention that $(X_-)_0 = 0$, we could have written the formula:

$$X_t Y_t = \int_{0+}^t X_{s-} dY_s + \int_{0+}^t Y_{s-} dX_s + [X,Y]_t.$$

Corollary 3. *All semimartingales on a given filtered probability space form an algebra.*

Proof. Since semimartingales form a vector space, Corollary 2 shows they form an algebra. □

A theorem analogous to Theorem 22 holds for $[X,Y]$ as well as $[X,X]$. It can be proved analogously to Theorem 22, or more simply by polarization. We omit the proof.

Theorem 23. *Let X and Y be two semimartingales. Then the bracket process $[X,Y]$ satisfies:*

(i) $[X,Y]_0 = X_0Y_0$; $\Delta[X,Y] = \Delta X \Delta Y$;

(ii) If σ_n is a sequence of random partitions tending to the identity, then

$$[X,Y] = X_0Y_0 + \lim_{n\to\infty} \sum_i (X^{T_{i+1}^n} - X^{T_i^n})(Y^{T_{i+1}^n} - Y^{T_i^n}),$$

where convergence is in ucp, and where σ_n is the sequence $0 = T_0^n \leq T_1^n \leq \cdots \leq T_i^n \leq \cdots \leq T_{k_n}^n$, with T_i^n stopping times.

(iii) If T is any stopping time, then $[X^T,Y] = [X,Y^T] = [X^T,Y^T] = [X,Y]^T$.

We next record a real analysis theorem from the Lebesgue-Stieltjes theory of integration. It can be proved via the monotone class theorem.

Theorem 24. *Let α, β, γ be functions mapping $[0,\infty)$ to $\mathbb{R}$ with $\alpha(0) = \beta(0) = \gamma(0) = 0$. Suppose α, β, γ are all right continuous, α is of finite variation, and β and γ are each increasing. Suppose further that for all s, t with $s \leq t$, we have*

$$|\int_s^t d\alpha_u| \leq (\int_s^t d\beta_u)^{\frac{1}{2}}(\int_s^t d\gamma_u)^{\frac{1}{2}}.$$

Then for any measurable functions f, g we have

$$\int_s^t |fg||d\alpha| \leq (\int_s^t f^2 d\beta)^{\frac{1}{2}}(\int_s^t g^2 d\gamma)^{\frac{1}{2}}.$$

In particular, the measure $d\alpha$ is absolutely continuous with respect to both $d\beta$ and $d\gamma$.

Note that $|d\alpha|$ denotes the total variation measure corresponding to the measure $d\alpha$, the Lebesgue-Stieltjes signed measure induced by α. We use this theorem to prove an important inequality concerning the quadratic variation and bracket processes.

Theorem 25 (The Kunita-Watanabe Inequality). *Let X and Y be two semimartingales, and let H and K be two measurable processes. Then one has a.s.*

$$\int_0^\infty |H_s||K_s||d[X,Y]_s| \leq (\int_0^\infty H_s^2 d[X,X]_s)^{\frac{1}{2}}(\int_0^\infty K_s^2 d[Y,Y]_s)^{\frac{1}{2}}.$$

Proof. By Theorem 24 we only need to show that there exists a null set N, such that for $\omega \notin N$, and (s,t) with $s \leq t$, we have:

$$(*) \qquad |\int_s^t d[X,Y]_u| \leq (\int_s^t d[X,X]_u)^{\frac{1}{2}}(\int_s^t d[Y,Y]_u)^{\frac{1}{2}}.$$

Let N be the null set such that if $\omega \notin N$, then $0 \leq \int_s^t d[X+rY,X+rY]_u$, for every r, s, t; $s \leq t$, with r, s, t all rational numbers. Then

$$\begin{aligned} 0 &\leq [X+rY,X+rY]_t - [X+rY,X+rY]_s \\ &= r^2([Y,Y]_t - [Y,Y]_s) + 2r([X,Y]_t - [X,Y]_s) + ([X,X]_t - [X,X]_s). \end{aligned}$$

The right side being positive for all rational r, it must be positive for all real r by continuity. Thus the discriminant of this quadratic equation in r must be nonnegative, which gives us exactly the inequality (*). Since we have, then, the inequality for all rational (s,t), it must hold for all real (s,t), by the right continuity of the paths of the processes. □

Corollary. *Let X and Y be two semimartingales, and let H and K be two measurable processes. Then*

$$E\{\int_0^\infty |H_s||K_s||d[X,Y]_s|\} \leq \|(\int_0^\infty H_s^2 d[X,X]_s)^{\frac{1}{2}}\|_{L^p} \|(\int_0^\infty K_s^2 d[Y,Y]_s)^{\frac{1}{2}}\|_{L^q}$$

if $\frac{1}{p} + \frac{1}{q} = 1$.

Proof. Apply Hölder's inequality to the Kunita-Watanabe inequality of Theorem 25. □

Since Theorem 25 and its Corollary are path-by-path Lebesgue-Stieltjes results, we *do not have to assume* that the integrand processes H and K be adapted.

Since the process $[X,X]$ is nondecreasing with right continuous paths, and since $\Delta[X,X]_t = (\Delta X_t)^2$ for all $t \geq 0$ (*with the convention that* $X_{0-} = 0$), we can decompose $[X,X]$ path by path into its continuous part and its pure jump part.

Definition. For a semimartingale X, the process $[X,X]^c$ denotes the path by path continuous part of $[X,X]$.

We can then write:

$$[X,X]_t = [X,X]_t^c + X_0^2 + \sum_{0<s\leq t} (\Delta X_s)^2$$

and

$$[X,X]_t = [X,X]_t^c + \sum_{0\leq s\leq t} (\Delta X_s)^2.$$

Observe that $[X,X]_0^c = 0$.

Comment. In Chap. IV, Sect. 5, we briefly discuss the *continuous local martingale part* of a semimartingale X satisfying an auxiliary hypothesis known as Hypothesis A. This (unique) continuous local martingale part of X is denoted X^c. It can be shown that a unique continuous local martingale part, X^c, exists for every semimartingale X.[1] (If X is an FV process, then $X^c \equiv 0$, as we shall see in Chap. III.) *It is always true that* $[X^c, X^c] = [X,X]^c$. Although we will have no need of this result in this book, this notation is often used in the literature.

We remark also that the conditional quadratic variation of a semimartingale X, denoted $\langle X,X\rangle$, is defined in Chap. III, Sect. 3. *It is also true that* $\langle X^c, X^c\rangle = [X^c, X^c] = [X,X]^c$, and these notations are used interchangeably in the literature. *If X is already continuous and $X_0 = 0$, then* $\langle X,X\rangle = [X,X] = [X,X]^c$.

Definition. A semimartingale X will be called **quadratic pure jump** if $[X,X]^c = 0$.

If X is quadratic pure jump, then $[X,X]_t = X_0^2 + \sum_{0<s\leq t}(\Delta X_s)^2$. The Poisson process N is an obvious example of a quadratic pure jump semimartingale. From the definition or immediately from Theorem 22, we see that $[N,N]_t = N_t$. More generally it can be shown that if X is a Lévy process with a Lévy decomposition $X_t = B_t + Y_t$ as in Theorem 42 of Chap. I, where B is a Brownian motion and

$$Y_t = \int_{\{|x|<1\}} x(N_t(\cdot, dx) - t\nu(dx)) + \alpha t + \sum_{0<s\leq t} \Delta X_s 1_{\{|\Delta X_s|\geq 1\}}$$

(using the notation of Theorem 42 of Chap. I), then Y is a quadratic pure jump semimartingale.

The next theorem gives a simple criterion for a process X to be a quadratic pure jump semimartingale.

Theorem 26. *If X is adapted, càdlàg, with paths of finite variation on compacts, then X is a quadratic pure jump semimartingale.*

Proof. Without loss of generality we assume $X_0 = 0$. We have already seen that such an X is a semimartingale (Theorem 7), and that the stochastic integral with respect to X is nothing more than a path by path Lebesgue-Stieltjes integral (Theorem 17). The integration by parts formula for Lebesgue-Stieltjes differentials applied to X times itself yields: $X^2 = \int X_- dX + \int X dX$, computed path by path. The semimartingale integration by parts formula (Corollary 2 of Theorem 22), on the other hand,

[1] See, for example, Dellacherie-Meyer [2], p. 355.

yields: $X^2 = 2\int X_- dX + [X,X]$. Moreover

$$\int X dX = \int (X_- + \Delta X) dX = \int X_- dX + \int \Delta X dX,$$

and

$$\int_0^t X_- dX_s + \int_0^t \Delta X_s dX_s = \int_0^t X_{s-} dX_s + \sum_{s \leq t} (\Delta X_s)^2.$$

Equating the two formulas we deduce $[X,X]_t = \sum_{s \leq t} (\Delta X_s)^2$, whence the theorem. □

Note in particular that if X is adapted with *continuous* paths of finite variation, then $[X,X]_t = X_0^2$, all $t \geq 0$.

Theorem 27. *Let X be a local martingale with continuous paths that are not everywhere constant. Then $[X,X]$ is not the constant process X_0^2, and $X^2 - [X,X]$ is a continuous local martingale. Moreover if $[X,X]_t = 0$ for all t then $X_t = 0$ for all t.*

Proof. Note that a continuous local martingale is a semimartingale (Corollary 2 of Theorem 8). We have $X^2 - [X,X] = 2\int X_- dX$, and by the martingale preservation property (Theorem 20) we have that $2\int X_- dX$ is a local martingale. Moreover $\Delta 2 \int X_- dX = 2(X_-)(\Delta X)$, and since X is continuous, $\Delta X = 0$, and thus $2\int X_- dX$ is a *continuous* local martingale, hence locally square integrable. Thus $X^2 - [X,X]$ is a locally square integrable local martingale.

By stopping, we can suppose X is a square integrable martingale. Assume further $X_0 = 0$. Next assume that $[X,X]$ actually were constant. Then $[X,X]_t = [X,X]_0 = X_0^2 = 0$, for all t. Since $X^2 - [X,X]$ is a local martingale, we conclude X^2 is a nonnegative local martingale, with $X_0^2 = 0$. Thus $X_t^2 = 0$, all t. This is a contradiction. If X_0 is not identically 0, we set $\hat{X}_t = X_t - X_0$ and the result follows. □

The following corollary is of fundamental importance in the theory of martingales.

Corollary 1. *Let X be a continuous local martingale, and $S \leq T \leq \infty$ be stopping times. If X has paths of finite variation on the stochastic interval (S,T), then X is constant on $[S,T]$. Moreover if $[X,X]$ is constant on $[S,T] \cap [0,\infty)$, then X is constant there too.*

Proof. $M = X^T - X^S$ is also a continuous local martingale, and M has finite variation on compacts. Moreover $[M,M] = [X^T - X^S, X^T - X^S] = [X,X]^T - [X,X]^S$, and therefore $[M,M]$ is constant everywhere, hence by Theorem 27 M must be constant everywhere; thus X is constant on $[S,T]$. □

Observe that if $t > 0$ is arbitrary and we take $S = 0$ and $T = t$ in Corollary 1, then we can conclude that *a continuous local martingale with paths of finite variations on compacts is a.s. constant.* While non-trivial continuous local martingales must therefore always have paths of infinite variation on compacts, they are not the only such local martingales. For example, the Lévy process martingale Z^d of Theorem 41 of Chap. I will have paths of infinite variation if the Lévy measure ν has infinite mass in a neighborhood of the origin. Another example is Azema's martingale, which is presented in detail in Sect. 6 of Chap. IV.

Corollary 2. *Let X and Y be two locally square-integrable local martingales. Then $[X,Y]$ is the unique adapted càdlàg process A with paths of finite variation on compacts satisfying the two properties:*

(i) $XY - A$ is a local martingale;
(ii) $\Delta A = \Delta X \Delta Y$, $A_0 = X_0 Y_0$.

Proof. Integration by parts yields:

$$XY = \int X_- dY + \int Y_- dX + [X,Y];$$

but the martingale preservation property tells us that both stochastic integrals are local martingales. Thus $XY - [X,Y]$ is a local martingale. Property (ii) is simply an application of Theorem 23. Thus it remains to show uniqueness. Suppose A, B both satisfy properties (i) and (ii). Then $A - B = (XY - B) - (XY - A)$, the difference of two local martingales which is again a local martingale. Moreover,

$$\Delta(A - B) = \Delta A - \Delta B = \Delta X \Delta Y - \Delta X \Delta Y = 0.$$

Thus $A - B$ is a continuous local martingale, $A_0 - B_0 = 0$, and it has paths of finite variation on compacts. Corollary 1 yields $A_t - B_t - A_0 + B_0 = 0$ and we have uniqueness. □

Corollary 2 can be useful in determining the process $[X,Y]$. For example if X and Y are locally square integrable martingales without common jumps such that XY is a martingale, then $[X,Y] = X_0 Y_0$. One can also easily verify (as a consequence of Theorem 23 and Corollaries 1 and 2 above, for example) that if X is a continuous square integrable martingale and Y is a square integrable martingale with paths of finite variation on compacts, then $[X,Y] = X_0 Y_0$, and hence XY is a martingale. (An example would be X a Brownian motion and Y a compensated Poisson process.) Corollary 2 is true as well for X, Y local martingales, however we need Theorem 17 of Chap. III to prove the general result.

In Chap. III we show that any local martingale is a semimartingale (see the Corollary of Theorem 14 of Chap. III), and therefore if M is a local martingale its quadratic variation $[M,M]_t$ always exists and is finite a.s. for every $t \geq 0$. We use this fact in the next corollary.

Corollary 3. *Let M be a local martingale. Then M is a martingale with $E\{M_t^2\} < \infty$, all $t \geq 0$, if and only if $E\{[M,M]_t\} < \infty$, all $t \geq 0$. If $E\{[M,M]_t\} < \infty$, then $E\{M_t^2\} = E\{[M,M]_t\}$.*

Proof. First assume that M is a martingale with $E\{M_t^2\} < \infty$ for all $t \geq 0$. Then M is clearly a locally square integrable martingale. Let $N_t = M_t^2 - [M,M]_t = 2\int_0^t M_{s-}dM_s$, which is a locally square integrable local martingale by Theorem 20. Let $(T^n)_{n\geq 1}$ be stopping times increasing to ∞ a.s. such that $N_t^{T^n}$ is a square integrable martingale. Then $E\{N_t^{T^n}\} = E\{N_0\} = 0$, all $t \geq 0$. Therefore

$$E\{M_{t\wedge T^n}^2\} = E\{[M,M]_{t\wedge T^n}\}.$$

Doob's maximal quadratic inequality gives $E\{(M_t^*)^2\} \leq 4E\{M_t^2\} < \infty$. Therefore by the dominated convergence theorem

$$\begin{aligned} E\{M_t^2\} &= \lim_{n\to\infty} E\{M_{t\wedge T^n}^2\} \\ &= \lim_{n\to\infty} E\{[M,M]_{t\wedge T^n}\} \\ &= E\{[M,M]_t\}, \end{aligned}$$

where the last result is by the monotone convergence theorem. In particular we have that $E\{[M,M]_t\} < \infty$.

For the converse, we now assume $E\{[M,M]_t\} < \infty$, all $t \geq 0$. Define stopping times by

$$T^n = \inf\{t > 0 : |M_t| > n\} \wedge n.$$

Then T^n increase to ∞ a.s. Furthermore $(M^{T^n})^* \leq n + |\Delta M_{T^n}| \leq n + [M,M]_n^{1/2}$, which is in L^2. By Theorem 47 of Chap. I M^{T^n} is a uniformly integrable martingale for each n. Also we have that

$$E\{(M_t^{T^n})^2\} \leq E\{((M^{T^n})^*)^2\} < \infty,$$

for all $t \geq 0$. Therefore M^{T^n} satisfies the hypotheses of the first half of this theorem, and $E\{(M_t^{T^n})^2\} = E\{[M^{T^n}, M^{T^n}]_t\}$. Using Doob's inequality:

$$\begin{aligned} E\{(M_{t\wedge T^n}^*)^2\} &\leq 4E\{(M_t^{T^n})^2\} \\ &= 4E\{[M^{T^n}, M^{T^n}]_t\} \\ &= 4E\{[M,M]_{T^n\wedge t}\} \\ &\leq 4E\{[M,M]_t\}. \end{aligned}$$

The monotone convergence theorem next gives

$$E\{(M_t^*)^2\} = \lim_{n\to\infty} E\{(M_{t\wedge T^n}^*)^2\}$$
$$\leq 4E\{[M,M]_t\} < \infty.$$

Therefore, again by Theorem 47 of Chap. I, we conclude that M is a martingale. The preceding gives $E\{M_t^2\} < \infty$. □

For emphasis we state as another corollary a special case of Corollary 3.

Corollary 4. *If M is a local martingale and $E\{[M,M]_\infty\} < \infty$, then M is a square integrable martingale (that is $\sup_t E\{M_t^2\} = E\{M_\infty^2\} < \infty$). Moreover $E\{M_t^2\} = E\{[M,M]_t\}$ for all t, $0 \leq t \leq \infty$.*

Example. Before continuing we consider again an example of a local martingale that exhibits many of the surprising pathologies of local martingales. Let B be a standard Brownian motion in $\mathbf{R}^3$ with $B_0 = (1,1,1)$. Let $M_t = \frac{1}{\|B_t\|}$, where $\|x\|$ is standard Euclidean norm in $\mathbf{R}^3$. (We previously considered this example in Sect. 5 of Chap. I.) As noted in Chap. I, the process M is a continuous local martingale; hence it is a locally square integrable local martingale. Moreover $E\{M_t^2\} < \infty$ for all t. However instead of $t \to E\{M_t^2\}$ being an increasing function as it would if M were a martingale, $\lim_{t\to\infty} E\{M_t^2\} = 0$. Moreover $E\{[M,M]_t\} \geq E\{[M,M]_0\} = 1$ since $[M,M]_t$ is increasing. Therefore we cannot have $E\{M_t^2\} = E\{[M,M]_t\}$ for all t. Indeed, by Corollary 3 and the preceding we see that we must have $E\{[M,M]_t\} = \infty$ for all $t > 0$. In conclusion, $M = \frac{1}{\|B\|}$ is a continuous local martingale with $E\{M_t^2\} < \infty$ for all t which is both not a true martingale and for which $E\{M_t^2\} < \infty$ while $E\{[M,M]_t\} = \infty$ for all $t > 0$.

Corollary 5. *Let X be a continuous local martingale. Then X and $[X,X]$ have the same intervals of constancy a.s.*

Proof. Let r be a positive rational, and define

$$T_r = \inf\{t \geq r : X_t \neq X_r\}.$$

Then $M = X^{T_r} - X^r$ is a local martingale which is constant. Hence $[M,M] = [X,X]^{T_r} - [X,X]^r$ is also constant. Since this is true for any rational r a.s., any interval of constancy of X is also one of $[X,X]$.

Since X is continuous, by stopping we can assume without loss of generality that X is a bounded martingale (and hence square integrable). For every positive, rational r we define

$$S_r = \inf\{t \geq r : [X,X]_t > [X,X]_r\}.$$

Then

$$E\{(X_{S_r} - X_r)^2\} = E\{X_{S_r}^2\} - E\{X_r^2\}$$

by Doob's Optional Sampling Theorem. Moreover

$$E\{X_{S_r}^2\} - E\{X_r^2\} = E\{[X,X]_{S_r} - [X,X]_r\} = 0,$$

by Corollary 3. Therefore $E\{(X_{S_r} - X_r)^2\} = 0$, and $X_{S_r} = X_r$ a.s. Moreover this implies $X_q = X_{S_q}$ a.s. on $\{S_q = S_r\}$ for each pair of rationals (r,q), and therefore we deduce that any interval of constancy of $[X,X]$ is also one of X. □

Note that the continuity of the local martingale X is essential in Corollary 5. Indeed, let N_t be a Poisson process, and let $M_t = N_t - t$. Then M is a martingale and $[M,M]_t = N_t$; clearly M has no intervals of constancy while N is constant except for jumps.

Theorem 28. *Let X be a quadratic pure jump semimartingale. Then for any semimartingale Y we have:*

$$[X,Y]_t = X_0Y_0 + \sum_{0<s\leq t} \Delta X_s \Delta Y_s.$$

Proof. The Kunita-Watanabe Inequality (Theorem 25) tells us $d[X,Y]_s$ is a.s. absolutely continuous with respect to $d[X,X]$ (path by path). Thus $[X,X]^c = 0$ implies $[X,Y]^c = 0$, and hence $[X,Y]$ is the sum of its jumps, and the result follows by Theorem 23. □

Theorem 29. *Let X and Y be two semimartingales, and let H, $K \in \mathbf{L}$. Then*

$$[H\cdot X, K\cdot Y]_t = \int_0^t H_sK_s d[X,Y]_s$$

and, in particular,

$$[H\cdot X, H\cdot X]_t = \int_0^t H_s^2 d[X,X]_s.$$

Proof. First assume (without loss of generality) that $X_0 = Y_0 = 0$. It suffices to establish the following result

$$(*) \qquad [H\cdot X, Y]_t = \int_0^t H_s d[X,Y]_s,$$

and then apply it again, by the symmetry of the form $[\cdot,\cdot]$.

First suppose H is the indicator of a stochastic interval. That is, $H = 1_{[0,T]}$, where T is a stopping time. Establishing (*) is equivalent in this case to showing $[X^T, Y] = [X,Y]^T$, a result that is an obvious consequence of Theorem 23, which approximates $[X,Y]$ by sums.

Next suppose $H = U1_{(S,T]}$, where S, T are stopping times, $S \leq T$ a.s., and $U \in \mathcal{F}_S$. Then $\int H_s dX_s = U(X^T - X^S)$, and by Theorem 23

$$[H \cdot X, Y] = U\{[X^T, Y] - [X^S, Y]\}$$
$$= U\{[X,Y]^T - [X,Y]^S\} = \int H_s d[X,Y]_s.$$

The result now follows for $H \in \mathbf{S}$ by linearity. Finally, suppose $H \in \mathbf{L}$ and let H^n be a sequence in $\mathbf{S}$ converging in ucp to H. Let $Z^n = H^n \cdot X$; $Z = H \cdot X$. We know Z^n, Z are all semimartingales. We have $\int H^n_s d[X,Y]_s = [Z^n, Y]$, since $H^n \in \mathbf{S}$, and using integration by parts:

$$[Z^n, Y] = YZ^n - \int Y_- dZ^n - \int Z^n_- dY$$
$$= YZ^n - \int Y_- H^n dX - \int Z^n_- dY.$$

By the definition of the stochastic integral, we know $Z^n \to Z$ in ucp, and since $H^n \to H(ucp)$, letting $n \to \infty$ we have

$$\lim_{n\to\infty} [Z^n, Y] = YZ - \int Y_- H dX - \int Z_- dY$$
$$= YZ - \int Y_- dZ - \int Z_- dY$$
$$= [Z, Y],$$

again by integration by parts. Since $\lim_{n\to\infty} \int H^n_s d[X,Y]_s = \int H_s d[X,Y]_s$, we have $[Z,Y] = [H \cdot X, Y] = \int H_s d[X,Y]_s$, and the proof is complete. □

Theorem 30. *Let H be a càdlàg, adapted process, and let X, Y be two semimartingales. Let σ_n be a sequence of random partitions tending to the identity. Then*

$$\sum H_{T_i^n}(X^{T_{i+1}^n} - X^{T_i^n})(Y^{T_{i+1}^n} - Y^{T_i^n})$$

converges in ucp to $\int H_{s-} d[X,Y]_s$ ($H_{0-} = 0$). Here $\sigma_n = (0 \leq T_0^n \leq T_1^n \leq \cdots \leq T_i^n \leq \ldots T_{k_n}^n)$.

Proof. By the definition of quadratic variation, $[X,Y] = XY - X_- \cdot Y - Y_- \cdot X$, where $X_- \cdot Y$ denotes the process $(\int_0^t X_{s-} dY_s)_{t \geq 0}$. By the associativity of the stochastic integral (Theorem 19)

$$H_- \cdot [X,Y] = H_- \cdot (XY) - H_- \cdot (X_- \cdot Y) - H_- \cdot (Y_- \cdot X)$$
$$= H_- \cdot (XY) - (H_- X_-) \cdot Y - (H_- Y_-) \cdot X$$
$$= H_- \cdot (XY) - (HX)_- \cdot Y - (HY)_- \cdot X.$$

By Theorem 21 the above is the limit of

$$\sum_i H_{T_i^n}(X^{T_{i+1}^n}Y^{T_{i+1}^n} - X^{T_i^n}Y^{T_i^n}) - \sum_i H_{T_i^n}X_{T_i^n}(Y^{T_{i+1}^n} - Y^{T_i^n})$$

$$- \sum_i H_{T_i^n}Y_{T_i^n}(X^{T_{i+1}^n} - X^{T_i^n})$$

$$= \sum_i H_{T_i^n}\{(X^{T_{i+1}^n}Y^{T_{i+1}^n} - X^{T_i^n}Y^{T_i^n}) - X^{T_i^n}(Y^{T_{i+1}^n} - Y^{T_i^n})$$

$$- Y^{T_i^n}(X^{T_{i+1}^n} - X^{T_i^n})\}$$

$$= \sum_i H_{T_i^n}(X^{T_{i+1}^n} - X^{T_i^n})(Y^{T_{i+1}^n} - Y^{T_i^n}).$$

□

Example. Let B_t be a standard Wiener process with $B_0 = 0$, (i.e., Brownian motion). $B_t^2 - t$ is a continuous martingale by Theorem 27 of Chap. I. Let $H \in \mathbb{L}$ be such that $E\{\int_0^t H_s^2 ds\} < \infty$, each $t \geq 0$. By Theorem 28 of Chap. I we have $[B,B]_t = t$, hence $[H \cdot B, H \cdot B]_t = \int_0^t H_s^2 ds$. By the martingale preservation property, $\int H_s dB_s$ is also a continuous local martingale, with $(H \cdot B)_0 = 0$. By Corollary 3 to Theorem 27:

$$E\{(\int_0^t H_s dB_s)^2\} = E\{[H \cdot B, H \cdot B]_t\}$$

$$= E\{\int_0^t H_s^2 ds\}.$$

It was this last equality:

$$E\{(\int_0^t H_s dB_s)^2\} = E\{\int_0^t H_s^2 ds\},$$

that was crucial in K. Itô's original treatment of a stochastic integral.

7. Ito's Formula (Change of Variables)

Let A be a process with continuous paths of finite variation on compacts. If $H \in \mathbb{L}$ we know by Theorem 17 that the stochastic integral $H \cdot A$ agrees a.s. with the path by path Lebesgue-Stieltjes integral $\int H_s dA_s$. In Sect. 6 of Chap. I (Theorem 50) we proved the change of variables formula for $f \in C^1$:

$$(*) \qquad f(A_t) - f(A_0) = \int_0^t f'(A_s)dA_s.$$

We also saw at the end of Sect. 4 of this chapter that for a standard Wiener process B with $B_0 = 0$,

$$(**)\qquad \int_0^t B_s dB_s = \frac{1}{2}B_t^2 - \frac{1}{2}t.$$

Taking $f(x) = \frac{x^2}{2}$, the above formula is equivalent to

$$f(B_t) - f(B_0) = \int_0^t f'(B_s)dB_s + \frac{1}{2}\int_0^t f''(B_s)ds,$$

which does not agree with the Lebesgue-Stieltjes change of variables formula (*). In this section we will state and prove a change of variables formula valid for all semimartingales.

We first mention, however, that the change of variables formula for continuous Stieltjes integrals given in Theorem 50 of Chap. I has an extension to right continuous processes of finite variation on compacts. We state this result as a theorem but we do not prove it here because it is merely a special case of Theorem 32.

Theorem 31 (Change of Variables). *Let V be an FV process with right continuous paths, and let f be such that f' exists and is continuous. Then $(f(V_t))_{t\geq 0}$ is an FV process and*

$$f(V_t) - f(V_0) = \int_{0+}^t f'(V_{s-})dV_s + \sum_{0<s\leq t}\{f(V_s) - f(V_{s-}) - f'(V_{s-})\Delta V_s\}.$$

Recall that the notation $\int_{0+}^t = \int_{(0,t]}$ denotes the integral over the half open interval $(0,t]$. We wish to establish a formula analogous to the above, but for the stochastic integral; that is, when the process is a semimartingale. The formula is different in this case, as we can see by comparing equation (*) with equation (**); we must add an extra term!

Theorem 32 (Itô's Formula). *Let X be a semimartingale and let f be a C^2 real function. Then $f(X)$ is again a semimartingale, and the following formula holds:*

$$\begin{aligned} f(X_t) - f(X_0) = &\int_{0+}^t f'(X_{s-})dX_s + \frac{1}{2}\int_{0+}^t f''(X_{s-})d[X,X]_s^c \\ &+ \sum_{0<s\leq t}\{f(X_s) - f(X_{s-}) - f'(X_{s-})\Delta X_s\} \end{aligned}$$

Proof. First note that the jump part of the stochastic integral $\int f''(X_{s-})d[X,X]_s$ is given by $\sum_{s\leq t} f''(X_{s-})(\Delta X_s)^2$, and this is a convergent series. By adding and subtracting $\frac{1}{2}$ of this series, we can rewrite Itô's

formula in the equivalent form:
(***)

$$f(X_t)-f(X_0) = \int_{0+}^t f'(X_{s-})dX_s + \frac{1}{2}\int_{0+}^t f''(X_{s-})d[X,X]_s$$
$$+ \sum_{0<s\leq t} \{f(X_s) - f(X_{s-}) - f'(X_{s-})\Delta X_s - \frac{1}{2}f''(X_{s-})(\Delta X_s)^2\}$$

which is perhaps less obviously a generalization of the "classical" case, but notationally simpler to prove. The proof rests, of course, on Taylor's theorem:

$$f(y) - f(x) = f'(x)(y-x) + \frac{1}{2}f''(x)(y-x)^2 + R(x,y)$$

where $R(x,y) \leq r(|y-x|)(y-x)^2$, such that $r : \mathbf{R}_+ \to \mathbf{R}_+$ is an increasing function with $\lim_{u\downarrow 0} r(u) = 0$, which is valid for $f \in C^2$ defined on a compact set.

Proof for the continuous case. We first restrict our attention to a continuous semimartingale X, since the proof is less complicated but nevertheless gives the basic idea. Without loss of generality we can take $X_0 = 0$. Define stopping times

$$R_m = \inf\{t : |X_t| \geq m\}.$$

Then the stopped process X^{R_m} is bounded by m, and if Itô's formula is valid for X^{R_m} for each m, it's valid for X as well. Therefore we assume that X takes its values in a compact set. We fix a $t > 0$, and let σ_n be a refining sequence of random partitions of $[0,t]$ tending to the identity $[\sigma_n = (0 = T_0^n \leq T_1^n \leq \cdots \leq T_{k_n}^n = t)]$. Then

$$\begin{aligned} f(X_t) - f(X_0) &= \sum_{i=0}^{k_n} \left\{ f\left(X_{T_{i+1}^n}\right) - f\left(X_{T_i^n}\right)\right\} \\ &= \sum_i f'\left(X_{T_i^n}\right)\left(X_{T_{i+1}^n} - X_{T_i^n}\right) \\ &\quad + \frac{1}{2}\sum_i f''\left(X_{T_i^n}\right)\left(X_{T_{i+1}^n} - X_{T_i^n}\right)^2 + \sum_i R\left(X_{T_i^n}, X_{T_{i+1}^n}\right) \end{aligned}$$

The first sum converges in probability to the stochastic integral $\int_0^t f'(X_{s-})dX_s$ by Theorem 21; the second sum converges in probability to $\frac{1}{2}\int_0^t f''(X_s)d[X,X]_s$ by Theorem 30. It remains to consider the third sum: $\sum_i R\left(X_{T_i^n}, X_{T_{i+1}^n}\right)$. But this sum is majorized, in absolute value, by $\sup_i r(|X_{T_{i+1}^n} - X_{T_i^n}|)\{\sum_i (X_{T_{i+1}^n} - X_{T_i^n})^2\}$, and since $\sum_i (X_{T_{i+1}^n} - X_{T_i^n})^2$ converges in probability to $[X,X]_t$ (Theorem 22), the last term will tend to 0 if $\lim_{n\to\infty}\sup_i r(|X_{T_{i+1}^n} - X_{T_i^n}|) = 0$. However $s \to X_s(\omega)$ is a continuous function on $[0,t]$, each fixed ω, and hence uniformly continuous. Since $\lim_{n\to\infty}\sup_i |T_{i+1}^n - T_i^n| = 0$ by hypothesis, we have the result. Thus, in the continuous case, $f(X_t) - f(X_0) = \int_0^t f'(X_{s-})dX_s + \frac{1}{2}\int_0^t f''(X_{s-})d[X,X_s]$, for each t, a.s. The continuity of the paths then permits us to remove the dependence of the null set on t, giving the complete result in the continuous case.

Proof for the general case: X is now given as a right continuous semimartingale. Once again we have a representation as in (***), but we need a closer analysis. For any $t > 0$ we have $\sum_{0<s\le t}(\Delta X_s)^2 \le [X,X]_t < \infty$ a.s., hence $\sum_{0<s\le t}(\Delta X_s)^2$ is convergent. Given $\varepsilon > 0$ and $t > 0$, let $A = A(\varepsilon, t)$ be a set of jumps of X that has a.s. a finite number of times s, and let $B = B(\varepsilon, t)$ be such that $\sum_{s\in B}(\Delta X_s)^2 \le \varepsilon^2$, where A and B are disjoint and $A \cup B$ exhaust the jumps of X on $(0,t]$. We write

$$\begin{aligned} f(X_t) - f(X_0) &= \sum_i \left\{ f(X_{T_{i+1}^n}) - f(X_{T_i^n}) \right\} \\ &= \sum_{i,A} \left\{ f(X_{T_{i+1}^n}) - f(X_{T_i^n}) \right\} + \sum_{i,B} \left\{ f(X_{T_{i+1}^n}) - f(X_{T_i^n}) \right\} \end{aligned}$$

where $\sum_{i,A}$ denotes $\sum_i 1_{\{A\cap(T_i^n, T_{i+1}^n]\neq\phi\}}$. Then

$$\lim_n \sum_{i,A} \left\{ f(X_{T_{i+1}^n}) - f(X_{T_i^n}) \right\} = \sum_{s\in A} \{ f(X_s) - f(X_{s-}) \},$$

and by Taylor's formula

$$\begin{aligned} \sum_{i,B} \left\{ f(X_{T_{i+1}^n}) - f(X_{T_i^n}) \right\} &= \sum_i f'(X_{T_i^n})(X_{T_{i+1}^n} - X_{T_i^n}) \\ &\quad + \frac{1}{2}\sum_i f''(X_{T_i^n})(X_{T_{i+1}^n} - X_{T_i^n})^2 \\ &\quad - \sum_{i,A} \Big\{ f'(X_{T_i^n})(X_{T_{i+1}^n} - X_{T_i^n}) \\ &\quad + \frac{1}{2} f''(X_{T_i^n})(X_{T_{i+1}^n} - X_{T_i^n})^2 \Big\} \\ &\quad + \sum_{i,B} R(X_{T_i^n}, X_{T_{i+1}^n}). \end{aligned}$$

As in the continuous case, the first two sums on the right side above converge respectively to $\int_{0+}^t f'(X_{s-})dX_s$ and $\frac{1}{2}\int_{0+}^t f''(X_{s-})d[X,X]_s$. The third sum converges to

$$-\sum_{s\in A} \left\{ f'(X_{s-})\Delta X_s + \frac{1}{2} f''(X_{s-})(\Delta X_s)^2 \right\}.$$

Assume temporarily that $|X_s| \le k$, some constant k all $s \le t$. Then f'' is uniformly continuous, and using the right continuity of X we have

$$\lim_n \sup \sum_{i,B} R(X_{T_i^n}, X_{T_{i+1}^n}) \le r(\varepsilon+)[X,X]_t,$$

where $r(\varepsilon+)$ is lim $\sup_{\delta\downarrow\varepsilon} r(\delta)$. Next let ε tend to 0. Then $r(\varepsilon+)[X,X]_t$ tends to 0, and

$$\sum_{s\in A(\epsilon,t)} \left\{ f(X_s) - f(X_{s-}) - f'(X_{s-})\Delta X_s - \frac{1}{2} f''(X_{s-})(\Delta X_s)^2 \right\}$$

tends to the series in (***), provided this series is absolutely convergent.

Let $V_k = \inf\{t > 0 : |X_t| \geq k\}$, with $X_0 = 0$. By first establishing (***) for $X1_{[0,V_k)}$, which is a semimartingale since it is the product of two semimartingales (Corollary 2 of Theorem 22), it suffices to consider semimartingales taking their values in intervals of the form $[-k,k]$. For f restricted to $[-k,k]$ we have $|f(y) - f(x) - (y-x)f'(x)| \leq C(y-x)^2$. Then

$$\sum_{0<s\leq t} |f(X_s) - f(X_{s-}) - f'(X_{s-})\Delta X_s| \leq C \sum_{0<s\leq t} (\Delta X_s)^2 \leq C[X,X]_t < \infty,$$

and $\sum_{0<s\leq t} |f''(X_{s-})|(\Delta X_s)^2 \leq d \sum_{0<s\leq t} (\Delta X_s)^2 \leq d[X,X]_t < \infty$ a.s. Thus the series is absolutely convergent and this completes the proof. □

Corollary (Itô's Formula). *Let X be a continuous semimartingale and let f be a C^2 real function. Then $f(X)$ is again a semimartingale and the following formula holds:*

$$f(X_t) - f(X_0) = \int_{0+}^{t} f'(X_s)dX_s + \frac{1}{2}\int_{0+}^{t} f''(X_s)d[X,X]_s.$$

Theorem 32 has a multi-dimensional analog. We omit the proof.

Theorem 33. *Let $X = (X^1, \ldots, X^n)$ be an n-tuple of semimartingales, and let $f : \mathbb{R}^n \to \mathbb{R}$ have continuous second order partial derivatives. Then $f(X)$ is a semimartingale and the following formula holds:*

$$\begin{aligned} f(X_t) - f(X_0) &= \sum_{i=1}^{n} \int_{0+}^{t} \frac{\partial f}{\partial x_i}(X_{s-})dX_s^i \\ &+ \frac{1}{2} \sum_{1\leq i,j\leq n} \int_{0+}^{t} \frac{\partial^2 f}{\partial x_i \partial x_j}(X_{s-})d[X^i,X^j]_s^c \\ &+ \sum_{0<s\leq t} \left\{ f(X_s) - f(X_{s-}) - \sum_{i=1}^{n} \frac{\partial f}{\partial x_i}(X_{s-})\Delta X_s^i \right\}. \end{aligned}$$

The stochastic integral calculus, as revealed by Theorems 32 and 33, is different from the classical Lebesgue-Stieltjes calculus. By restricting the class of integrands to semimartingales made left continuous (instead of $\mathbb{L}$), *one can define a stochastic integral that obeys the traditional rules of the Lebesgue-Stieltjes calculus.*

Definition. Let X, Y be semimartingales. Define the **Fisk-Stratonovich integral of Y with respect to X**, denoted $\int_0^t Y_{s-} \circ dX_s$, by:

$$\int_0^t Y_{s-} \circ dX_s \equiv \int_0^t Y_{s-} dX_s + \frac{1}{2}[Y, X]_t^c.$$

The Fisk-Stratonovich integral is often referred to as simply the **Stratonovich integral**. The notation "$\circ$" is called **Itô's circle**. Note that we have defined the Fisk-Stratonovich integral in terms of the semimartingale integral. With some work one can slightly enlarge the domain of the definition and we do so in Sect. 5 of Chap. V. In particular, Theorem 34 below is proved with the weaker hypothesis that $f \in C^2$ (Theorem 20 of Chap. V). We will write **the F-S integral** as an abbreviation for the Fisk-Stratonovich integral.

Theorem 34. *Let X be a semimartingale and let f be C^3. Then*

$$f(X_t) - f(X_0) = \int_{0+}^t f'(X_{s-}) \circ dX_s + \sum_{0<s\leq t} \{f(X_s) - f(X_{s-}) - f'(X_{s-})\Delta X_s\}.$$

Proof. Note that f' is C^2, so that $f'(X)$ is a semimartingale by Theorem 32 and in the domain of the $F-S$ integral. By Theorem 32 and the definition, it suffices to establish $\frac{1}{2}[f'(X), X]^c = \frac{1}{2}\int_0^t f''(X_{s-}) d[X, X]_s^c$. However

$$\begin{aligned} f'(X_t) - f'(X_0) = &\int_{0+}^t f''(X_{s-}) dX_s + \frac{1}{2}\int_{0+}^t f^{(3)}(X_{s-}) d[X, X]_s^c \\ &+ \sum_{0<s\leq t} \{f'(X_s) - f'(X_{s-}) - f''(X_{s-})\Delta X_s\}. \end{aligned}$$

Thus

$$[f'(X), X]^c = [f''(X_-) \cdot X, X]^c + [\frac{1}{2} f^{(3)}(X_-) \cdot [X, X], X]^c.$$

The first term on the right side above is $\int_0^t f''(X_{s-}) d[X, X]_s^c$ by Theorem 28; the second term can easily be seen, as a consequence of Theorem 22 and the fact that $[X, X]$ has paths of finite variation, to be $(\sum_{0<s\leq \cdot} f^{(3)}(X_{s-})(\Delta X_s)^3)^c$; that is, zero, and the theorem is proved. □

Note that if X is a semimartingale with *continuous paths*, then Theorem 34 reduces to the classical Riemann-Stieltjes formula: $f(X_t) - f(X_0) = \int_0^t f'(X_s) \circ dX_s$; this is, of course, the main attraction of the Fisk-Stratonovich integral.

Corollary (Integration by Parts). *Let X, Y be semimartingales, with at least one of X and Y continuous. Then*

$$X_tY_t - X_0Y_0 = \int_{0+}^t X_{s-} \circ dY_s + \int_{0+}^t Y_{s-} \circ dX_s.$$

Proof. The standard integration by parts formula is

$$X_tY_t = \int_0^t X_{s-}dY_s + \int_0^t Y_{s-}dX_s + [X,Y]_t.$$

However $[X,Y]_t = [X,Y]_t^c + X_0Y_0$ if one of X or Y is continuous. Thus adding $\frac{1}{2}[X,Y]_t^c$ to each integral on the right side yields the result. □

One can extend the stochastic calculus to complex valued semimartingales. We give a simple example.

Theorem 35. *Let X, Y be continuous semimartingales, let $Z_t = X_t + iY_t$, and let f be analytic. Then*

$$f(Z_t) = f(Z_0) + \int_{0+}^t f'(Z_s)dZ_s + \frac{1}{2}\int_{0+}^t f''(Z_s)d[Z,Z]_s.$$

Proof. Using Itô's formula for the real and imaginary parts of f yields

$$\begin{aligned}
f(Z_t) = f(Z_0) &+ \int_{0+}^t \frac{\partial f}{\partial x}(Z_s)dX_s + \int_{0+}^t \frac{\partial f}{\partial y}(Z_s)dY_s \\
&+ \frac{1}{2}\int_{0+}^t \frac{\partial^2 f}{\partial x^2}(Z_s)d[X,X]_s + \int_{0+}^t \frac{\partial^2 f}{\partial x \partial y}(Z_s)d[X,Y]_s \\
&+ \frac{1}{2}\int_{0+}^t \frac{\partial^2 f}{\partial y^2}(Z_s)d[Y,Y]_s.
\end{aligned}$$

Since f is analytic by the Cauchy-Riemann equations $\frac{\partial f}{\partial x} = f'$ and $\frac{\partial f}{\partial y} = if'$. Differentiating again gives $\frac{\partial^2 f}{\partial x^2} = \frac{\partial f'}{\partial x} = f''$ and $\frac{\partial^2 f}{\partial x \partial y} = if''$ and $\frac{\partial^2 f}{\partial y^2} = -f''$. The result now follows by collecting terms and observing that $dZ_s = dX_s + idY_s$ and $[Z,Z] = [X+iY, X+iY] = [X,X] + 2i[X,Y] - [Y,Y]$. □

Observe that for a complex valued semimartingale *the process* $[Z,Z]$ *is in general a complex valued process.* For many applications, it is more appropriate to use the nonnegative increasing process

$$[Z,\overline{Z}] = [X,X] + [Y,Y]$$

to play the role of the quadratic variation.

8. Applications of Itô's Formula

As an application of the change of variables formula, we investigate a simple, yet important and non-trivial, stochastic differential equation. We treat it, of course, in integral form.

Theorem 36. *Let X be a semimartingale, $X_0 = 0$. Then there exists a (unique) semimartingale Z that satisfies the equation: $Z_t = 1 + \int_0^t Z_{s-} dX_s$; Z is given by*

$$Z_t = \exp\left(X_t - \frac{1}{2}[X,X]_t\right) \prod_{0<s\leq t} (1 + \Delta X_s) \exp\left(-\Delta X_s + \frac{1}{2}(\Delta X_s)^2\right)$$

where the infinite product converges.

Proof. We will not prove the uniqueness here, since it is a trivial consequence of the general theory to be established in Chap. V. (For example see Theorem 7 of Chap. V.) Note that the formula for Z_t is equivalent to the formula:

$$Z_t = \exp\left(X_t - \frac{1}{2}[X,X]_t^c\right) \prod_{s\leq t} (1 + \Delta X_s) \exp(-\Delta X_s),$$

and since $X_t - \frac{1}{2}[X,X]_t^c$ is a semimartingale, $\exp(x)$ is C^2, we need only show that $\prod_{s\leq t}(1 + \Delta X_s)\exp(-\Delta X_s)$ is càdlàg, adapted, and of finite variation and it will be a semimartingale, too; thus Z will be a semimartingale. The product is clearly càdlàg, adapted; it thus suffices to show the product converges and is of finite variation.

Since X has càdlàg paths, there are only a finite number of s such that $|\Delta X_s| \geq 1/2$ on each compact interval (fixed ω). Thus it suffices to show

$$V_t = \prod_{0<s\leq t} \left(1 + \Delta X_s 1_{\{|\Delta X_s|<1/2\}}\right) \exp\left(-\Delta X_s 1_{\{|\Delta X_s|<1/2\}}\right)$$

converges and is of finite variation. Let $U_s = \Delta X_s 1_{\{|\Delta X_s|\leq 1/2\}}$. Then we have $\log V_t = \sum_{s\leq t}\{\log(1 + U_s) - U_s\}$, which is an absolutely convergent series a.s., since $\sum_{0<s\leq t}(U_s)^2 \leq [X,X]_t < \infty$ a.s., because $|\log(1+x) - x| \leq x^2$ when $|x| < 1/2$. Thus $\log(V_t)$ is a process with paths of finite variation, and hence so also is $\exp(\log V_t) = V_t$.

To show that Z is a solution, we set $K_t = X_t - \frac{1}{2}[X,X]_t^c$, and let $f(x,y) = ye^x$. Then $Z_t = f(K_t, S_t)$, where $S_t = \prod_{0\leq s\leq t}(1 + \Delta X_s)\exp(-\Delta X_s)$. By

the change of variables formula we have
(*)

$$Z_t - 1 = \int_{0+}^t Z_{s-} dK_s + \int_{0+}^t e^{K_{s-}} dS_s + \frac{1}{2}\int_{0+}^t Z_{s-} d[K,K]_s^c + \sum_{0<s\leq t}(Z_s - Z_{s-} - Z_{s-}\Delta K_s - e^{K_{s-}}\Delta S_s)$$

$$= \int_{0+}^t Z_{s-} dX_s - \frac{1}{2}\int_{0+}^t Z_{s-} d[X,X]_s^c + \int_{0+}^t e^{K_{s-}} dS_s + \frac{1}{2}\int_{0+}^t Z_{s-} d[X,X]_s^c + \sum_{0<s\leq t}(Z_s - Z_{s-} - Z_{s-}\Delta K_s - e^{K_{s-}}\Delta S_s),$$

since $[K,S]^c = [S,S]^c = 0$. Note that S is a pure jump process; hence $\int_{0+}^t e^{K_{s-}} dS_s = \sum_{0<s\leq t} e^{K_{s-}}\Delta S_s$; also $Z_s = Z_{s-}(1+\Delta X_s)$, and $Z_{s-}\Delta K_s = Z_{s-}\Delta X_s$, so the last sum on the right side of equation (*) becomes:

$$\sum_{0<s\leq t}(Z_{s-}(1+\Delta X_s) - Z_{s-} - Z_{s-}\Delta X_s - e^{K_{s-}}\Delta S_s) = \sum_{0<s\leq t} - e^{K_{s-}}\Delta S_s.$$

Thus equation (*) simplifies due to cancellation: $Z_t - 1 = \int_0^t Z_{s-} dX_s$, and we have the result. □

Definition. For a semimartingale X, $X_0 = 0$ **the stochastic exponential of** X, written $\mathcal{E}(X)$, is the (unique) semimartingale Z that is a solution of: $Z_t = 1 + \int_0^t Z_{s-} dX_s$.

The stochastic exponential is also known as the **Doléans-Dade exponential**. Theorem 36 gives a general formula for $\mathcal{E}(X)$. This formula simplifies considerably when X is continuous. Indeed, let X be a continuous semimartingale with $X_0 = 0$. Then

$$\mathcal{E}(X)_t = \exp(X_t - \frac{1}{2}[X,X]_t).$$

An important special case is when the semimartingale X is a multiple λ of a standard Brownian motion $B = (B_t)_{t\geq 0}$. Since λB has no jumps we have

$$\mathcal{E}(\lambda B)_t = \exp\left(\lambda B_t - \frac{\lambda^2}{2}[B,B]_t\right) = \exp\left(\lambda B_t - \frac{\lambda^2}{2}t\right).$$

Moreover, since $\mathcal{E}(\lambda B)_t = 1 + \lambda\int_0^t \mathcal{E}(\lambda B)_{s-} dB_s$ we see that $\mathcal{E}(\lambda B)_t = e^{\lambda B_t - \frac{\lambda^2}{2}t}$ is a continuous martingale. The process $\mathcal{E}(\lambda B)$ is sometimes referred to as *geometric Brownian motion*. Note that the previous theorem gives us $\mathcal{E}(X)$ in closed form. We also have the following pretty result.

Theorem 37. *Let X and Y be two semimartingales with $X_0 = Y_0 = 0$. Then $\mathcal{E}(X)\mathcal{E}(Y) = \mathcal{E}(X + Y + [X,Y])$.*

Proof. Let $U_t = \mathcal{E}(X)_t$ and $V_t = \mathcal{E}(Y)_t$. Then the integration by parts formula gives that $U_tV_t - 1 = \int_{0+}^t U_{s-}dV_s + \int_{0+}^t V_{s-}dU_s + [U,V]_t$. Using that U and V are exponentials, this is equivalent to:

$$= \int_{0+}^t U_{s-}V_{s-}dY_s + \int_{0+}^t U_{s-}V_{s-}dX_s + \int_{0+}^t U_{s-}V_{s-}d[X,Y]_s;$$

letting $W_t = U_tV_t$, we deduce:

$W_t = 1 + \int_0^t W_{s-}d(X + Y + [X,Y])_s$, and so $W = \mathcal{E}(X + Y + [X,Y])$, which was to be shown. □

Corollary. *Let X be a continuous semimartingale, $X_0 = 0$. Then $\mathcal{E}(X)^{-1} = \mathcal{E}(-X + [X,X])$.*

Proof. By Theorem 37,

$$\mathcal{E}(X)\mathcal{E}(-X + [X,X]) = \mathcal{E}(X + (-X + [X,X]) + [X,-X]),$$

since $[-X,[X,X]] = 0$. However $\mathcal{E}(0) = 1$, and we are done. □

In Sect. 9 of Chap. V we consider general linear equations. In particular, we obtain an explicit formula for the solution of the equation

$$X_t = H_t + \int_0^t X_s dZ_s,$$

where Z is a continuous semimartingale. We also consider more general inverses of stochastic exponentials. See, for example, Theorem 63 of Chap. V.

Another application of the change of variables theorem (and indeed of the stochastic exponential) is a proof of Lévy's characterization of Brownian motion in terms of its quadratic variation.

Theorem 38 (Lévy). *A stochastic process $X = (X_t)_{t\geq 0}$ is a standard Brownian motion if and only if it is a continuous local martingale with $[X,X]_t = t$.*

Proof. We have already observed that a Brownian motion B is a continuous local martingale and that $[B,B]_t = t$ (see the remark following Theorem 22). Thus it remains to show sufficiency. Fix $u \in \mathbf{R}$ and set $F(x,t) = \exp(iux + \frac{u^2}{2}t)$. Let $Z_t = F(X_t,t) = \exp(iuX_t + \left(\frac{u^2}{2}\right)t)$. Since $F \in C^2$ we can apply Itô's formula (Theorem 33) to obtain:

$$\begin{aligned} Z_t &= 1 + iu\int_0^t Z_s dX_s + \frac{u^2}{2}\int_0^t Z_s ds - \frac{u^2}{2}\int_0^t Z_s d[X,X]_s \\ &= 1 + iu\int_0^t Z_s dX_s, \end{aligned}$$

which is the exponential equation. Since X is a continuous local martingale, we now have that Z is also one (complex valued, of course) by the martingale preservation property. Moreover stopping Z at a fixed time t_0, Z^{t_0}, we have that Z^{t_0} is bounded and hence a martingale. It then follows for $0 \leq s < t$ that

$$E\{\exp(iu(X_t - X_s))|\mathcal{F}_s\} = \exp\left(-\frac{u^2}{2}(t-s)\right).$$

Since this holds for any $u \in \mathbb{R}$ we conclude that $X_t - X_s$ is independent of $\mathcal{F}_s$ and that it is normally distributed with mean zero and variance $(t-s)$. Therefore X is a Brownian motion. □

Observe that if M and N are two continuous martingales such that MN is a martingale, then $[M,N] = 0$ by Corollary 2 of Theorem 27. Therefore if $\mathbf{B}_t = (B_t^1, \ldots, B_t^n)$ is an n-dimensional standard Brownian motion, $B_t^i B_t^j$ is a martingale for $i \neq j$, and we have that

$$[B^i, B^j]_t = \begin{cases} t & \text{if } i = j \\ 0 & \text{if } i \neq j \end{cases}.$$

Theorem 38 then has a multi-dimensional version, which has an equally simple proof.

Theorem 39 (Lévy). *Let* $\mathbf{X} = (X^1, \ldots, X^n)$ *be continuous local martingales such that*

$$[X^i, X^j]_t = \begin{cases} t & \text{if } i = j \\ 0 & \text{if } i \neq j \end{cases}.$$

Then $\mathbf{X}$ *is a standard* n*-dimensional Brownian motion.*

As another application of Itô's formula, we exhibit the relationship between harmonic and subharmonic functions and martingales.

Theorem 40. *Let* $\mathbf{X} = (X^1, \ldots, X^n)$ *be an* n*-dimensional continuous local martingale with values in an open subset* D *of* $\mathbb{R}^n$. *Suppose that* $[X^i, X^j] = 0$ *if* $i \neq j$, *and* $[X^i, X^i] = A$, $1 \leq i \leq n$. *Let* $u : D \to \mathbb{R}$ *be harmonic (respectively subharmonic). Then* $u(\mathbf{X})$ *is a local martingale (resp. submartingale).*

Proof. By Itô's formula (Theorem 33) we have $u(\mathbf{X}_t) - u(\mathbf{X}_0) = \int_{0+}^t \operatorname{grad} u(\mathbf{X}_s) \cdot d\mathbf{X}_s + \frac{1}{2}\int_{0+}^t \Delta u(\mathbf{X}_s) dA_s$ where the "dot" denotes Euclidean inner product, and Δ denotes the Laplacian. If u is harmonic (subharmonic), then $\Delta u = 0 (\Delta u \geq 0)$ and the result follows. □

If $\mathbf{X}$ is a standard n-dimensional Brownian motion, then $\mathbf{X}$ satisfies the hypotheses of Theorem 40 with the process $A_t = t$. That $u(\mathbf{X}_t)$ is a submartingale (respectively supermartingale) when u is subharmonic (resp. superharmonic) is the motivation for the terminology submartingale and supermartingale. The relationship between stochastic calculus and potential theory suggested by Theorem 40 has proven fruitful (see, for example, Doob [2]).

Lévy's characterization of Brownian motion (Theorem 38) allows us to prove a useful "change of time" result.

Theorem 41. *Let $M = (M_t)_{t\geq 0}$ be a continuous local martingale with $M_0 = 0$ and such that $\lim_{t\to\infty}[M,M]_t = \infty$ a.s. Let*

$$T_s = \inf\{t > 0 : [M,M]_t > s\}.$$

Define $\mathcal{G}_s = \mathcal{F}_{T_s}$ and $B_s = M_{T_s}$. Then $(B_s, \mathcal{G}_s)_{s\geq 0}$ is a standard Brownian motion. Moreover $([M,M]_t)_{t\geq 0}$ are stopping times for $(\mathcal{G}_s)_{s\geq 0}$ and

$$M_t = B_{[M,M]_t} \quad a.s.\ 0 \leq t < \infty.$$

That is, M can be represented as a time change of a Brownian motion.

Proof. The $(T_s)_{s\geq 0}$ are stopping times by Theorem 3 of Chap. I. Each T_s is finite a.s. by the hypothesis that $\lim_{t\to\infty}[M,M]_t = \infty$ a.s. Therefore the σ-fields $\mathcal{G}_s = \mathcal{F}_{T_s}$ are well defined. The filtration $(\mathcal{G}_s)_{s\geq 0}$ need not be right continuous, but one can take $\mathcal{H}_s = \mathcal{G}_{s+} = \mathcal{F}_{T_s+}$ to obtain one. Note further that $\{[M,M]_t \leq s\} = \{T_s \geq t\}$, hence $([M,M]_t)_{t\geq 0}$ are stopping times for the filtration $\mathcal{G} = (\mathcal{G}_s)_{s\geq 0}$.

By Corollary 3 of Theorem 27 we have $E\{M_{T_s}^2\} = E\{[M,M]_{T_s}\} = s < \infty$, since $[M,M]_{T_s} = s$ identically because $[M,M]$ is continuous. Thus the time changed process is square integrable. Moreover

$$E\{B_u - B_s | \mathcal{G}_s\} = E\{M_{T_u} - M_{T_s} | \mathcal{F}_{T_s}\} = 0$$

by the Optional Sampling Theorem. Also

$$\begin{aligned}
E\{B_u^2 - B_s^2 | \mathcal{G}_s\} &= E\{(B_u - B_s)^2 | \mathcal{G}_s\} \\
&= E\{(M_{T_u} - M_{T_s})^2 | \mathcal{F}_{T_s}\} \\
&= E\{[M,M]_{T_u} - [M,M]_{T_s} | \mathcal{F}_{T_s}\} \\
&= u - s.
\end{aligned}$$

Therefore $B_s^2 - s$ is a martingale, whence $[B,B]_s = s$ provided B has continuous paths.

We want to show that $B_s = M_{T_s}$ has continuous paths. However by Corollary 5 of Theorem 27 almost surely all intervals of constancy of $[M,M]$ are also intervals of constancy of M. It follows easily that B is continuous. It remains to show that $M_t = B_{[M,M]_t}$. Since $B_s = M_{T_s}$, we have that $B_{[M,M]_t} = M_{T_{[M,M]_t}}$, a.s. Since $(T_s)_{s\geq 0}$ is the right continuous inverse of $[M,M]$, we have that $T_{[M,M]_t} \geq t$, with equality holding if and only if t is a point of right increase of $[M,M]$. (If $(T_s)_{s\geq 0}$ were continuous, then we would always have that $T_{[M,M]_t} = t$.) However $T_{[M,M]_t} > t$ implies that $t \to [M,M]_t$ is constant on the interval $(t, T_{[M,M]_t})$; thus by Corollary 4 of Theorem 27 we conclude M is constant on $(t, T_{[M,M]_t})$. Therefore $B_{[M,M]_t} = M_{T_{[M,M]_t}} = M_t$ a.s., and we are done. □

Another application of the change of variables formula is the determination of the distribution of *Lévy's stochastic area process*. Let $\mathbf{B}_t = (X_t, Y_t)$ be an $\mathbf{R}^2$-valued Brownian motion with $(X_0, Y_0) = (0,0)$. Then during the times s to $s + ds$ the chord from the origin to $\mathbf{B}$ sweeps out a triangular region of area $\frac{1}{2}R_s dN_s$, where

$$R_s = \sqrt{X_s^2 + Y_s^2}$$

and

$$dN_s = -\frac{Y_s}{R_s}dX_s + \frac{X_s}{R_s}dY_s.$$

Therefore the integral $A_t = \int_0^t R_s dN_s = \int_0^t (-Y_s dX_s + X_s dY_s)$ is equal to twice the area swept out from time 0 until time t. Paul Lévy found the characteristic function of A_t and therefore determined its distribution. Theorem 42 is known as **Lévy's stochastic area formula.**

Theorem 42. *Let $\mathbf{B}_t = (X_t, Y_t)$ be an $\mathbf{R}^2$-valued Brownian motion, $\mathbf{B}_0 = (0,0)$, $u \in \mathbf{R}$. Let $A_t = \int_0^t X_s dY_s - \int_0^t Y_s dX_s$. Then*

$$E\{e^{iuA_t}\} = \frac{1}{\cosh(ut)}, \quad 0 \le t < \infty, -\infty < u < \infty.$$

Proof. Let $\alpha(t)$, $\beta(t)$ be C^1 functions, and set

$$V_t = iuA_t - \frac{\alpha(t)}{2}(X_t^2 + Y_t^2) + \beta(t).$$

Then

$$\begin{aligned} dV_t &= iudA_t - \frac{\alpha'(t)}{2}(X_t^2 + Y_t^2)dt \\ &\quad - \alpha(t)\{X_t dX_t + Y_t dY_t + dt\} + \beta'(t)dt \\ &= (-iuY_t - \alpha(t)X_t)dX_t + (iuX_t - \alpha(t)Y_t)dY_t \\ &\quad - \frac{1}{2}dt\{\alpha'(t)X_t^2 + \alpha'(t)Y_t^2 + 2\alpha(t) - 2\beta'(t)\}. \end{aligned}$$

Next observe that, from the above calculation:

$$\begin{aligned} d[V,V]_t &= (-iuY_t - \alpha(t)X_t)^2 dt + (iuX_t - \alpha(t)Y_t)^2 dt \\ &= (\alpha(t)^2 - u^2)(X_t^2 + Y_t^2)dt. \end{aligned}$$

Always using the change of variables formula and the preceding calculations:

$$
\begin{aligned}
de^{V_t} &= e^{V_t}(dV_t + \frac{1}{2}d[V,V]_t) \\
&= e^{V_t}(-iuY_t - \alpha(t)X_t)dX_t + e^{V_t}(iuX_t - \alpha(t)Y_t)dY_t \\
&+ \frac{1}{2}e^{V_t}dt\{(\alpha(t)^2 - u^2 - \alpha'(t))(X_t^2 + Y_t^2) + 2\beta'(t) - 2\alpha(t)\}.
\end{aligned}
$$

Therefore e^{V_t} is a local martingale provided

$$
\begin{aligned}
\alpha'(t) &= \alpha(t)^2 - u^2 \\
\beta'(t) &= \alpha(t).
\end{aligned}
$$

Next we fix $t_0 > 0$ and solve the above ordinary differential equations with $\alpha(t_0) = \beta(t_0) = 0$. The solution is:

$$
\begin{aligned}
\alpha(t) &= u \tanh(u(t_0 - t)) \\
\beta(t) &= -\log \cosh(u(t_0 - t)),
\end{aligned}
$$

where tanh and cosh are hyperbolic tangent and hyperbolic cosine. Note that for $0 \leq t \leq t_0$,

$$
|e^{V_t}| = \exp(-\alpha(t)\left\{\frac{X_t^2 + Y_t^2}{2}\right\} + \beta(t)) \leq e^{\beta(t)} \leq 1.
$$

Thus e^{V_t}, $0 \leq t \leq t_0$ is bounded and hence it is a true martingale, not just a local martingale, by Theorem 47 of Chap. I. Therefore

$$
E\{e^{V_{t_0}}\} = E\{e^{V_0}\}.
$$

However $V_{t_0} = iuA_{t_0}$, since $\alpha(t_0) = \beta(t_0) = 0$; and since $A_0 = X_0 = Y_0 = 0$, also $V_0 = -\log\ \cosh(ut_0)$. We conclude that

$$
\begin{aligned}
E\{e^{iuA_{t_0}}\} &= e^{-\log\ \cosh(ut_0)} \\
&= \frac{1}{\cosh(ut_0)},
\end{aligned}
$$

and the proof is complete. □

There are of course other proofs of Lévy's stochastic area formula (e.g., Yor [**5**], or Lévy [**2**]). As a corollary to Theorem 42 we obtain the density for the distribution of A_t.

Corollary. *Let* $\mathbf{B}_t = (X_t, Y_t)$ *be an* $\mathbf{R}^2$*-valued Brownian motion,* $\mathbf{B}_0 = (0,0)$*, and set* $A_t = \int_0^t X_s dY_s - \int_0^t Y_s dX_s$*. Then the density function for the distribution of* A_t *is*

$$f_{A_t}(x) = \frac{1}{2t\cosh(\pi x/2t)}, \quad -\infty < x < \infty.$$

Proof. By Theorem 42 we have the Fourier transform (or characteristic function) of A_t : $E\{e^{iuA_t}\} = \frac{1}{\cosh(ut)}$. Thus we need only to calculate $\frac{1}{2\pi}\int \frac{e^{-iux}}{\cosh(ut)}du$. The integrand is of the form $f(z) = \frac{P(z)}{Q(z)} = \frac{e^{-izx}}{\cosh(zt)}$. Since $\cosh(zt)$ has a pole at $z_0 = \frac{i\pi}{2t}$, we have

$$\mathrm{Res}(f, z_0) = \frac{P(z_0)}{Q'(z_0)} = \frac{e^{\pi x/2t}}{it}.$$

Next we integrate along the closed curve C_r given by

$$C_r : \begin{cases} y = 0 & -r \le x \le r & C_r^1 \\ x = r & 0 \le y \le \frac{\pi}{t} & C_r^2 \\ y = \frac{2\pi}{t} & r \ge x \ge -r & C_r^3 \\ x = -r & \frac{\pi}{t} \ge y \ge 0 & C_r^4 \end{cases}$$

and therefore $\int_{C_r} f(z)dz = 2\pi i\ \mathrm{Res}(f, z_0)$. Along C_r^3 the integral is

$$-\int_r^{-r} \frac{e^{-i(u+\frac{\pi i}{t})x}}{\cosh(ut)}du = e^{\frac{\pi x}{t}} \int_{-r}^{r} \frac{e^{-iux}}{\cosh(ut)}du.$$

The integrands on C_r^2 and C_r^4 are dominated by $2e^{-r}$, and therefore

$$\begin{aligned} \lim_{r\to\infty} \int_{C_r} f(z)dz &= (1 + e^{\pi x/t}) \int_{-\infty}^{\infty} \frac{e^{-iux}}{\cosh(ut)}du \\ &= 2\pi i\ \mathrm{Res}(f, z_0) \\ &= 2\pi \frac{e^{\pi x/2t}}{t}. \end{aligned}$$

Finally we can conclude:

$$\begin{aligned} \frac{1}{2\pi}\int_{-\infty}^{\infty} \frac{e^{-iux}}{\cosh(ut)}du &= \frac{1}{2\pi}\left(\frac{2\pi e^{\frac{\pi x}{2t}}}{t(1 + e^{\frac{\pi x}{t}})}\right) \\ &= \frac{1}{t(e^{\frac{\pi x}{2t}} + e^{\frac{-\pi x}{2t}})} \\ &= \frac{1}{2t\cosh(\frac{\pi x}{2t})}. \end{aligned}$$

□

The stochastic area process A shares some of the properties of Brownian motion, as is seen by recalling that $A_t = \int_0^t R_s dN_s$, where N is a Brownian motion by Lévy's theorem (Theorem 38), and N and R are independent (this must be proven, of course). For example A satisfies a *reflection principle*: If one changes the sign of the increments of A after a stopping time, the process obtained thereby has the same distribution as that of A. One can use this fact to show, for example, that if $S_t = \sup_{0 \leq s \leq t} A_s$, then S_t has the same distribution as $|A_t|$, for $t > 0$.

Bibliographic Notes

The definition of semimartingale and the treatment of stochastic integration as a Riemann-type limit of sums is in essence new. It has its origins in the fundamental theorem of Bichteler [**1**,**2**], and Dellacherie [**2**]. The pedagogic approach used here was first suggested by Meyer [**13**], and it was then outlined by Dellacherie [**2**]. Dellacherie's outline was further expanded by Lenglart [**3**] and Protter [**6**,**7**]. A similar idea was developed by Letta [**1**].

We will not attempt to give a comprehensive history of stochastic integration here, but rather just a sketch. The important early work was that of Wiener [**1**,**2**], and then of course Itô [**1**–**5**]. Doob stressed the martingale nature of the Itô integral in his book [**1**] and proposed a general martingale integral. Doob's proposed development depended on a decomposition theorem (the Doob-Meyer decomposition, Theorem 6 of Chap. III) which did not yet exist. Meyer proved this decomposition theorem in [**1**,**2**], and commented that a theory of stochastic integration was now possible. This was begun by Courrège [**1**], and extended by Kunita and Watanabe [**1**], who revealed an elegant structure of square-integrable martingales and established a general change of variables formula. Meyer [**4**–**7**] extended Kunita and Watanabe's work, realizing that the restriction of integrands to predictable processes is essential. He also extended the integrals to local martingales, which had been introduced earlier by Itô and Watanabe [**1**]. Up to this point, stochastic integration was tied indirectly to Markov processes, by the assumption that the underlying filtration of σ-algebras be "quasi-left continuous." This hypothesis was removed by Doléans-Dade and Meyer [**1**], thereby making stochastic integration a purely martingale theory. It was also in this article that semimartingales were first proposed in the form we refer to as classical semimartingales in Chap. III.

A different theory of stochastic integration was developed independently by McShane [**1**,**2**], which was close in spirit to the approach given here. However it was technically complicated and not very general. It was shown in Protter [**5**] (building on the work of Pop-Stojanovic [**1**]) that the theory of McShane could for practical purposes be viewed as a special case of the semimartingale theory.

The subject of stochastic integration essentially lay dormant for six years until Meyer [**8**] published a seminal "course" on stochastic integration. It was here that the importance of semimartingales was made clear, but it was not until the late 1970's that the theorem of Bichteler [**1**,**2**], and Dellacherie [**2**] gave an *a posteriori* justification of semimartingales: The seemingly *ad hoc* definition of a semimartingale as a process having a decomposition into the sum of a local martingale and an FV process was shown to be the most general reasonable stochastic differential possible. (See also Kussmaul [**1**] in this regard, and the bibliographic notes in Chap. III.)

Most of the results of this chapter can be found in Meyer [**8**], though they are proven for classical semimartingales and hence of necessity the proofs are much more complicated. Theorem 4 (Stricker's Theorem) is (of course) due to Stricker [**1**]; see also Meyer [**10**]. Theorem 5 is due to Meyer [**11**]. There are many other methods of expanding a filtration and still preserving the semimartingale property. The initial result is that of Itô [**8**], while a general reference is Jeulin [**1**]. A simple proof of Itô's original result (with an extension) is in Jacod-Protter [**1**]. A pretty, general result is in Jacod [**2**].

Theorem 14 is originally due to Lenglart [**1**]. Theorem 16 is known to be true only in the case of integrands in $\mathbb{L}$. The local behavior of the integral (Theorems 17 and 18) is due to Meyer [**8**] (see also McShane [**1**]). The a.s. Kunita-Watanabe inequality, Theorem 25, is due to Meyer [**8**], while the expected version (the Corollary to Theorem 25) is due to Kunita-Watanabe [**1**].

That continuous martingales have paths of infinite variation or are constant a.s. was first published by Fisk [**1**] (Corollary 1 of Theorem 27). The proof given here of Corollary 4 of Theorem 27 (that a continuous local martingale X and its quadratic variation $[X, X]$ have the same intervals of constancy) is due to Maisonneuve [**1**]. The proof of Itô's formula (Theorem 32) is by now classic; however we benefited from Föllmer's presentation of it [**1**].

The Fisk-Stratonovich integral was developed independently by Fisk [**1**] and Stratonovich [**1**], and it was extended to general semimartingales by Meyer [**8**]. Theorem 35 is inspired by the work of Getoor and Sharpe [**1**].

The stochastic exponential of Theorem 36 is due to Doléans-Dade [**2**]. It has become extraordinarily important. See, for example, Jacod-Shiryaev [**1**]. The pretty formula of Theorem 37 is due to Yor [**1**]. Exponentials have of course a long history in analysis. For an insightful discussion of exponentials see Gill-Johansen [**1**]. That every continuous local martingale is the time change of a Brownian motion is originally due to Dubins-Schwarz [**1**] and Dambis [**1**]. The proof of Lévy's stochastic area formula (Theorem 42) is new and is due to S. Janson. See Janson-Wichura [**1**] for related results. The original result is in Lévy [**2**], and another proof can be found in Yor [**5**].

CHAPTER III
Semimartingales and Decomposable Processes

1. Introduction

In Chap. II we defined a semimartingale as a good integrator and we developed a theory of stochastic integration for integrands in $\mathbb{L}$, the space of adapted processes with left continuous, right-limited paths. Such a space of integrands suffices to establish a change of variables formula (or "Itô's formula"), and it also suffices for many applications, such as the study of stochastic differential equations. Nevertheless the space $\mathbb{L}$ is not general enough for the consideration of such important topics as local times and martingale representation theorems. We need a space of integrands analogous to measurable functions in the theory of Lebesgue integration; thus defining an integral as a limit of sums – which requires a degree of smoothness on the sample paths – is inadequate. In this chapter we lay the groundwork necessary for an extension of our space of integrands, and the stochastic integral is then extended in Chap. IV.

Historically the stochastic integral was first proposed for Brownian motion, then for continuous martingales, then for square-integrable martingales and finally for processes which can be written as the sum of a locally square integrable local martingale and an adapted, càdlàg processes with paths of finite variation on compacts; that is, a decomposable process. Later Doléans-Dade and Meyer [**1**] showed that the local square integrability hypothesis could be removed, which led to the traditional definition of a semimartingale (what we call a *classical semimartingale*). More formally, let us recall two definitions from Chaps. I and II and then define classical semimartingales.

Definition. An adapted, càdlàg process A is a **finite variation process (FV)** if almost surely the paths of A are of finite variation on each compact interval of $[0,\infty)$. We write $\int_0^\infty |dA_s|$ or $|A|_\infty$ for the random variable which is the total variation of the paths of A.

Definition. An adapted, càdlàg process X is **decomposable** if there exist processes M, A such that

$$X_t = X_0 + M_t + A_t$$

with $M_0 = A_0 = 0$, M a locally square integrable local martingale, and A an FV process.

Definition. An adapted, càdlàg process Y is a **classical semimartingale** if there exist processes N, B with $N_0 = B_0 = 0$ such that

$$Y_t = Y_0 + N_t + B_t$$

where N is a local martingale and B is an FV process.

Clearly an FV process is decomposable, and both FV processes and decomposable processes are semimartingales (Theorems 7 and 9 of Chap. II). *The goal of this chapter is to show that a process X is a classical semimartingale if and only if it is a semimartingale.* To do this we have to develop a small amount of "the general theory of processes". The key result is Theorem 13 which states that any local martingale M can be written

$$M = N + A$$

where N is a local martingale with bounded jumps (and hence locally square integrable), and A is an FV process. An immediate consequence is that a classical semimartingale is decomposable and hence a semimartingale by Theorem 9 of Chap. II. The theorem of Bichteler and Dellacherie (Theorem 22) gives the converse: a semimartingale is decomposable.

We summarize the results of this chapter, that are important to our treatment, in Theorems 1 and 3.

Theorem 1. *Let X be an adapted, càdlàg process. The following are equivalent:*

(i) X is a semimartingale;
(ii) X is decomposable;
(iii) given $\beta > 0$, there exist M, A with $M_0 = A_0 = 0$, M a local martingale with jumps bounded by β, A an FV process, such that $X_t = X_0 + M_t + A_t$;
(iv) X is a classical semimartingale.

Before stating Theorem 2 we give several definitions.

Definition. An FV process A is of **integrable variation** if $E\{\int_0^\infty |dA_s|\} < \infty$. The FV process A is of **locally integrable variation** if there exists a sequence of stopping times $(T^n)_{n\geq 1}$ increasing to ∞ a.s. such that $E\{\int_0^{T^n}|dA_s|\}<\infty$, for each n.

Definition. Let A be an (adapted) FV process, $A_0=0$, of integrable variation. Then A is a **natural process** if

$$E\{[M,A]_\infty\}=0$$

for all bounded martingales M.

Definition. Let A be an (adapted) FV process, $A_0=0$, which is of locally integrable variation. A is **locally a natural process** if

$$E\{[M,A^T]_\infty\}=0$$

for any stopping time T such that $E\{\int_0^T|dA_s|\}<\infty$, and for all bounded martingales M.

Note that in the last definition above, by hypothesis there exist stopping times T^n increasing to ∞ a.s. such that A^{T^n} is of integrable variation, each n. Also, if A is natural, then A^T is also natural for any stopping time T, since $E\{[M,A^T]_\infty\}=E\{[M^T,A]_\infty\}$ by Theorem 23 of Chap. II.

Remark. In Sect. 8 we define the predictable σ-algebra on $\mathbf{R}_+\times\Omega$ and we show that an FV process A of locally integrable variation and $A_0=0$ is *natural if it is predictable*. We also show that if A is *bounded and natural, then it is predictable.*

Theorem 2. *Let A be an FV process, $A_0=0$, and $E\{|A|_\infty\}<\infty$. Then A is natural if and only if*

$$E\left\{\int_0^\infty M_{s-}dA_s\right\}=E\{M_\infty A_\infty\}$$

for any bounded martingale M.

Proof. By integration by parts we have

$$\int_0^\infty M_{s-}dA_s=M_\infty A_\infty-M_0A_0-\int_0^\infty A_{s-}dM_s-[M,A]_\infty$$

Then $M_0A_0=0$ and letting $N_t=\int_0^t A_{s-}dM_s$, we know that N is a local martingale (Theorem 20 of Chap. II). However using integration by parts we see that $E\{N_\infty^*\}<\infty$, hence N is a true martingale (Theorem 47 of Chap. I). Therefore $E\left\{\int_0^\infty A_{s-}dM_s\right\}=E\{N_\infty\}-E\{N_0\}=0$, since N is a martingale. Therefore the equality holds if and only if $E\{[M,A]_\infty\}=0$. □

Note that if A is an FV process with $A_0 = 0$ then it is a quadratic pure jump semimartingale, whence $[M, A]_\infty = \sum_{0<s<\infty} \Delta M_s \Delta A_s$. In particular *if A is continuous then $[M, A] = 0$, and A is locally a natural process.*

Theorem 3. *Let X be a semimartingale. If X has a decomposition $X_t = X_0 + M_t + A_t$ with M a local martingale and A a locally natural FV process, $M_0 = A_0 = 0$, then such a decomposition is unique.*

In Theorem 1, clearly (ii) or (iii) each imply (iv), and (iii) implies (ii), and (ii) implies (i). That (iv) imples (iii) is an immediate consequence of Theorem 12. While Theorem 12 (and Theorems 10 and 11) is quite deep, nevertheless the heart of Theorem 1 is the implication (i) implies (ii), essentially the theorem of K. Bichteler and C. Dellacherie, which itself uses the Doob-Meyer decomposition theorem, Rao's theorem on quasimartingales, and the Girsanov-Meyer theorem on changes of probability laws. Theorem 3 is essentially Theorem 18.

We have tried to present this succession of deep theorems in the most direct and elementary manner possible. This contrasts with their usual treatment, which is customarily part of a general pedagogic presentation of the "general theory of processes". In our effort to keep to a low technical level, we have used some old notions (e.g., "naturality") as well as some old but intuitive proofs.

2. The Doob-Meyer Decompositions

We begin with a definition. Let $\mathbb{N}$ denote the natural numbers.

Definition. An adapted, càdlàg process X is a **potential** if it is a nonnegative supermartingale such that $\lim_{t\to\infty} E(X_t) = 0$. A process $(X_n)_{n\in\mathbb{N}}$ is also called a **potential** if it is a nonnegative supermartingale for $\mathbb{N}$ and $\lim_{n\to\infty} E(X_n) = 0$.

Theorem 4 (The Doob Decomposition). *A potential $(X_n)_{n\in\mathbb{N}}$ has a decomposition $X_n = M_n - A_n$, where $A_{n+1} \geq A_n$ a.s., $A_0 = 0$, $A_n \in \mathcal{F}_{n-1}$, and $M_n = E(A_\infty | \mathcal{F}_n)$. Such a decomposition is unique.*

Proof. Let $M_0 = X_0$ and $A_0 = 0$. Define $M_1 = M_0 + (X_1 - E(X_1|\mathcal{F}_0))$; $A_1 = X_0 - E\{X_1|\mathcal{F}_0\}$. Define M_n, A_n inductively as follows:

$$M_n = M_{n-1} + (X_n - E\{X_n|\mathcal{F}_{n-1}\})$$
$$A_n = A_{n-1} + (X_{n-1} - E\{X_n|\mathcal{F}_{n-1}\}).$$

Note that $E\{A_n\} = E\{X_0\} - E\{X_n\} \leq E\{X_0\} < \infty$, as is easily checked by induction. It is then simple to check that M_n and A_n so defined satisfy the hypotheses.

Next suppose $X_n = N_n - B_n$ is another such representation. Then $M_n - N_n = A_n - B_n$ and in particular $M_1 - N_1 = A_1 - B_1 \in \mathcal{F}_0$; thus $M_1 - N_1 = E\{M_1 - N_1|\mathcal{F}_0\} = M_0 - N_0 = X_0 - X_0 = 0$, hence $M_1 = N_1$. Continuing inductively shows $M_n = N_n$, all n. □

We wish to extend Theorem 4 to continuous time potentials. First note that if A is a process with $A_0 = 0$, $A_n \leq A_{n+1}$ a.s. and $A_n \in \mathcal{F}_n$, then it is simple to check that $A_{n+1} \in \mathcal{F}_n$ if and only if for all bounded martingales Y we have

$$E\left\{\sum_{k=1}^{\infty} Y_{k-1}(A_k - A_{k-1})\right\} = E\{Y_\infty A_\infty\}.$$

If A is indexed by $[0, \infty)$ we can sample A at $t = \frac{i}{2^n}$ and use the discrete parameter example to obtain Riemann-type sums of the form

$$E\left\{\sum_{i=1}^{\infty} Y_{\frac{(i-1)}{2^n}}(A_{\frac{i}{2^n}} - A_{\frac{i-1}{2^n}})\right\}$$

for Y a bounded martingale. Letting n tend to ∞ we see that a condition in the continuous parameter case analogous to $A_n \in \mathcal{F}_{n-1}$ is

$$E\left\{\int_0^\infty Y_{s-} dA_s\right\} = E\{Y_\infty A_\infty\}.$$

This is, of course, precisely the characterization for A to be natural given by Theorem 2 in Sect. 1.

The next theorem is known as the Riesz decomposition, due to its potential theory analogue.

Theorem 5 (Riesz Decomposition). *Let $(X_t)_{t\in\mathbf{R}_+}$ be a positive supermartingale. Then there exists a unique decomposition of X into a martingale and a potential Z such that $X = M + Z$.*

Proof. As is well known $\lim_{t\to\infty} X_t = Y$ exists a.s., and moreover $Y \in L^1$. Let $M_t = E\{Y|\mathcal{F}_t\}$, and let $Z_t = X_t - M_t$. One easily verifies that Z is a potential.

Let $X = N + W$ be another such decomposition. Then $N - M = Z - W$ is a martingale with $Z_\infty - W_\infty = 0$. Therefore $Z = W$ and $N = M$. □

Before stating our next decomposition theorem we establish a preliminary result. Recall that in Chap. II we saw that a continuous local martingale with paths of finite variation on compacts was, in fact, constant (Corollary 1 of Theorem 27 of Chap. II). The concept of a natural process allows us to extend this result, which we do in the following lemma.

Lemma. *Let A be an FV process with $A_0 = 0$, of locally integrable variation, and locally natural. If A is a local martingale then A is identically zero.*

Proof. By stopping we can assume A is of integrable variation and natural. Let T be a finite stopping time and let H be any bounded, nonnegative martingale. Then $E\{\int_0^T H_{s-}dA_s\} = 0$, as is easily seen by approximating sums and the dominated convergence theorem, since $\int_0^T |dA_s| \in L^1$ and $E\{A_T\} = 0$. Using the naturality of A, $E\{H_T A_T\} = E\{\int_0^T H_{s-}dA_s\} = 0$, and letting $H_t = E\{1_{\{A_T>0\}}|\mathcal{F}_t\}$ then shows that $P(A_T > 0) = 0$. Since $E\{A_T\} = 0$, we conclude $A_T \equiv 0$ a.s., hence $A \equiv 0$. □

The next theorem is our first version of the Doob-Meyer decomposition theorem.

Theorem 6. *Let X be a potential such that the collection $\mathcal{H} = \{X_T;\ T$ a stopping time$\}$ is uniformly integrable. Then X has a decomposition $X = M - A$, where M is a uniformly integrable martingale and A is a right continuous, increasing process, $A_0 = 0$, and A is natural. Such a decomposition is unique.*

Proof. For each $n \in \mathbb{N}$, define $Y_i^n = (X_{\frac{i}{2^n}})_{i\in\mathbb{N}}$. Each process $(Y_i^n)_{i\in\mathbb{N}}$ is a discrete potential. By the Doob decomposition (Theorem 4), there exist A_i^n such that $Y_i^n = E\{A_\infty^n|\mathcal{F}_{\frac{i}{2^n}}\} - A^n_{\frac{i}{2^n}}$, where $A^n_{\frac{i}{2^n}} \in \mathcal{F}_{\frac{(i-1)}{2^n}}$ and $A_\infty^n = \lim_{i\to\infty} A^n_{\frac{i}{2^n}}$. Suppose we know $(A_\infty^n)_{n\in\mathbb{N}}$ is a uniformly integrable collection. Then by the Dunford-Pettis theorem there exists a r.v. A_∞ and a subsequence n_k such that $A_\infty^{n_k}$ tends to A_∞ in $\sigma(L^1, L^\infty)$.[1] Let M_t be the right continuous version of $E(A_\infty|\mathcal{F}_t)$. Then for $r \le s$ dyadic rationals and n sufficiently large, $A_r^n \le A_s^n$ a.s., hence $E\{A_\infty^n|\mathcal{F}_r\} - X_r \le E\{A_\infty^n|\mathcal{F}_s\} - X_s$, and it follows that $M_r - X_r \le M_s - X_s$ a.s. Therefore $A_t = M_t - X_t$ is right continuous and a.s. increasing on the dyadic rationals; hence we can take A right continuous, everywhere increasing, and $\lim_{t\to\infty} A_t = A_\infty$ since $\lim_{t\to\infty} X_t = 0$. Moreover $M_0 = X_0$ and $A_0 = 0$ because $E\{A_\infty^n|\mathcal{F}_0\} = X_0$ and the operator $E\{\cdot|\mathcal{F}_0\}$ is weakly continuous.

Now let N be a bounded martingale. By the dominated convergence theorem we have

$$E\left\{\int_0^\infty N_{s-}dA_s\right\} = \lim_{n\to\infty}\sum_{i=0}^\infty E\left\{N_{\frac{i}{2^n}}(A_{\frac{i+1}{2^n}} - A_{\frac{i}{2^n}})\right\}$$

[1] Random variables X^n converge to X in $\sigma(L^1, L^\infty)$ if $X^n, X \in L^1$ and for any bounded r.v. Z, $E\{X^n Z\} \to E\{XZ\}$. For the Dunford-Pettis theorem the reader can consult Meyer [3], p.20, or Dellacherie-Meyer [1], for example.

and since $N_{\frac{i}{2^n}} \in \mathcal{F}_{\frac{i}{2^n}}$, this implies:

$$\begin{aligned} E\left\{\int_0^\infty N_{s-}dA_s\right\} &= \lim_n \sum_i E\left\{N_{\frac{i}{2^n}} E\left\{A_{\frac{i+1}{2^n}} - A_{\frac{i}{2^n}} | \mathcal{F}_{\frac{i}{2^n}}\right\}\right\} \\ &= -\lim_n \sum_i E\left\{N_{\frac{i}{2^n}} E\left\{X_{\frac{i+1}{2^n}} - X_{\frac{i}{2^n}} | \mathcal{F}_{\frac{i}{2^n}}\right\}\right\} \\ &= \lim_n \sum_i E\left\{N_{\frac{i}{2^n}} (A^n_{\frac{i+1}{2^n}} - A^n_{\frac{i}{2^n}})\right\}. \end{aligned}$$

Moreover $E\{N_{\frac{i}{2^n}} A^n_{\frac{i+1}{2^n}}\} = E\{N_{\frac{i+1}{2^n}} A^n_{\frac{i+1}{2^n}}\}$, hence the above implies

$$E\left\{\int_0^\infty N_{s-}dA_s\right\} = \lim_{n\to\infty} \sum_i E\left\{N_{\frac{i+1}{2^n}} A^n_{\frac{i+1}{2^n}} - N_{\frac{i}{2^n}} A^n_{\frac{i}{2^n}}\right\}$$

and we have a telescoping sum. Therefore

$$E\left\{\int_0^\infty N_{s-}dA_s\right\} = \lim_{n\to\infty} E\left\{A^n_\infty N_\infty\right\},$$

and taking the limit along the subsequence n_k shows that the limit is $E\{A_\infty N_\infty\}$, and therefore A is natural by Theorem 2.

As for uniqueness, let $X = L - B$ be another decomposition. Subtracting yields $M - L = A - B$, and therefore $A - B$ is an FV process of integrable total variation, 0 at zero, natural, and a local martingale because it is equal to $M - L$. By the lemma preceding this theorem we conclude $A - B$ is identically zero, which gives uniqueness.

It remains only to show that the collection (A^n_∞) is uniformly integrable. For each $\lambda > 0$ define the stopping times

$$T^\lambda_n = \inf\{\frac{i}{2^n} : A^n_{\frac{i+1}{2^n}} > \lambda\}.$$

Then $A^n_\infty > \lambda$ if and only if $T^\lambda_n < \infty$. Moreover $X_{T^\lambda_n} = E\left\{A^n_\infty | \mathcal{F}_{T^\lambda_n}\right\} - A^n_{T^\lambda_n}$. Therefore

$$(*) \qquad \begin{aligned} E\left\{A^n_\infty 1_{(A^n_\infty > \lambda)}\right\} &= E\left\{A^n_{T^\lambda_n} 1_{(T^\lambda_n < \infty)}\right\} + E\left\{X_{T^\lambda_n} 1_{(T^\lambda_n < \infty)}\right\} \\ &\leq \lambda P(A^n_\infty > \lambda) + E\left\{X_{T^\lambda_n} 1_{(T^\lambda_n < \infty)}\right\}, \end{aligned}$$

and

$$E\left\{(A^n_\infty - \lambda) 1_{(A^n_\infty > 2\lambda)}\right\} \leq E\left\{X^n_{T^\lambda_n} 1_{(T^\lambda_n < \infty)}\right\}$$

which implies

$$2\lambda P(A^n_\infty > 2\lambda) \leq 2E\left\{X_{T^\lambda_n} 1_{(T^\lambda_n < \infty)}\right\} \leq c,$$

a finite constant, by the uniform integrability assumption. Thus since

$$P(T^\lambda_n < \infty) = P(A^n_\infty > \lambda),$$

we have $\lim_{\lambda\to\infty} P(T_n^\lambda < \infty) = 0$, *uniformly in* n. Replacing λ by 2λ in (*) yields

$$\begin{aligned} E\left\{A_\infty^n 1_{(A_\infty^n > 2\lambda)}\right\} &\le 2\lambda P(A_\infty^n > 2\lambda) + E\left\{X_{T_n^{2\lambda}} 1_{(T_n^{2\lambda} < \infty}\right\} \\ &\le 2E\left\{X_{T_n^\lambda} 1_{(T_n^\lambda < \infty)}\right\} + E\left\{X_{T_n^{2\lambda}} 1_{(T_n^{2\lambda} < \infty)}\right\} \end{aligned}$$

and this tends to 0 uniformly in n as λ tends to ∞ by the hypothesis that $\mathcal{H} = \{X_T;\ T \text{ a stopping time}\}$ is uniformly integrable. □

Corollary (Doob-Meyer Decomposition). *Let X be a positive supermartingale, and suppose $\mathcal{H} = \{X_T;\ T$ a stopping time$\}$ is uniformly integrable. Then X has a unique decomposition $X = M - A$ where M is a martingale and A is a right continuous, increasing, natural process with $A_0 = 0$.*

Proof. This is a combination of Theorems 5 and 6. □

The next theorem can also be considered a Doob-Meyer decomposition theorem. It exchanges the uniform integrability for a weakening of the conclusion that M be a martingale to that of M being a *local martingale.* Note that the supermartingale need not be positive.

Theorem 7. *Let Z be a supermartingale. Then Z has a decomposition $Z = Z_0 + M - A$ where M is a local martingale and A is an increasing process which is locally natural, and $M_0 = A_0 = 0$. Such a decomposition is unique. Moreover if $\lim_{t\to\infty} E\{Z_t\} > -\infty$, then $E\{A_\infty\} < \infty$, and A is natural.*

Proof. First consider uniqueness. Let $Z = Z_0 + M - A$ and $Z = Z_0 + N - C$ be two decompositions. Then $M - N = A - C$ by subtraction, hence $A - C$ is a local martingale. Let M^{T^n} be a uniformly integrable martingale. Then

$$E\{Z_0 - Z_{t\wedge T^n}\} = E\{-M_t^{T^n}\} + E\{A_t^{T^n}\} = E\{A_t^{T^n}\}$$

and therefore $E\{A_t^{T^n}\} \le E\{Z_0 - Z_t\}$, using Theorem 17 of Chap. I. Letting n tend to ∞ yields $E\{A_t\} \le E\{Z_0 - Z_t\}$. Thus A is integrable on $[0, t_0]$, each t_0, as is C. Therefore $A - C$ is of locally integrable variation, locally natural, and a local martingale. Since $A_0 - C_0 = 0$, by the Lemma preceding Theorem 6, $A - C = 0$; that is, $A = C$, and hence $M = N$ as well and we have uniqueness.

Next we turn to existence. Let $T^m = \inf\{t : |Z_t| \ge m\} \wedge m$. Then T^m increase to ∞ a.s. and since they are bounded stopping times $Z_{T^m} \in L^1$ each m (Theorem 17 of Chap. I). Moreover the stopped process Z^{T^m} is dominated by the integrable random variable $\max(|Z_{T^m}|, m)$. Hence if $X = Z^{T^m}$ for fixed m, then $\mathcal{H} = \{X_T; T \text{ a stopping time}\}$ is uniformly integrable. Let us implicitly stop Z at the stopping time T^m and for $n > 0$ define $Y_t^n = Z_t - E\{Z_n|\mathcal{F}_t\}$, with $Y_t^n = Y_{t\wedge n}^n = 0$ when $t \geq n$. Then Y^n is a positive supermartingale satisfying the hypotheses of the Corollary of Theorem 6. Thus $Y_t^n = Y_0^n + M_t^n - A_t^n$. Letting $N_t^n = E\{Z_n|\mathcal{F}_t\}$, a martingale, we have on $[0,n]$ that

$$Z_t = Z_0 + M_t^n + N_t^n - A_t^n.$$

To conclude, therefore, it suffices to show that $A_t^m = A_t^n$ on $[0,n]$, for $m \geq n$. This is a consequence of the uniqueness already established. The uniqueness also allows us to remove the assumption that Z is stopped at the time T^m.

Finally, note that since $E\{A_t\} \leq E\{Z_0 - Z_t\}$, and since A is increasing, we have by the monotone convergence theorem

$$E\{A_\infty\} = \lim_{t\to\infty} E\{A_t\} \leq \lim_{t\to\infty} E\{Z_0 - Z_t\}$$

which is finite if $\lim_{t\to\infty} E\{Z_t\} > -\infty$. □

3. Quasimartingales

Let X be a càdlàg, adapted process defined on $[0,\infty]$.[2]

Definition. A finite tuple of points $\tau = (t_0, t_1, \ldots, t_{n+1})$ such that $0 = t_0 < t_1 < \cdots < t_{n+1} = \infty$ is a **partition** of $[0,\infty]$.

Definition. Suppose that τ is a partition of $[0,\infty]$ and that $X_{t_i} \in L^1$, each $t_i \in \tau$. Define

$$C(X,\tau) = \sum_{i=0}^{n} |E\{X_{t_i} - X_{t_{i+1}}|\mathcal{F}_{t_i}\}|;$$

the **variation of** X **along** τ is defined to be

$$\mathrm{Var}_\tau(X) = E\{C(X,\tau)\}.$$

The **variation of** X is defined to be

$$\mathrm{Var}(X) = \sup_\tau \mathrm{Var}_\tau(X),$$

where the supremum is taken over all such partitions.

Definition. An adapted, càdlàg process X is a **quasimartingale** on $[0,\infty]$ if $E\{|X_t|\} < \infty$, for each t, and if $\mathrm{Var}(X) < \infty$.

[2] It is convenient when discussing quasimartingales to include ∞ in the index set, thus making it homeomorphic to $[0,t]$ for $0 < t \leq \infty$. If a process X is defined only on $[0,\infty)$ we extend it to $[0,\infty]$ by setting $X_\infty = 0$.

Before stating the next theorem we recall the notational convention: if X is a random variable then

$$X^+ = \max(X,0)$$
$$X^- = -\min(X,0).$$

Also recall that by convention if X is defined only on $[0,\infty)$, we set $X_\infty = 0$.

Theorem 8. *Let X be a process indexed by $[0,\infty)$. Then X is a quasimartingale if and only if X has a decomposition $X = Y - Z$ where Y and Z are each positive right continuous supermartingales.*

Proof. For given $s \geq 0$, let $\sum(s)$ denote the set of finite subdivisions of $[s,\infty]$. For each $\tau \in \sum(s)$, set

$$Y_s^\tau = E\{C(X,\tau)^+|\mathcal{F}_s\} : \ Z_s^\tau = E\{C(X,\tau)^-|\mathcal{F}_s\}$$

where $C(X,\tau)^+$ denotes $\sum_{t_i\in\tau} E\{X_{t_i} - X_{t_{i+1}}|\mathcal{F}_{t_i}\}^+$, and analogously for $C(X,\tau)^-$. Also let $\prec$ denote the ordering of set containment. Suppose $\sigma,\tau \in \sum(s)$ with $\sigma \prec \tau$. We claim $Y_s^\sigma \leq Y_s^\tau$ a.s. To see this let $\sigma = (t_0,\dots,t_n)$, and it suffices to consider what happens upon adding a subdivision point t before t_0, after t_n, or between t_i and t_{i+1}. The first two situations being clear, let us consider the third. Set

$$A = E\{X_{t_i} - X_t|\mathcal{F}_{t_i}\}; \ B = E\{X_t - X_{t_{i+1}}|\mathcal{F}_t\}$$
$$C = E\{X_{t_i} - X_{t_{i+1}}|\mathcal{F}_{t_i}\};$$

then $C = A + E\{B|\mathcal{F}_{t_i}\}$, hence

$$\begin{aligned} C^+ &\leq A^+ + E\{B|\mathcal{F}_{t_i}\}^+ \\ &\leq A^+ + E\{B^+|\mathcal{F}_{t_i}\}, \end{aligned}$$

by Jensen's inequality. Therefore

$$E\{C^+|\mathcal{F}_s\} \leq E\{A^+|\mathcal{F}_s\} + E\{B^+|\mathcal{F}_s\}$$

and we conclude $Y_s^\sigma \leq Y_s^\tau$. Since $E\{Y_s^\tau\}$ is bounded by $\mathrm{Var}(X)$, taking limits in L^1 along the directed ordered set $\sum(s)$ we define

$$\hat{Y}_s = \lim_\tau Y_s^\tau,$$

and we can define $\hat{Z}_s$ analogously. Taking a subdivision with $t_0 = s$ and $t_{n+1} = \infty$ we see $Y_s^\tau - Z_s^\tau = E\{C^+ - C^-|\mathcal{F}_s\} = X_s$, and we deduce that $\hat{Y}_s - \hat{Z}_s = X_s$. Moreover if $s < t$ it is easily checked that $\hat{Y}_s \geq E\{\hat{Y}_t|\mathcal{F}_s\}$ and $\hat{Z}_s \geq E\{\hat{Z}_t|\mathcal{F}_s\}$. Define the right continuous processes $Y_t \equiv \hat{Y}_{t+}$, $Z_t \equiv \hat{Z}_{t+}$, with the right limits taken through the rationals. Then Y and Z are positive supermartingales and $Y_s - Z_s = X_s$.

For the converse, suppose $X = Y - Z$, where Y and Z are each positive supermartingales. Then for a partition τ of $[0,t]$

$$
\begin{aligned}
E\{\sum_{t_i\in\tau}|E\{X_{t_i}-X_{t_{i+1}}|\mathcal{F}_{t_i}\}|\} &\le E\{\sum_{t_i\in\tau} E\{Y_{t_i}-Y_{t_{i+1}}|\mathcal{F}_{t_i}\}\} \\
&\quad + E\{\sum_{t_i\in\tau} E\{Z_{t_i}-Z_{t_{i+1}}|\mathcal{F}_{t_i}\}\} \\
&= \sum_{t_i\in\tau}(E\{Y_{t_i}\}-E\{Y_{t_{i+1}}\}+E\{Z_{t_i}\}-E\{Z_{t_{i+1}}\}) \\
&= E\{Y_0\}+E\{Z_0\}-(E\{Y_t\}+E\{Z_t\}).
\end{aligned}
$$

Thus X is a quasimartingale on $[0,t]$, each $t>0$. □

Theorem 9 (Rao). *A quasimartingale X has a unique decomposition $X = M + A$, where M is a local martingale and A is a locally natural process with paths of finite variation on compacts and $A_0 = 0$.*

Proof. This theorem is a combination of Theorems 7 and 8. □

If A is of locally integrable variation it is then locally a quasimartingale, and hence by Rao's theorem (Theorem 9), there exists a *unique* decomposition

$$A = M + \widetilde{A}$$

where $\widetilde{A}$ is a locally natural FV process. That is, there exists a unique, locally natural (predictable) FV process $\widetilde{A}$ such that $A - \widetilde{A}$ is a local martingale.

Definition. Let A be an FV process with $A_0 = 0$, with locally integrable total variation. The unique FV locally natural process $\widetilde{A}$ such that $A - \widetilde{A}$ is a local martingale is called the **compensator** of A.

We have commented that a bounded FV process A is locally natural if and only if it is predictable. Actually it is known that A is locally natural if and only if it is predictable. Therefore we could have equivalently defined $\widetilde{A}$ to be the unique *predictable* FV process such that $A-\widetilde{A}$ is a local martingale.

As an example consider the Poisson process $N = (N_t)_{t\ge 0}$. Recall that $N_t - t$ is a martingale. Since the process $V_t = t$ is continuous and obviously adapted, it is natural (predictable). Therefore $\widetilde{N}_t = t$, $t \ge 0$.

Compensators have elementary properties that are very useful. For example *if the FV process A is actually increasing and integrable, then its compensator $\widetilde{A}$ is also increasing and integrable*: Indeed, since A is increasing and $E\{A_\infty\} < \infty$, we have $-A$ is a supermartingale and hence by Theorem 7, $-A = M - C$ where M is a local martingale and C is increasing, natural, and $E\{C_\infty\} < \infty$. Obviously $\widetilde{A} = C$, and moreover $E\{A_\infty\} = E\{\widetilde{A}_\infty\}$. The above clearly extends to: If A is of locally integrable variation and increasing, then $\widetilde{A}$ is (locally integrable and) increasing.

As we saw in Chap. II, processes of fundamental importance to the theory of stochastic integration are the quadratic variation processes $[X,X] = ([X,X]_t)_{t\geq 0}$, where X is a semimartingale.

Definition. Let X be a semimartingale such that its quadratic variation process $[X,X]$ is locally integrable. Then **the conditional quadratic variation of X, denoted** $\langle X,X\rangle = (\langle X,X\rangle_t)_{t\geq 0}$, *exists and it is defined to be the compensator of* $[X,X]$. That is $\langle X,X\rangle = \widetilde{[X,X]}$.

If X is a continuous semimartingale then $[X,X]$ is also continuous and hence already locally natural (predictable); *thus* $[X,X] = \langle X,X\rangle$ *when* X *is continuous.* In particular for a standard Brownian motion B, $[B,B]_t = \langle B,B\rangle_t = t$, all $t \geq 0$. The conditional quadratic variation is also known in the literature by its notation: It is sometimes called the "*sharp bracket*", the "*angle bracket,*" or the "*oblique bracket*". It has properties analogous to that of the quadratic variation processes. For example if X and Y are two semimartingales such that $\langle X,X\rangle$ and $\langle Y,Y\rangle$ both exist, then $\langle X,Y\rangle$ exists and can be defined by polarization

$$\langle X,Y\rangle = \frac{1}{2}(\langle X+Y,X+Y\rangle - \langle X,X\rangle - \langle Y,Y\rangle).$$

Also, $\langle X,X\rangle$ is also a nondecreasing process by the preceding discussion, since $[X,X]$ is nondecreasing. The conditional quadratic variation is inconvenient since unlike the quadratic variation it doesn't always exist. Although it is ubiquitous in the literature it is sometimes unnecessary as one could have often used the quadratic variation instead, and indeed *we will have no need of the conditional quadratic variation* $\langle X,X\rangle$ *of a semimartingale* X *in this book.*

4. The Fundamental Theorem of Local Martingales

We begin by defining two special types of stopping times.

Definition. A stopping time T is **predictable** if there exists a sequence of stopping times $(T_n)_{n\geq 1}$ such that T_n is increasing, $T_n < T$ on $\{T > 0\}$, all n, and $\lim_{n\to\infty} T_n = T$ a.s. such a sequence (T_n) is said to **announce** T.

If X is a continuous, adapted process with $X_0 = 0$, and $T = \inf\{t : |X_t| \geq c\}$, for some $c > 0$, then T is predictable; indeed, the sequence $T_n = \inf\{t : |X_t| \geq c - \frac{1}{n}\} \wedge n$ is an announcing sequence. Fixed times are also predictable.

Definition. A stopping time T is **accessible** if there exists a sequence $(T_k)_{k\geq 1}$ of predictable times such that

$$P(\bigcup_{k=1}^{\infty}\{\omega : T_k(\omega) = T(\omega) < +\infty\}) = P(\{T < \infty\}).$$

Such a sequence $(T_k)_{k\geq 1}$ is said to **envelop** T.

Any stopping time that takes on a countable number of values is clearly accessible. The first jump time of a Poisson process is not an accessible stopping time (indeed, any jump time of a Lévy process is not accessible).

Definition. A stopping time T is **totally inaccessible** if for every predictable stopping time S,

$$P(\{\omega : T(\omega) = S(\omega) < \infty\}) = 0.$$

Let T be a stopping time and $\Lambda \in \mathcal{F}_T$. We define

$$T_\Lambda(\omega) = \begin{cases} T(\omega) & \text{if } \omega \in \Lambda \\ \infty & \text{if } \omega \notin \Lambda \end{cases}$$

It is simple to check that since $\Lambda \in \mathcal{F}_T$, T_Λ is a stopping time. Note further that $T = \min(T_\Lambda, T_{\Lambda^c}) = T_\Lambda \wedge T_{\Lambda^c}$.

Theorem 10. *Let T be a stopping time. There exist disjoint events A, B such that $A \cup B = \{T < \infty\}$ a.s., such that T_A is accessible and T_B is totally inaccessible; and $T = T_A \wedge T_B$ a.s. Such a decomposition is a.s. unique.*

Proof. Let $\mathcal{H} = \{\Lambda \in \mathcal{F} : \Lambda = \bigcup_{n=1}^{\infty}\{S_n = T < \infty\}\}$, where S_n is any sequence of predictable stopping times. $\mathcal{H}$ is stable under countable unions, and $\Lambda \in \mathcal{H}$ implies $\Lambda \in \mathcal{F}_T$. Let

$$H = \text{ess sup } \mathcal{H},$$

and set $A = H \cap \{T < \infty\}$, $B = H^c \cap \{T < \infty\}$. Then T_A is accessible, and T_B is totally inaccessible, as is easily seen. The uniqueness of such a decomposition is left to the reader. □

Theorem 11. *Let T be a totally inaccessible stopping time. Then there exists a martingale M with paths of finite variation and with exactly one jump, of size one, occurring at time T (that is, $M_T \neq M_{T-}$ on $\{T < \infty\}$).*

Proof. Define

$$U_t = 1_{\{t\geq T\}}.$$

and let $A = \widetilde{U}$, the compensator of U. Since U is a bounded submartingale we know that A is increasing, $E\{A_\infty\} < \infty$, and $M = U - A$ is a martingale.

Note that since M is the difference of two increasing processes, it has paths of finite variation.

The theorem will be proved if we can show that A is continuous. We first show that $E\{A_\infty^2\} < \infty$ by establishing a special case of an **energy inequality**. Let $X_t = E\{A_\infty|\mathcal{F}_t\} - A_t$. Then X is a potential and moreover $X_t = E\{U_\infty|\mathcal{F}_t\} - U_t$, hence X is *bounded*, because U is. Since X is bounded we have $E\{\int_0^\infty X_{s-}dA_s\} < \infty$, and for large enough n this implies

$$E\{\sum_{i=0}^{\infty} X_{i/2^n}(A_{i+1/2^n} - A_{i/2^n})\} < \infty.$$

This reduces the problem to the discrete case: Let $X_n = E\{A_\infty|\mathcal{F}_n\} - A_n$ and suppose $\sum_{n=0}^{\infty} E\{X_n(A_{n+1} - A_n)\} < \infty$. We will show $E\{A_\infty^2\} < \infty$. Since $E\{X_{n+1}|\mathcal{F}_n\} \leq X_n$,

$$\sum_n E\{X_{n+1}(A_{n+1} - A_n)\} \leq \sum_n E\{X_n(A_{n+1} - A_n)\} < \infty,$$

and for a process Y letting $\hat{Y}_n$ denote $\min(Y_n, N)$ we have since $A_n \in \mathcal{F}_{n-1}$,

$$\begin{aligned} E\{\hat{A}_\infty^2\} &= \sum_n E\{(\hat{X}_{n+1} + \hat{X}_n)(\hat{A}_{n+1} - \hat{A}_n)\} \\ &\leq \sum_n E\{(X_{n+1} + X_n)(\hat{A}_{n+1} - \hat{A}_n)\} \\ &\leq \sum_n E\{(X_{n+1} + X_n)(A_{n+1} - A_n)\} < \infty, \end{aligned}$$

and therefore $E\{A_\infty^2\} < \infty$ by letting N tend to ∞.

We next show that $A = \widetilde{U}$ is continuous. First, let $(S_n)_{n\geq 1}$ be an increasing sequence of stopping times with limit S. Then

$$\{\lim_{n\to\infty} U_{S_n} \neq U_S\} = \{\lim_{n\to\infty} S_n \wedge T = T, S_n \wedge T < T, \text{ all } n\}.$$

Since T is totally inaccessible we conclude $P(\{\lim_n U_{S_n} \neq U_S\}) = 0$. This in turn implies $\lim_{n\to\infty} E\{U_{S_n}\} = E\{U_S\}$ which implies $\lim_n E\{X_{S_n}\} = E\{X_S\}$. Such a potential is said to regular: That is, a potential Z is **regular** if for any sequence of stopping times S_n increasing to S we have $\lim_{n\to\infty} E\{Z_{S_n}\} = E\{Z_S\}$. Define $A_t^n = E\{A_{k+1/2^n}|\mathcal{F}_t\}$ for $\frac{k}{2^n} \leq t < \frac{k+1}{2^n}$, taking the càdlàg version on $[\frac{k}{2^n}, \frac{k+1}{2^n})$, each k, n. Then A^n decreases to A a.s. for all t. Define

$$T^{n,\epsilon} = \inf\{t > 0 : A_t^n - A_t \geq \epsilon\}$$

and let $T^\epsilon = \lim_{n\to\infty} T^{n,\epsilon}$. Then $E\{A_{T^\epsilon} - A_{T^{n,\epsilon}}\} \geq \epsilon P(\{T_\epsilon < \infty\})$ for all n. Since X is a regular potential we have

$$\lim_{n\to\infty} E\{A_{T^{n,\epsilon}}\} = E\{A_{T^\epsilon}\},$$

and therefore $P(\{T_\epsilon < \infty\}) = 0$. Also, since A^n is a martingale on the intervals $I_{k,n} = [\frac{k}{2^n}, \frac{k+1}{2^n})$, we have by the naturality of A that

$$E\{\int_{I_{k,n}} A^n_{s-} dA_s\} = E\{A_{k+1/2^n}(A_{k+1/2^n} - A_{k/2^n})\}$$

and summing over k with n fixed yields

$$E\{\int_0^\infty A^n_{s-} dA_s\} = \sum_{k=0}^\infty E\{A_{k+1/2^n}(A_{k+1/2^n} - A_{k/2^n})\}$$

which in turn implies

$$(*) \qquad E\{\int_0^\infty (A_s - A_{s-}) dA_s\} = \lim_{n\to\infty} E\{\int_0^\infty (A^n_{s-} - A_{s-}) dA_s\}.$$

However

$$\begin{aligned} E\{\int_0^\infty (A^n_{s-} - A_{s-}) dA_s\} =& E\{\int_0^{T^{n,\epsilon}} (A^n_{s-} - A_{s-}) dA_s\} \\ &+ E\{\int_{T^{n,\epsilon}}^\infty (A^n_{s-} - A_{s-}) dA_s\} \\ &\le \epsilon E\{A_{T^{n,\epsilon}}\} + E\{\int_{T^{n,\epsilon}}^\infty A^n_{s-} dA_s\} \\ &\le \epsilon E\{A_{T^\epsilon}\} + E\{A_\infty(A_\infty - A_{T^{n,\epsilon}})\}. \end{aligned}$$

Combining the above with (*) gives

$$\begin{aligned} E\{\int_0^\infty (A_s - A_{s-}) dA_s\} &= \lim_{n\to\infty} \epsilon E\{A_{T^{n,\epsilon}}\} + E\{A_\infty(A_\infty - A_{T^{n,\epsilon}})\} \\ &\le \epsilon E\{A_{T^\epsilon}\} + E\{A_\infty(A_\infty - A_{T^\epsilon})\} \\ &= \epsilon E\{A_\infty\} \end{aligned}$$

because $E\{A^2_\infty\} < \infty$ and $P(\{T^\epsilon = \infty\}) = 1$. The choice of $\epsilon > 0$ was arbitrary, whence

$$E\{\sum_s (\Delta A_s)^2\} = E\{\int_0^\infty (A_s - A_{s-}) dA_s\} = 0,$$

hence A must be continuous. □

Theorem 12. *Let A be an adapted, càdlàg, nondecreasing locally natural process, and let T be a stopping time. If $P(A_T \neq A_{T-}) > 0$ then T is not totally inaccessible.*

Proof. By stopping we may assume A is of integrable variation. Suppose T is totally inaccessible. By Theorem 11 we know there exists a martingale M which jumps only at T and has jump size one. Without loss of generality take $M_0 = 0$ and let $R_n = \inf\{t : |M_t| > n\}$. Then M^{R_n} is bounded by $n+1$, and since A is natural,

$$0 = E\{[M^{R_n}, A]\} = E\{\sum_t \Delta M_t^{R_n} \Delta A_t\} = E\{\Delta A_T 1_{\{T \le R_n\}}\}.$$

Letting n tend to ∞ yields $E\{A_T - A_{T-}\} = 0$, and since A is nondecreasing we conclude $A_T = A_{T-}$ a.s., which is a contradiction. Therefore T is not totally inaccessible. □

Theorem 13 (Fundamental Theorem of Local Martingales). *Let M be a local martingale and let $\beta > 0$. Then there exist local martingales N, A such that A is an FV process, the jumps of N are bounded by 2β, and $M = N + A$.*

Proof. We first set

$$C_t = \sum_{0<s\le t} |\Delta M_s| 1_{\{|\Delta M_s| \ge \beta\}},$$

and we want to show C is locally integrable. By stopping *we can assume without loss that M is a uniformly integrable martingale and that $M_0 = 0$.* For each ω, $C_t(\omega)$ is a finite sum. Define:

$$R_n = \inf\{t : C_t \ge n \text{ or } |M_t| \ge n\}.$$

Then $|\Delta M_{R_n}| \le |M_{R_n}| + |M_{R_n-}| \le |M_{R_n}| + n$. Moreover $C_{R_n} \le C_{R_n-} + |\Delta M_{R_n}| \le |M_{R_n}| + 2n$, which is in L^1. Since R_n increases to ∞, we have C is locally integrable. By stopping we further assume that $E\{C_\infty\} < \infty$.

Next we define

$$A_t = \sum_{0<s\le t} \Delta M_s 1_{\{|\Delta M_s| \ge \beta\}},$$

and since C is assumed integrable we deduce that A is a quasimartingale. By Theorems 7 and 8 we know that A decomposes as follows:

$$A = L + B_t^1 - B_t^2,$$

where L is a local martingale and B^1, B^2 are nondecreasing, *natural* processes, such that $E\{B_\infty^1\} < \infty$, $E\{B_\infty^2\} < \infty$. Let

$$\widetilde{A}_t = B_t^1 - B_t^2.$$

Then $|A_t - \widetilde{A}_t| \le C_\infty + B_\infty^1 + B_\infty^2 \in L^1$, and hence the martingales $A - \widetilde{A}$ and

$$N = M - (A - \widetilde{A})$$

are uniformly integrable. Here we denote $\widetilde{A}$ for the compensator of A.

It remains only to show that the jumps of N are bounded by 2β. To this end, let T be a stopping time. If $\Delta\widetilde{A}_T = 0$ a.s., then

$$\begin{aligned}\Delta N_T &= \Delta M_T - \Delta A_T \\ &= \Delta M_t 1_{\{|\Delta M_T| \leq \beta\}}.\end{aligned}$$

This implies $|\Delta N_T| \leq \beta$, and we are done by Theorem 7 of Chap. I.

We therefore assume if $\Lambda = \{|\Delta\widetilde{A}_T| > 0\}$, then $P(\Lambda) > 0$. Let $R = T_\Lambda$. By Theorem 10 we can decompose R as $R = R_A \wedge R_B$, where R_A is accessible and R_B is totally inaccessible. Then $P(R_B < \infty) = 0$ by Theorem 12, whence $R = R_A$. Let $(T_k)_{k\geq 1}$ be a sequence of predictable times enveloping R. It will suffice to show $|\Delta N_{T_k}| \leq 2\beta$, for each T_k. Thus without loss of generality we can take $R = T$ to be predictable. By convention, we set $\Delta N_T = 0$ on $\{T = \infty\}$. Let $(S_n)_{n\geq 1}$ be a sequence of stopping times announcing T. We want to show that $\Delta\widetilde{A}_T$ is $\bigvee_n \mathcal{F}_{S_n}$ measurable. Since $\widetilde{A}_{T-} = \lim_{n\to\infty} \widetilde{A}_{S_n}$, it suffices to show $\widetilde{A}_T \in \bigvee_n \mathcal{F}_{S_n}$. Let $A_t^k = E\{\widetilde{A}_{i+1/2^k}|\mathcal{F}_t\}$ for $\frac{i}{2^k} \leq t < \frac{i+1}{2^k}$, the same approximations that were used in the proof of Theorem 11. Then

$$E\{A_\infty^k|\mathcal{F}_t\} = E\{\widetilde{A}_\infty - \widetilde{A}_{i+1/2^k}|\mathcal{F}_t\} + A_{i+1/2^n}^k$$

for $\frac{i}{2^k} \leq t < \frac{i+1}{2^k}$. Next define the stopping time T^k by $T^k = \frac{i+1}{2^k}$ if $\frac{i}{2^k} \leq T < \frac{i+1}{2^k}$. Then

$$E\{A_\infty^k|\mathcal{F}_T\} = E\{\widetilde{A}_\infty - \widetilde{A}_{T^k}|\mathcal{F}_T\} + A_{T^k}^k.$$

However A_∞^k converges to $\widetilde{A}_\infty$ in $\sigma(L^1, L^\infty)$,[3] hence $E\{A_\infty^k|\mathcal{F}_T\}$ converges to $E\{\widetilde{A}_\infty|\mathcal{F}_T\}$ in $\sigma(L^1, L^\infty)$ as well. Since $\widetilde{A}$ is right continuous, $E\{\widetilde{A}_{T^k}|\mathcal{F}_T\}$ converges to $\widetilde{A}_T$ in L^1. Thus $A_{T^k}^k$ converges to $\widetilde{A}_T$ in $\sigma(L^1, L^\infty)$. However if S^n increases to T, then since $A_{i+1/2^k}^k \in \mathcal{F}_{i/2^k}$, we conclude $A_T^k \in \bigvee_n \mathcal{F}_{S^n}$; hence $\widetilde{A}_T \in \bigvee_n \mathcal{F}_{S^n}$ as well.

Also observe that since S_n increases to T and since $\mathcal{F}_{S_n} \supset \mathcal{F}_{S_m}$ for $n \geq m$, we have

$$\begin{aligned}E\{N_T|\bigvee_n \mathcal{F}_{S_n}\} &= \lim_n E\{N_T|\mathcal{F}_{S_n}\} \\ &= \lim_n N_{S_n} \\ &= N_{T-},\end{aligned}$$

implying

$$E\{\Delta N_T|\bigvee_n \mathcal{F}_{S_n}\} = 0.$$

[3] See footnote one of this chapter for the definition of convergence in $\sigma(L^1, L^\infty)$.

Since $A = L + \widetilde{A}$, the above implies

$$E\{\Delta A_T | \bigvee_n \mathcal{F}_{S_n}\} = E\{\Delta \widetilde{A}_T | \bigvee_n \mathcal{F}_{S_n}\}$$
$$= \Delta \widetilde{A}_T.$$

Finally, we need only observe that

$$\begin{aligned}\Delta N_T &= \Delta N_T - E\{\Delta N_T | \bigvee \mathcal{F}_{S_n}\} \\ &= \Delta(M-A)_T + \Delta \widetilde{A}_T - E\{\Delta(M-A)_T | \bigvee \mathcal{F}_{S_n}\} - E\{\Delta \widetilde{A}_T | \bigvee \mathcal{F}_{S_n}\} \\ &= \Delta(M-A)_T - E\{\Delta(M-A)_T | \bigvee \mathcal{F}_{S_n}\}.\end{aligned}$$

Since $|\Delta(M-A)_T| \le \beta$, the result follows. □

If the jumps of a local martingale M with $M_0 = 0$ are bounded by a constant β, then M itself is locally bounded: let $T_n = \inf\{t : |M_t| \ge n\}$. Then $|M_{t \wedge T_n}| \le n + \beta$. Therefore M is *a fortiori* locally square integrable. We thus have as a corollary:

Corollary. *A local martingale is decomposable.*

Of course if all local martingales were locally square integrable, they would then be trivially decomposable, and we would need not have gone to the trouble to prove Theorems 11, 12 and 13. The next example shows that there are martingales that are not locally square integrable (a more complex example is published in Doléans-Dade [**3**]).

Example. Let $(\Omega, \mathcal{F}, P)$ be complete probability space and let X be a random variable such that $X \in L^1$, but $X \notin L^2$. Define the filtration

$$\mathcal{F}_t^0 = \{\emptyset, \Omega\} \qquad 0 \le t < 1$$
$$\mathcal{F}_t^0 = \mathcal{F} \qquad t \ge 1,$$

where $\mathcal{F}_t = \mathcal{F}_t^0 \vee \mathcal{N}$, with $\mathcal{N}$ all the P-null sets of $\mathcal{F}$. Let $M_t = E\{X|\mathcal{F}_t\}$, the right continuous version. Then M is not locally square integrable.

The next example shows another way in which local martingales differ from martingales: A local martingale need not remain a local martingale under a shrinkage of the filtration. They do, however, remain semimartingales and thus they still have an interpretation as a differential.

Example. Let Y be a symmetric random variable with a continuous distribution and such that $E\{|Y|\} = \infty$. Let $X_t = Y 1_{\{t \ge 1\}}$, and define

$$\mathcal{G}_t^0 = \begin{cases} \sigma\{|Y|\} & 0 \le t < 1 \\ \sigma\{Y\} & t \ge 1 \end{cases}$$

where $\mathcal{G}_t$ is the completed filtration. Define stopping times T^n by

$$T^n = \begin{cases} 0 & \text{if } |Y| \geq n \\ \infty & \text{otherwise.} \end{cases}$$

Then T^n reduce X and show that it is a local martingale. However X is not a local martingale relative to its completed minimal filtration. Note that X is still a semimartingale however.

The full power of Theorem 13 will become apparent in Sect. 5.

5. Classical Semimartingales

We have seen that a decomposable process is a semimartingale (Theorem 9 of Chap. II). We can now show that a classical semimartingale is indeed a semimartingale as well.

Theorem 14. *A classical semimartingale is a semimartingale.*

Proof. Let X be a classical semimartingale. Then $X_t = M_t + A_t$ where M is a local martingale and A is an FV process. The process A is a semimartingale by Theorem 7 of Chap. II, and M is decomposable by the Corollary of Theorem 13, hence also a semimartingale (Theorem 9 of Chap. II). Since semimartingales form a vector space (Theorem 1 of Chap. II) we conclude X is a semimartingale. □

Corollary. *A càdlàg local martingale is a semimartingale.*

Proof. A local martingale is a classical semimartingale. □

Theorem 15. *A càdlàg quasimartingale is a semimartingale.*

Proof. By Theorem 9 a quasimartingale is a classical semimartingale. Hence it is a semimartingale by Theorem 14. □

Theorem 16. *A càdlàg supermartingale is a semimartingale.*

Proof. Since a local semimartingale is a semimartingale (Corollary to Theorem 6 of Chap. II), it suffices to show that for a supermartingale X, the stopped process X^t is a semimartingale. However for a partition τ of $[0, t]$,

$$\begin{aligned} E\{\sum_{t_i \in \tau} |E\{X_{t_i} - X_{t_{i+1}} | \mathcal{F}_{t_i}\}|\} &= E\{\sum_{t_i \in \tau} E\{X_{t_i} - X_{t_{i+1}} | \mathcal{F}_{t_i}\}\} \\ &= \sum_{t_i \in \tau} (E\{X_{t_i}\} - E\{X_{t_{i+1}}\}) \\ &= E\{X_0\} - E\{X_t\}. \end{aligned}$$

Therefore X^t is a quasimartingale, hence a semimartingale by Theorem 15. □

Corollary. *A submartingale is a semimartingale.*

We saw in Chap. II that if X is a locally square integrable local martingale and $H \in \mathbb{L}$, then the stochastic integral $H \cdot X$ is also a locally square integrable local martingale (Theorem 20 of Chap. II). Because of the Corollary of Theorem 13 we can now improve this result.

Theorem 17. *Let M be a local martingale and let $H \in \mathbb{L}$. Then the stochastic integral $H \cdot M$ is again a local martingale.*

Proof. A local martingale is a semimartingale by the Corollary of Theorem 13 and Theorem 9 of Chap. II; thus $H \cdot M$ is defined. By the Fundamental Theorem of Local Martingales (Theorem 13) for $\beta > 0$ we can write $M = N + A$ where N, A are local martingales, the jumps of N are bounded by β, and A has paths of finite variation on compacts. Since N has bounded jumps by stopping we can assume N is bounded. Define T by

$$T = \inf\{t > 0 : \int_0^t |dA_s| > m\}.$$

Then $E\{\int_0^{t\wedge T} |dA_s|\} \leq m+\beta+E\{|\Delta M_T|\} < \infty$, and thus by stopping A can be assumed to be of integrable variation: Also by replacing H by $H^S 1_{\{S>0\}}$ for an appropriate stopping time S we can assume without loss of generality that H is bounded, since H is left continuous. We also assume without loss that $M_0 - N_0 = A_0 = 0$. We know $H \cdot N$ is a local martingale by Theorem 20 of Chap. II, thus we need show only that $H \cdot A$ is a local martingale.

Let σ_n be a sequence of random partitions of $[0, t]$ tending to the identity. Then $\sum H_{T_i^n}(A^{T_i^{n+1}} - A^{T_i^n})$ tends to $(H \cdot A)_t$ in ucp, where σ_n is the sequence $0 = T_0^n \leq T_1^n \leq \cdots \leq T_i^n \leq \ldots$. Let (n_k) be a subsequence such that the sums converge uniformly a.s. on $[0, t]$. Then

$$\begin{aligned}
E\left\{\int_0^t H_u dA_u | \mathcal{F}_s\right\} &= E\left\{\lim_{n_k} \sum_i H_{T_i^{n_k}}(A_t^{T_{i+1}^{n_k}} - A_t^{T_i^{n_k}}) | \mathcal{F}_s\right\} \\
&= \lim_{n_k} E\left\{\sum_i H_{T_i^{n_k}}(A_t^{T_{i+1}^{n_k}} - A_t^{T_i^{n_k}}) | \mathcal{F}_s\right\} \\
&= \lim_{n_k} \sum_i H_{T_i^{n_k}}(A_s^{T_{i+1}^{n_k}} - A_s^{T_i^{n_k}})
\end{aligned}$$

by Lebesgue's dominated convergence theorem. Since the last limit above equals $(H \cdot A)_s$, we conclude that $H \cdot A$ is indeed a local martingale. □

Let X be a classical semimartingale, and let $X_t = X_0 + M_t + A_t$ be a decomposition where $M_0 = A_0 = 0$, M is a local martingale, and A is an

FV process. Then if the space $(\Omega, \mathcal{F}, (\mathcal{F}_t)_{t\geq 0}, P)$ supports a Poisson process N, we can write

$$X_t = X_0 + \{M_t + N_t - t\} + \{A_t - N_t + t\}$$

as another decomposition of X. In other words, the decomposition of a classical semimartingale need not be unique. This problem can often be solved by choosing a certain canonical decomposition which is unique.

Definition. Let X be a semimartingale. If X has a decomposition $X_t = X_0 + M_t + A_t$ with $M_0 = A_0 = 0$, M a local martingale, A an FV process, and with A *locally natural*, then X is said to be a **special semimartingale**.

To simplify notation we henceforth assume $X_0 = 0$.

Theorem 18. *If X is a special semimartingale, then its decomposition $X = M + A$ with A locally natural is unique.*

Proof. Let $X = N+B$ be another such decomposition. Then $M-N = B-A$, hence $B - A$ is an FV process which is a local martingale. Moreover $B - A$ is locally natural, hence constant by the Lemma preceding Theorem 6. Since $B_0 - A_0 = 0$, we conclude $B = A$. □

Definition. If X is a special semimartingale, then the unique decomposition $X = M + A$ with A locally natural is called the **canonical decomposition**.

Theorem 9 shows that any quasimartingale is special. A useful sufficient condition for a semimartingale X to be special is that X be a classical semimartingale, or equivalently decomposable, *and also have bounded jumps*.

Theorem 19. *Let X be a classical semimartingale with bounded jumps. Then X is a special semimartingale.*

Proof. Let $X_t = X_0 + M_t + A_t$ be a decomposition of X with $M_0 = A_0 = 0$, M a local martingale, and A an FV process. By Theorem 13 we can then also write

$$X_t = X_0 + N_t + B_t$$

where N is a local martingale with bounded jumps and B is an FV process. Since X and N each have bounded jumps, so also does B. It is therefore locally a quasimartingale and therefore decomposes

$$B = L + \widetilde{B}$$

where L is a local martingale and $\widetilde{B}$ is a locally natural FV process (Theorem 9). Therefore

$$X_t = X_0 + \{N_t + L_t\} + \widetilde{B}_t$$

is the canonical decomposition of X and hence X is special. □

Corollary. *Let X be a classical semimartingale with continuous paths. Then X is special and in its canonical decomposition*

$$X_t = X_0 + M_t + A_t,$$

the local martingale M and the FV process A have continuous paths.

Proof. X is continuous hence trivially has bounded jumps, so it is special by Theorem 19. Since X is continuous we must have

$$\Delta M_T = -\Delta A_T$$

for any stopping time T ($\Delta A_T = 0$ by convention on $\{T = \infty\}$). Suppose A jumps at a stopping time T. By Theorem 10, $T = T_A \wedge T_B$, where T_A is accessible and T_B is totally inaccessible. By Theorem 12 it follows that $P(|\Delta A_{T_B}| > 0) = 0$; hence without loss of generality we can assume T is accessible. It then suffices to consider T predictable (as in the proof of Theorem 13). Let S_n be a sequence of stopping times announcing T. Since A is natural we saw in the proof of Theorem 13 that ΔA_T is $\bigvee_n \mathcal{F}_{S_n}$-measurable. Therefore ΔM_T is also $\bigvee_n \mathcal{F}_{S_n}$ measurable. Stop M so that it is a uniformly integrable martingale. Then

$$\begin{aligned}\Delta M_T &= E\{\Delta M_T | \bigvee_n \mathcal{F}_{S_n}\}\\ &= 0,\end{aligned}$$

and M, and hence A, are continuous, using Theorem 7 of Chap. I. □

A class of examples of special semimartingales is Lévy processes with bounded jumps. By Theorem 19 we need only show that a Lévy process Z with bounded jumps is a classical semimartingale. This is merely Theorem 40 of Chap. I.

For examples of semimartingales which are *not* special, we need only to construct semimartingales with finite variation terms that are not locally integrable. For example, let X be a compound Poisson process with non-integrable, nonnegative jumps. Or, more generally, let X be a Lévy process with Lévy measure ν such that $\int_1^\infty x\nu(dx) = \infty$.

6. Girsanov's Theorem

Let X be a semimartingale on a space $(\Omega, \mathcal{F}, (\mathcal{F}_t)_{t\geq 0}, P)$ satisfying the usual hypotheses. We saw in Chap. II that if Q is another probability law on $(\Omega, \mathcal{F})$, and $Q \ll P$, then X is a Q-semimartingale as well. If X is a classical semimartingale (or equivalently, if X is decomposable) and has a decompostion $X = M + A$, M a local martingale and A an FV process, then it is often useful to be able to calculate the analogous decomposition $X = N + B$, if it exists, under Q. This is rather tricky unless we make a simplifying assumption which usually holds in practice: that both $Q \ll P$ *and* $P \ll Q$.

Definition. Two probability laws P, Q on $(\Omega, \mathcal{F})$ are said to be **equivalent** if $P \ll Q$ and $Q \ll P$. (Recall that $P \ll Q$ denotes that P is absolutely continuous with respect to Q.) We write $Q \sim P$ to denote equivalence.

If $Q \ll P$, then there exists a random variable Z in $L^1(dP)$ such that $\frac{dQ}{dP} = Z$ and $E_P\{Z\} = 1$, where E_P denotes expectation with respect to the law P. We let

$$Z_t = E_P\left\{\frac{dQ}{dP}|\mathcal{F}_t\right\}$$

be the right continuous version. Then Z is a uniformly integrable martingale and hence a semimartingale (by the Corollary of Theorem 14). Note that if Q is equivalent to P, then $\frac{dP}{dQ} \in L^1(dQ)$ and $\frac{dP}{dQ} = \left(\frac{dQ}{dP}\right)^{-1}$.

We begin with a simple lemma, the proof of which we leave to the reader.

Lemma. *Let $Q \sim P$, and $Z_t = E_P\left\{\frac{dQ}{dP}|\mathcal{F}_t\right\}$. An adapted, càdlàg process M is a Q-local martingale if and only if MZ is a P-local martingale.*

Theorem 20 (Girsanov-Meyer). *Let P and Q be equivalent. Let X be a classical semimartingale under P with decomposition $X = M+A$. Then X is also a classical semimartingale under Q and has a decomposition $X = L+C$, where*

$$L_t = M_t - \int_0^t \frac{1}{Z_s} d[Z, M]_s$$

is a Q-local martingale, and $C = X - L$ is a Q - FV process.

Proof. Recall that by Theorem 2 of Chap. II it is trivial that X is a Q-semimartingale. We need to show it is a classical semimartingale, with the above decomposition being valid.

Since M and Z are P-local martingales, they are semimartingales (Corollary of Theorem 14) and

$$\int Z_- dM + \int M_- dZ$$

is a local martingale as well (Theorem 17). Using integration by parts we have

$$(*) \qquad ZM - [Z, M] = \int Z_- dM + \int M_- dZ$$

is also a P-local martingale. Since Z is a version of $E_P\{\frac{dQ}{dP}|\mathcal{F}_t\}$, we have $\frac{1}{Z}$ is a càdlàg version of $E_Q\{\frac{dP}{dQ}|\mathcal{F}_t\}$; therefore $\frac{1}{Z}$ is a Q-semimartingale. Since $P \ll Q$, it is also a P-semimartingale. Multiplying equation (*) by $\frac{1}{Z}$ we have

$$(**) \qquad M - \left(\frac{1}{Z}\right)[Z,M] = \frac{1}{Z}\left(\int Z_- dM + \int M_- dZ\right).$$

Note that if we were to multiply the right side of (**) by Z we would obtain a P-local martingale. We conclude by the Lemma preceding this theorem that $M - (\frac{1}{Z})[Z,M]$ is a Q-local martingale. We next use integration by parts (under Q):

$$\left(\frac{1}{Z}\right)[Z,M] = \int \frac{1}{Z_-} d[Z,M] + \int [Z,M]_- d\left(\frac{1}{Z}\right) + [[Z,M],\frac{1}{Z}].$$

Let $N = \int [Z,M]_- d\left(\frac{1}{Z}\right)$. Since $\frac{1}{Z}$ is a Q-local martingale, so also is N (Theorem 17). The above becomes:

$$\begin{aligned}\left(\frac{1}{Z_t}\right)[Z,M]_t &= \int_0^t \frac{1}{Z_{s-}} d[Z,M]_s + N_t + [[Z,M],\frac{1}{Z}]_t \\ &= \int_0^t \frac{1}{Z_{s-}} d[Z,M]_s + N_t + \sum_{0<s\le t} \Delta\left(\frac{1}{Z_s}\right)\Delta[Z,M]_s \\ &= \int_0^t \frac{1}{Z_s} d[Z,M]_s + N_t.\end{aligned}$$

Adding our two Q-local martingales yields:

$$\begin{aligned}N + M - \left(\frac{1}{Z}\right)[Z,M] &= N + M - \int \frac{1}{Z} d[Z,M] - N \\ &= M - \int \frac{1}{Z} d[Z,M]\end{aligned}$$

which, being the sum of two Q-local martingales, is itself one. This establishes the theorem. □

The Girsanov-Meyer theorem has a similar version when the hypothesis that Q and P be equivalent is relaxed to $Q \ll P$ only. See Lenglart [**1**].

The Girsanov-Meyer theorem is extremely useful, both in applications to the theory of stochastic processes as well as in more applied fields. It often gives an explicit expression for a Radon-Nikodym derivative, which can be interpreted as a likelihood ratio. We give an example from Statistical Communication Theory.

Suppose we are receiving a signal corrupted by noise, and we wish to determine if there is indeed a signal, or if we are just receiving noise (e.g., we could be searching for signs of intelligent life in our galaxy with a radio telescope). Let $x(t)$ be the received signal, $\xi(t)$ the noise, and $s(t)$ the actual (transmitted) signal. Then

$$x(t) = s(t) + \xi(t).$$

A frequent assumption is that the noise is "white". A **white noise** is usually described as a second order wide sense stationary process with a constant spectral density function (that is, $E(\xi_t) = 0$, and $V(\tau) = E\{X_t X_{t+\tau}\} = \delta_\tau S_0$, where δ_τ is the Dirac delta function at τ, and S_0 is the constant spectral density; one then has $S(\nu) = \int_{-\infty}^{\infty} e^{i2\pi\nu\tau} V(\tau) d\tau$). Such a process does not exist in a rigorous mathematical sense. Indeed it can be interpreted as the derivative of the Wiener process, in a generalized function sense. (See Arnold [**1**], p.53, for example.) This suggests that we consider the integrated version of our signal-noise equation:

$$\begin{aligned} X_t &= \int_0^t s(u)du + W_t \\ &= S_t + W_t \end{aligned}$$

where W is a standard Wiener process and where X_t is thought of as "$\int_0^t x(s)ds$", the cumulative received signal.

The key step in our analysis is the following consequence of the Girsanov-Meyer theorem.

Theorem 21. *Let W be a standard Brownian motion on $(\Omega, \mathcal{F}, (\mathcal{F}_t)_{t\geq 0}, P)$, and let $H \in \mathbb{L}$ be bounded. Let*

$$X_t = \int_0^t H_s ds + W_t$$

and define Q by $\frac{dQ}{dP} = \exp(\int_0^T -H_s dW_s - \frac{1}{2}\int_0^T H_s^2 ds)$, for some $T > 0$. Then under Q, X is a standard Brownian motion for $0 \leq t \leq T$.

Proof. Let $Z_T = \exp(\int_0^T -H_s dW_s - \frac{1}{2}\int_0^T H_s^2 ds)$. Then if $Z_t = E\{Z_T | \mathcal{F}_t\}$ we know by Theorem 36 of Chap. II that Z satisfies the equation

$$Z_t = 1 - \int_0^t Z_{s-} H_s dW_s.$$

By the Girsanov-Meyer Theorem (Theorem 20), we know that

$$N_t = W_t - \int_0^t \frac{1}{Z_s} d[Z, W]_s$$

is a Q-local martingale. However

$$\begin{aligned} [Z, W]_t = [-ZH \cdot W, W]_t &= \int_0^t -Z_s H_s d[W, W]_s \\ &= -\int_0^t Z_s H_s ds, \end{aligned}$$

since $[W,W]_t = t$ for Brownian motion. Therefore

$$\begin{aligned} N_t &= W_t - \int_0^t -\frac{1}{Z_s} Z_s H_s ds \\ &= W_t + \int_0^t H_s ds \\ &= X_t, \end{aligned}$$

and therefore X is a Q-local martingale. Since $(\int_0^t H_s ds)_{t\geq 0}$ is a continuous FV process we have that $[X,X]_t = [W,W]_t = t$, and therefore by Lévy's Theorem (Theorem 38 of Chap. II) we conclude that X is a standard Brownian motion. □

Corollary. *Let W be a standard Brownian motion and $H \in \mathbf{L}$ be bounded. Then the law of*

$$X_t = \int_0^t H_s ds + W_t,$$

$0 \leq t \leq T < \infty$, *is equivalent to Wiener measure.*

Proof. Let $C[0,T]$ be the space of continuous functions on $[0,T]$ with values in $\mathbf{R}$ (such a space is called a *path space*). If $W = (W_t)_{0\leq t\leq T}$ is a standard Brownian motion, it induces a measure μ_W on $C[0,T]$:

$$\mu_W(A) = P(\{\omega : t \mapsto W_t(\omega) \in A\}).$$

Let μ_X be the analogous measure induced by X. Then by Theorem 21 we have $\mu_X \sim \mu_W$ and further we have

$$\frac{d\mu_W}{d\mu_X} = \exp\left(\int_0^T -H_s dW_s - \frac{1}{2}\int_0^T H_s^2 ds\right).$$

□

Remark. We have not tried for maximum generality here. For example the hypothesis that H be bounded can be weakened. It is also desirable to weaken the restriction that $H \in \mathbf{L}$. Indeed we only needed that hypothesis to be able to form the integral $(\int_0^t H_s dW_s)$. This is one example to indicate why we need a space of integrands more general than $\mathbf{L}$.

We are now in a position to consider the problem posed earlier: is there a signal corrupted by noise, or is there just noise (that is, does $s(t) = 0$ a.e., a.s.)? In terms of hypothesis testing, let $\mathbf{H}_0$ denote the null hypothesis, $\mathbf{H}_1$ the alternative. We have:

$$\mathbf{H}_0 : X_t = W_t$$

$$\mathbf{H}_1 : X_t = \int_0^t H_s ds + W_t.$$

We then have

$$\frac{d\mu_W}{d\mu_X} = \exp\left(\int_0^t -H_s dW_s - \frac{1}{2}\int_0^t H_s^2 ds\right),$$

by the preceding Corollary. This leads to a likelihood ratio test:

$$\text{if} \quad \frac{d\mu_W}{d\mu_X}(\omega) \leq \lambda, \quad \text{reject} \quad \mathbf{H}_0;$$

$$\text{if} \quad \frac{d\mu_W}{d\mu_X}(\omega) > \lambda, \quad \text{fail to reject} \quad \mathbf{H}_0;$$

where the threshold level λ is chosen so that the fixed Type I error is achieved.

To indicate another use of the Girsanov-Meyer theorem let us consider stochastic differential equations. Since stochastic differential equations[4] are treated systematically in Chap. V we are free here to restrict our attention to a simple but illustrative situation. Let W be a standard Brownian motion on a space $(\Omega, \mathcal{F}, (\mathcal{F}_t)_{t\geq 0}, P)$ satisfying the usual hypotheses. Let $f_i(\omega, s, x)$ be functions satisfying $(i = 1, 2)$:

(i) $|f_i(\omega, s, x) - f_i(\omega, s, y)| \leq K|x - y|$ for fixed (ω, s)
(ii) $f_i(\cdot, s, x) \in \mathcal{F}_s$ for fixed (s, x)
(iii) $f_i(\omega, \cdot, x)$ is left continuous with right limits for fixed (ω, x).

By a Picard-type iteration procedure one can show there exists a unique solution (with continuous paths) of:

$$(***) \qquad X_t = X_0 + \int_0^t f_1(\cdot, s, X_s) dW_s + \int_0^t f_2(\cdot, s, X_s) ds.$$

The Girsanov-Meyer theorem allows us to establish the existence of solutions of analogous equations where the Lipschitz hypothesis on the "drift" coefficient f_2 is removed. Indeed if X is the solution of (***), let γ be any bounded, measurable function such that $\gamma(\omega, s, X_s) \in \mathsf{L}$. Define

$$g(\omega, s, x) = f_2(\omega, s, x) + f_1(\omega, s, x)\gamma(s, \omega, x).$$

We will see that we can find a solution of

$$Y_t = Y_0 + \int_0^t f_1(\cdot, s, Y_s) dB_s + \int_0^t g(\cdot, s, Y_s) ds$$

provided we choose a new Brownian motion B appropriately.

We define a new probability law Q by

$$\frac{dQ}{dP} = \exp\left(\int_0^T \gamma(s, X_s) dW_s - \frac{1}{2}\int_0^T \gamma(s, X_s)^2 ds\right).$$

[4] Stochastic "differential" equations have meaning only if they are interpreted as stochastic integral equations.

By Theorem 21 we have that

$$B_t = W_t - \int_0^t \gamma(s, X_s)ds$$

is a standard Brownian motion under Q. We then have that the solution X of $(***)$ also satisfies

$$\begin{aligned} X_t &= X_0 + \int_0^t f_1(\cdot, s, X_s)dB_s + \int_0^t (f_2 + f_1\gamma)(\cdot, s, X_s)ds \\ &= X_0 + \int_0^t f_1(\cdot, s, X_s)dB_s + \int_0^t g(\cdot, s, X_s)ds, \end{aligned}$$

which is a solution of a stochastic differential equation driven by a Brownian motion, under the law Q.

7. The Bichteler-Dellacherie Theorem

In Sect. 5 we saw that a classical semimartingale is a semimartingale. In this section we will show the converse.

Theorem 22 (Bichteler-Dellacherie). *An adapted, càdlàg process X is a semimartingale if and only if it is a classical semimartingale. That is, X is a semimartingale if and only if it can be written $X = M + A$, where M is a local martingale and A is an FV process.*

Proof. The sufficiency is exactly Theorem 14. We therefore establish here only the necessity.

Since X is càdlàg, the process $J_t = \sum_{0<s\leq t} \Delta X_s 1_{\{|\Delta X_s|\geq 1\}}$ has paths of finite variation on compacts; hence it is an FV process. Let $Y = X - J$; then Y has bounded jumps. If we show that X is a classical semimartingale on $[0, t_0]$, some t_0, then $Y = X - J$ is one as well; moreover Y is special by Theorem 19. Let $Y = Y_0 + \overline{M} + \overline{A}$ be the canonical decomposition of Y. Suppose $t_1 > t_0$ and X is also shown to be a classical semimartingale on $[0, t_1]$. Then let $Y = Y_0 + \overline{N} + \overline{B}$ be the canonical decomposition of Y on $[0, t_1]$. We have that, for $t \leq t_0$,

$$\overline{M} - \overline{N} = \overline{B} - \overline{A}$$

by subtraction. Thus $\overline{B} - \overline{A}$ is a locally natural, FV process which is a local martingale; hence by the Lemma preceding Theorem 6 we have $\overline{B} = \overline{A}$. Thus if we take t_n increasing to ∞ and show that X is a classical semimartingale on $[0, t_n]$, each n, we have that X is a classical semimartingale on $[0, \infty)$.

We now choose $u_0 > 0$, and by the above it suffices to show X is a classical semimartingale on $[0, u_0]$. Thus it is no loss to assume X is a total semimartingale on $[0, u_0]$. We will show that X is a quasimartingale, under an equivalent probability Q. Rao's theorem (Theorem 9) shows that X is a classical semimartingale under Q, and the Girsanov-Meyer theorem (Theorem 20) then shows that X is a classical semimartingale under P.

Let us take $H \in \mathbf{S}$ of the special form:

$$(*) \qquad H_t = \sum_{i=0}^{n-1} H_i 1_{(T_i, T_{i+1}]}$$

where $0 = T_0 \leq T_1 \leq \cdots \leq T_{n-1} < T_n = u_0$. In this case the mapping I_X is given by:

$$I_X(H) = (H \cdot X)_{u_0} = H_0(X_{T_1} - X_0) + \cdots + H_{n-1}(X_{u_0} - X_{T_{n-1}}).$$

The mapping $I_X : \mathbf{S}_u \to L^0$ is continuous, where L^0 is endowed with the topology of convergence in probability, by the hypothesis that X is a total semimartingale. Let

$$\mathcal{B} = \{H \in \mathbf{S} : H \text{ has a representation } (*) \text{ and } |H| \leq 1\}.$$

Let $\beta = I_X(\mathcal{B})$, the image of $\mathcal{B}$ under I_X. It will now suffice to find a probability Q equivalent to P such that $X_t \in L^1(dQ)$, $0 \leq t \leq u_0$ and such that $\sup_{U \in \beta} E_Q(U) = c < \infty$. The reason this suffices is that if we take, for a given $0 = t_0 < t_1 < \cdots < t_n = u_0$, the random variables $H_0 = \operatorname{sign}(E_Q\{X_{t_1} - X_0 | \mathcal{F}_0\})$, $H_1 = \operatorname{sign}(E_Q\{X_{t_2} - X_{t_1} | \mathcal{F}_{t_1}\}), \ldots$, we have that for this $H \in \mathcal{B}$,

$$E_Q(I_X(H)) = E_Q\{|E_Q\{X_{t_0} - X_{t_1} | \mathcal{F}_0\}| + \cdots + |E_Q\{X_{t_{n-1}} - X_{t_n} | \mathcal{F}_{t_{n-1}}\}|\}.$$

Since this partition τ was arbitrary, we have $\operatorname{Var}(X) = \sup_\tau \operatorname{Var}_\tau(X) \leq \sup_{U \in \beta} E_Q(U) = c < \infty$, and so X is a Q-quasimartingale.

Lemma 1. $\lim_{c \to \infty} \sup_{Y \in \beta} P(|Y| > c) = 0$.

Proof. Suppose $\lim_{c \to \infty} \sup_{Y \in \beta} P(|Y| > c) > 0$. Then there exists a sequence c_n tending to ∞, $Y_n \in \beta$, and $a > 0$ such that $P(|Y_n| > c_n) \geq a$, all n. This is equivalent to

$$(*) \qquad P\left(\frac{|Y_n|}{c_n} > 1\right) \geq a > 0.$$

Since $Y_n \in \beta$, there exists $H^n \in \mathcal{B}$ such that $I_X(H^n) = Y_n$. Then $I_X(\frac{1}{c_n} H^n) = \frac{1}{c_n} I_X(H^n) = \frac{1}{c_n} Y_n \in \beta$, if $c_n \geq 1$. But $\frac{1}{c_n} H^n$ tends to 0 uniformly a.s. which implies that $I_X(\frac{1}{c_n} H^n) = \frac{1}{c_n} Y_n$ tends to 0 in probability. This contradicts (*). $\square$

Lemma 2. *There exists a law Q equivalent to P such that $X_t \in L^1(dQ)$, $0 \leq t \leq u_0$.*

Proof. Let $Y = \sup_{0\leq t\leq u_0}|X_t|$. Since X has càdlàg paths, $Y < \infty$ a.s., and moreover if D is a countable dense subset of $[0, u_0]$ then $Y = \sup_{t\in D}|X_t|$; hence Y is a random variable. Let $A_m = \{m \leq Y < m+1\}$, and set $Z = \sum_{m=0}^{\infty} 2^{-m} 1_{A_m}$. Then Z is bounded, strictly positive, and $YZ \in L^1(dP)$. Define Q by setting $\frac{dQ}{dP} = \frac{1}{E_P(Z)} Z$. Then $E_Q\{|X_t|\} \leq E_Q\{Y\} = E_P\{YZ\}/E_P\{Z\} < \infty$. Hence $E_Q\{|X_t|\} < \infty$, $0 \leq t \leq u_0$. □

Observe that $\beta \subset L^1(dQ)$ for the law Q constructed in Lemma 2. Lemma 3 below next implies that X is an R-quasimartingale for $R \sim Q \sim P$. Hence by Rao's Theorem (Theorem 9) it is a classical semimartingale (R), and by the Girsanov-Meyer theorem (Theorem 20) it is a classical semimartingale for the equivalent law P as well. Thus Lemma 3 completes the proof of Theorem 22. We follow Yan ([1]).

Lemma 3. *Let β be a convex subset of $L^1(dQ)$, $0 \in \beta$, that is bounded in probability: that is, for any $\epsilon > 0$ there exists a $c > 0$ such that $Q(\zeta > c) \leq \epsilon$, for any $\zeta \in \beta$. Then there exists a probability R equivalent to Q, with a bounded density, such that $\sup_{U\in\beta} E_R(U) < \infty$.*

Proof. First note that the hypotheses imply that $\beta \subset L^1(dR)$. What we must show is that there exists a bounded random variable Z, such that $P(Z > 0) = 1$, and such that $\sup_{\zeta\in\beta} E_Q(Z\zeta) < \infty$.

Let $A \in \mathcal{F}$ such that $Q(A) > 0$. Then there exists a constant d such that $Q(\zeta > d) \leq Q(A)/2$, for all $\zeta \in \beta$, by assumption. Using this constant d, let $c = 2d$, and we have that $0 \leq c1_A \notin \beta$, and moreover if B_+ denotes all bounded, positive r.v., then $c1_A$ is not in the $L^1(dQ)$ closure of $\beta - B_+$, denoted $\overline{\beta - B_+}$. That is, $c1_A \notin \overline{\beta - B_+}$. Since the dual of L^1 is L^∞, and $\overline{\beta - B_+}$ is convex, by a version of the Hahn-Banach theorem (see, e.g., Treves [1, p.190]) there exists a bounded random variable Y such that

$$(*) \qquad \sup_{\zeta\in\beta, \eta\in B_+} E_Q\{Y(\zeta - \eta)\} < cE_Q\{Y1_A\}.$$

Replacing η by $a1_{\{Y<0\}}$ and letting a tend to ∞ shows that $Y \geq 0$ a.s., since otherwise the expectation on the left side above would get arbitrarily large. Next suppose $\eta = 0$. Then the inequality above gives

$$\sup_{\zeta\in\beta} E_Q\{Y\zeta\} \leq cE_Q\{Y1_A\} < +\infty.$$

Now set $\mathcal{H} = \{Y \in B_+ : \sup_{\zeta\in\beta} E_Q\{Y\zeta\} < \infty\}$. Since $0 \in B_+$, we know $\mathcal{H}$ is not empty. Let $\mathcal{A} = \{$all sets of the form $\{Z = 0\}, Z \in \mathcal{H}\}$. We wish to show that there exists a $Z \in \mathcal{H}$ such that $Q(\{Z = 0\}) = \inf_{A\in\mathcal{A}} Q(A)$. Suppose, then, that Z_n is a sequence of elements of $\mathcal{H}$. Let $c_n = \sup_{\zeta\in\beta} E\{Z_n\zeta\}$ and $d_n = \|Z_n\|_{L^\infty}$. (Since $0 \in \beta$, we have $c_n \geq 0$). Choose b_n such that

$\sum b_n c_n < \infty$ and $\sum b_n d_n < \infty$, and set $Z = \sum b_n Z_n$. Then clearly $Z \in \mathcal{H}$. Moreover, $\{Z = 0\} = \cap_n \{Z_n = 0\}$. Thus $\mathcal{A}$ is stable under countable intersections, and so there exists a Z such that $Q(\{Z = 0\}) = \inf_{A \in \mathcal{A}} Q(A)$.

We now wish to show $Z > 0$ a.s. Suppose not. That is, suppose $Q(\{Z = 0\}) > 0$. Let Y satisfy $(*)$ (we have seen that there exists such a Y and that it hence is in $\mathcal{H}$). Further we take for our set A in $(*)$ the set $A = \{Z = 0\}$, for which we are assuming $Q(A) > 0$. Since $0 \in \beta$ and $0 \in B_+$, we have from Lemma 2 that

$$0 < E\{Y 1_A\} = E\{Y 1_{\{Z=0\}}\}.$$

Since each of Y and Z are in $\mathcal{H}$, their sum is in $\mathcal{H}$ as well. But then the above implies

$$Q\{Y + Z = 0\} = Q\{Z = 0\} - Q(\{Z = 0\} \cap \{Y > 0\}) < Q(\{Z = 0\}).$$

This, then, is a contradiction, since $Q(\{Z = 0\})$ is minimal for $Z \in \mathcal{H}$. Therefore we conclude $Z > 0$ a.s., and since $Z \in B_+$, it is bounded as well, and Lemma 3 is proved; thus also, Theorem 22 is proved. □

We state again, for emphasis, that Theorems 14 and 22 together allows us to conclude that semimartingales (as we have defined them) and classical semimartingales are the same.

8. Natural Versus Predictable Processes

Recall that $\mathbb{L}$ is the space of adapted processes having left continuous paths with right limits.

Definition. The **predictable σ-algebra** $\mathcal{P}$ on $\mathbb{R}_+ \times \Omega$ is the smallest σ-algebra making all processes in $\mathbb{L}$ measurable. We also let $(\mathbf{b}\mathcal{P})$ $\mathcal{P}$ denote the (bounded) processes that are predictably measurable.

In this section we show that if A is a bounded FV process of locally integrable variation with $A_0 = 0$, then A is locally natural if and only if A is predictable (Theorem 27). This result is not needed for our treatment,[5] and this section could be viewed as an appendix. We include it because in the literature the concept of a natural process has largely been replaced with that of a predictable process, and it is relatively simple to see that the two notions are equivalent. Indeed, the sufficiency is extremely simple if one accepts a result of Chap. IV.

[5] It is however used once in the proof of Theorem 24 of Chap. IV. (Theorem 26 is not used to prove Theorem 24 of Chap. IV.)

Theorem 23. *Let A be an FV process of integrable variation with $A_0 = 0$. If A is predictable, then A is natural.*

Proof. Let M be a bounded martingale. First assume A is bounded. Then the stochastic integral $\int_0^\infty A_s dM_s$ exists, and it is a martingale by Theorem 29 in Chap. IV combined with, for example, Corollary 3 of Theorem 29 of Chap. II. Therefore $E\{\int_0^\infty A_s dM_s\} = E\{A_0 M_0\} = 0$, since $A_0 = 0$. However $E\{\int_0^\infty A_{s-} dM_s\} = E\{A_{0-} M_0\} = 0$ as well, since $A_{0-} = 0$. However

$$\begin{aligned}\int_0^\infty A_s dM_s - \int_0^\infty A_{s-} dM_s &= \int_0^\infty (A_s - A_{s-}) dM_s \\ &= \int_0^\infty \Delta A_s dM_s \\ &= \sum_{0<s<\infty} \Delta A_s \Delta M_s \\ &= [A, M]_\infty,\end{aligned}$$

since A is a quadratic pure jump semimartingale. Therefore

$$E\{[A, M]_\infty\} = E\{\int_0^\infty A_s dM_s - \int_0^\infty A_{s-} dM_s\} = 0.$$

Since M was an arbitrary bounded martingale, A is natural by definition. Finally we remove the assumption that A is bounded: Let $A^n = n \wedge (A \vee (-n))$. Then A^n is bounded and still predictable, hence it is natural. For a bounded martingale M

$$E\{[M, A]_\infty\} = \lim_n E\{[M, A^n]_\infty\} = 0,$$

by the dominated convergence theorem. Therefore A is natural. □

Corollary 1. *Let A be an FV process of locally integrable variation with $A_0 = 0$. If A is predictable, then A is natural.*

Proof. Let T^n increase to ∞ a.s. such that A^{T^n} is of integrable variation. Then $A^{T^n} = A 1_{[0,T^n]} + A_{T^n} 1_{(T^n,\infty)}$ hence A^{T^n} is still predictable; therefore A^{T^n} is natural by Theorem 23. Thus A is locally natural. □

Corollary 2. *Let A be an FV process of locally integrable variation with $A_0 = 0$. If A is predictable and a local martingale, then A is identically zero.*

Proof. Since A is predictable it is also natural by Theorem 23. The result then follows by the lemma preceding Theorem 6. □

To prove the converse we need the existence of a predictable projection. This entails another definition.

Definition. Let T be a stopping time. **The σ-field $\mathcal{F}_{T-}$** is the smallest σ-field containing $\mathcal{F}_0$ and all sets of the form $A \cap \{t < T\}$, $t > 0$ and $A \in \mathcal{F}_t$. Observe that $\mathcal{F}_{T-} \subset \mathcal{F}_T$, and also the stopping time T is $\mathcal{F}_{T-}$ measurable.

Theorem 24. *Let H be a bounded, measurable process. There exists a unique predictable process $\dot{H}$, bounded by* $\sup|H|$, *such that*

$$\dot{H}_T = E\{H_T|\mathcal{F}_{T-}\} \quad on\ \{T < \infty\}$$

for all predictable stopping times T.

Proof. Uniqueness is interpreted as follows: if J is another such process, then $\dot{H}$ and J are indistinguishable. The proof of uniqueness involves deep results known as the section theorems, and we refer the reader to Jacod-Shiryaev [**1**], p.23. We will prove existence.

Let $\mathcal{H}$ be the collection of bounded $\mathcal{F} \otimes \mathcal{B}_+$ measurable processes H such that such an $\dot{H}$ exists. Then $\mathcal{H}$ is a monotone vector space: Indeed, if $H^n \in \mathcal{H}$ converges monotonely to H, where H is bounded, then $J = \lim_n \sup \dot{H}^n$ satisfies that $J_T = E\{H_T|\mathcal{F}_{T-}\}$, and we can take $\dot{H}$ to be J. Therefore by the monotone class theorem it suffices to establish the result for the multiplicative family $1_A 1_{[a,b)}$, where $A \in \mathcal{F}$, $0 \leq a < b$.

Let

$$H_t(\omega) = 1_A(\omega)1_{[a,b)}(t),$$

and let $M_t = E\{1_A|\mathcal{F}_t\}$, the càdlàg version. M is a bounded martingale. Let $J_t = M_{t-}1_{[a,b)}$. The process J is the product of two predictable processes ($1_{[a,b)}(t)$ is non-random and hence predictable) and hence it is predictable. However for a predictable time T, we know that there exist stopping times S^n increasing to T with $S^n < T$ on $\{T > 0\}$. Therefore

$$\begin{aligned} M_{T-} = \lim_{n\to\infty} M_{S^n} &= \lim_{n\to\infty} E\{M_T|\mathcal{F}_{S^n}\} \\ &= E\{M_T|\mathcal{F}_{T-}\}, \end{aligned}$$

since $\mathcal{F}_{T-} = \bigvee_n \mathcal{F}_{S^n}$ for a predictable time T. (This is because $\mathcal{F}_{S^n} \subset \mathcal{F}_{T-}$ since $S^n < T$ on $\{T > 0\}$, whence $\bigvee \mathcal{F}_{S^n} \subset \mathcal{F}_{T-}$; also $A \cap \{t < T\} = \bigcup_n (A \cap \{t \leq S^n\})$ a.s. on $\{T > 0\}$, and therefore $\mathcal{F}_{T-} \subset \bigvee_n \mathcal{F}_{S^n}$).

We can now conclude that for a predictable stopping time T,

$$\begin{aligned} J_T = M_{T-}1_{\{a \leq T < b\}} &= E\{M_T|\mathcal{F}_{T-}\}1_{\{a \leq T < b\}} \\ &= E\{1_A|\mathcal{F}_{T-}\}1_{\{a \leq T < b\}} \\ &= E\{H_T|\mathcal{F}_{T-}\}, \end{aligned}$$

since T is $\mathcal{F}_{T-}$ measurable. Therefore $J = \dot{H}$ and H is an element of $\mathcal{H}$, so we are done by the monotone class theorem. □

Definition. Let H be a bounded, measurable process. The **predictable projection** of H, written $\dot{H}$, is the unique predictable process such that

$$\dot{H}_T = E\{H_T|\mathcal{F}_{T-}\} \quad \text{a.s. on } \{T<\infty\}$$

for all predictable stopping times T.

Before proving the converse of Theorem 23, we need a preliminary result.

Theorem 25. *Let V be a natural process (of integrable variation), and let H be bounded, measurable. Then*

$$E\{\int_0^\infty H_s dV_s\} = E\{\int_0^\infty \dot{H}_s dV_s\},$$

where the stochastic integrals are taken in the Stieltjes sense, path by path.

Proof. Let $\mathcal{H}$ be the collection of all bounded $\mathcal{F}\otimes\mathcal{B}_+$ measurable processes such that the conclusion of the theorem holds. Then $\mathcal{H}$ is a monotone vector space. The result is true for the multiplicative class $\mathcal{M} = \{H : H_t = 1_\Lambda 1_{[a,b)}\}$, where $\Lambda \in \mathcal{F}$. Indeed,

$$E\{\int_0^\infty H_s dV_s\} = E\{1_\Lambda(V_{b-} - V_{a-})\}$$

whereas

$$E\{\int_0^\infty \dot{H}_s dV_s\} = E\{\int_0^\infty M_{s-}1_{[a,b)}(s)dV_s\}$$

since $\dot{H}_s = M_{s-}1_{[a,b)}$ as we saw in the proof of Theorem 24. Continuing we have

$$\begin{aligned} E\{\int_0^\infty M_{s-}1_{[a,b)}(s)dV_s\} &= E\{\int_0^\infty M_{s-}(dV_s^{b-} - dV_s^{a-})\} \\ &= E\{\int_0^\infty M_{s-}dV_s^{b-}\} - E\{\int_0^\infty M_{s-}dV_s^{a-})\}. \end{aligned}$$

Now

$$\begin{aligned} E\{\int_0^\infty M_{s-}dV_s^{b-}\} &= \lim_{\substack{\beta\to b \\ \beta<b}} E\{\int_0^\infty M_{s-}dV_s^{\beta}\} \\ &= \lim_{\beta\uparrow b} E\{M_\infty V_\infty^{\beta}\} \\ &= E\{M_\infty V_\infty^{b-}\}. \end{aligned}$$

The case for dV_s^{a-} is analogous, whence

$$\begin{aligned} E\{\int_0^\infty M_{s-}1_{[a,b)}(s)dV_s\} &= E\{M_\infty V_\infty^{b-}\} - E\{M_\infty V_\infty^{a-}\} \\ &= E\{M_\infty(V_{b-} - V_{a-})\} \\ &= E\{1_\Lambda(V_{b-} - V_{a-})\}, \end{aligned}$$

where $M_t = E\{1_\Lambda | \mathcal{F}_t\}$, and where we have used that V is natural. The result now follows by the monotone class theorem. □

Theorem 26. *Let A be a bounded natural process. Then A is predictable.*

Proof. Let M be a bounded martingale of integrable variation, and let $\int_0^\infty A_s dM_s$ be the Stieltjes integral, computed path by path. Then

$$\int_0^\infty A_s dM_s = \int_0^\infty A_{s-} dM_s + [A, M]_\infty,$$

hence $E\{\int_0^\infty A_s dM_s\} = 0$ because $E\{\int_0^\infty A_{s-} dM_s\} = 0$ (since it is a martingale and $A_{0-}M_0 = 0$) and $E\{[A,M]_\infty\} = 0$, because A is natural.

Let V be the bounded, increasing process $V_t = 1_\Lambda 1_{\{t \geq T\}}$, where T is a finite stopping time and $\Lambda \in \mathcal{F}_T$. Let $\widetilde{V}$ be the compensator of V: i.e., the unique natural process such that $V - \widetilde{V}$ is a local martingale. Letting $\dot{A}$ denote the predictable projection of A, we have that $E\{\int_0^\infty (A_s - \dot{A}_s) d\widetilde{V}_s\} = 0$ by Theorem 25. $E\{(\widetilde{V}_\infty)^2\} < \infty$ by the energy inequality established in the proof of Theorem 11. Since $V - \widetilde{V}$ is then a square integrable martingale, $E\{[V - \widetilde{V}, V - \widetilde{V}]_\infty\} < \infty$ by Corollary 3 of Theorem 27 of Chap. II. Since $\dot{A}$ is bounded, by the lemma preceding Theorem 28 of Chap. IV, $\int \dot{A}_s d(V_s - \widetilde{V}_s)$ is a square integrable martingale, and we have $E\{\int_0^\infty \dot{A}_s d(V_s - \widetilde{V}_s)\} = 0$. However since $E\{\int_0^\infty A_s d(V_s - \widetilde{V}_s)\} = 0$ because A is natural, we conclude $E\{\int_0^\infty (A_s - \dot{A}_s) d(V_s - \widetilde{V}_s)\} = 0$. Therefore $E\{\int_0^\infty (A_s - \dot{A}_s) dV_s\} = 0$ as well.

From the preceding we conclude $E\{(A_T - \dot{A}_T) 1_\Lambda 1_{\{T < \infty\}}\} = 0$, hence $A_T = \dot{A}_T$ a.s., since both A_T and $\dot{A}_T$ are $\mathcal{F}_T$ measurable. Therefore A and $\dot{A}$ are indistinguishable, hence A is predictable. □

Combining Theorems 23 and 26 we have the following.

Theorem 27. *Let A be a bounded FV process of locally integrable variation with $A_0 = 0$. Then A is predictable if and only if A is locally natural.*

Actually Theorem 27 is true without the simplifying assumption that A is bounded, but the proofs become more technical. One method to extend Theorem 27 to the unbounded case is to use the extended predictable projection (see Jacod-Shiryaev [1], p. 23).

Bibliographic Notes

The material of Chap. III comprises a large part of the core of the "general theory of processes" as is presented, for example, in Dellacherie [**1**]. We have, however, presented only the minimum required for our treatment of general stochastic integration and stochastic differential equations in Chaps. IV and V. We have also tried to keep the proofs as non-technical and as intuitive as possible. To this end we have presented K.M. Rao's proof of the Doob-Meyer decomposition (Theorem 6), see Rao [**1**, **2**], rather than the usual Doléans-Dade measure approach. This involved the use of natural processes, although our definition is a new one which is equivalent to the original one of Meyer for increasing processes. In Sect. 8 we give an elementary proof that a bounded process of integrable variation is natural if and only if it is predictable, thus showing the equivalence of the approaches.

The Doob decomposition is from Doob [**1**], and the Doob-Meyer decomposition (Theorem 6) is due to Meyer [**1**, **2**]. Our proof is taken from K.M. Rao [**1**, **2**]. The theory of quasimartingales was developed by Fisk [**1**], Orey [**1**], K.M. Rao [**2**], Stricker [**1**], and Métivier-Pellaumail [**1**].

The Fundamental Theorem of Local Martingales is due to Jia-an Yan and appears in an article of Meyer [**9**]; it was also proved independently by Doléans-Dade [**4**].

The notion of special semimartingales and canonical decompositions is due to Meyer [**8**]. The Girsanov-Meyer theorem (Theorem 20) dates back to the 1954 work of Maruyama [**1**], though we don't try to change its name here. The version presented here is due to Meyer [**8**] who extended the work of Girsanov [**1**]; see also Lenglart [**1**]. For more on how the stochastic calculus can be used in signal detection theory, see (for example) Wong [**1**].

The Bichteler-Dellacherie theorem (Theorem 22) is due independently to Bichteler [**1**, **2**] and Dellacherie [**2**]. It was proved in the late 1970's but it didn't appear in print until 1979 in the book of Jacod [**1**]. Many people have made contributions to this theorem, which had at least some of its origins in the work of Métivier-Pellaumail [**3**], Mokobodzki, Nikishin, Letta, and Lenglart. Our treatment was inspired by Meyer [**13**] and by Yan [**1**].

Our treatment of natural versus predictable processes is new, though the idea comes from Lenglart [**2**]. For a more customary treatment which includes the unbounded case, see (for example) Doob [**2**], pp. 483–487.

CHAPTER IV

General Stochastic Integration and Local Times

1. Introduction

We defined a semimartingale as a "good integrator" in Chap. II, and this led naturally to defining the stochastic integral as a limit of sums. To express an integral as a limit of sums requires some path smoothness of the integrands and we limited our attention to processes in $\mathbb{L}$: the space of adapted processes with paths that are left continuous and have right limits. The space $\mathbb{L}$ is sufficient to prove Itô's formula, the Girsanov-Meyer theorem, and it also suffices in some applications such as stochastic differential equations. But other uses, such as martingale representation theory or local times, require a larger space of integrands.

In this chapter we define stochastic integration for predictable processes. Our extension from Chap. II is very roughly analogous to how the Lebesgue integral extends the Riemann integral. We first define stochastic integration for bounded, predictable processes and a subclass of semimartingales known as $\mathcal{H}^2$. We then extend the definition to arbitrary semimartingales and to locally bounded predictable integrands.

2. Stochastic Integration for Predictable Integrands

In this section, we will weaken the restriction that an integrand H must be in $\mathbb{L}$; we will show our definition of stochastic integrals can be extended to a class of *predictably measurable* integrands.

Throughout this section X will denote a semimartingale such that $X_0 = 0$. This is a convenience involving no loss of generality: If Y is any semimartingale we can set $\hat{Y}_t = Y_t - Y_0$, and if we have defined stochastic integrals for semimartingales that are zero at 0, we can next define:

$$\int_0^t H_s dY_s \equiv \int_0^t H_s d\hat{Y}_s + H_0 Y_0.$$

When $Y_0 \neq 0$, recall that we write $\int_{0+}^t H_s dY_s$ to denote integration on $(0,t]$, and $\int_0^t H_s dY_s$ denotes integration on the closed interval $[0,t]$.

Let $X = M + A$ be a decomposition of a semimartingale X, with $X_0 = M_0 = A_0 = 0$. Here M is a local martingale and A is an FV process (such a decomposition exists by Theorem 22 of Chap. III). We will first consider special semimartingales. Recall that a semimartingale X is called *special* if it has a decomposition

$$X = \overline{N} + \overline{A}$$

where $\overline{N}$ is a local martingale and $\overline{A}$ is a *locally natural* (or equivalently *predictable*) FV process. This decomposition is unique by Theorem 18 in Chap. III and it is called the *canonical decomposition.*

Definitions. Let X be a special semimartingale with canonical decomposition $X = \overline{N} + \overline{A}$. **The $\mathcal{H}^2$ norm of X** is defined to be

$$\|X\|_{\mathcal{H}^2} = \|[\overline{N},\overline{N}]_\infty^{1/2}\|_{L^2} + \|\int_0^\infty |d\overline{A}_s|\|_{L^2}.$$

The space $\mathcal{H}^2$ of semimartingales consists of all special semimartingales with finite $\mathcal{H}^2$ norm.

In Chap. V we define an equivalent norm which we denote $\|\cdot\|_{\underline{\underline{H}}^2}$.

Theorem 1. *The space of $\mathcal{H}^2$ semimartingales is a Banach space.*

Proof. The space is clearly a normed linear space and it is easy to check that $\|\cdot\|_{\mathcal{H}^2}$ is a norm (recall that $E\{\overline{N}_\infty^2\} = E\{[\overline{N},\overline{N}]_\infty\}$, and therefore $\|X\|_{\mathcal{H}^2} = 0$ implies that $E\{\overline{N}_\infty^2\} = 0$ which implies, since $\overline{N}$ is a martingale, that $\overline{N} \equiv 0$).

To show completeness we treat the terms $\overline{N}$ and $\overline{A}$ separately. Consider first $\overline{N}$. Since $E\{\overline{N}_\infty^2\} = \|[\overline{N},\overline{N}]_\infty^{\frac{1}{2}}\|_{L^2}^2$, it suffices to show that the space of L^2 martingales is complete. However an L^2 martingale M can be identified with $M_\infty \in L^2$, and thus the space is complete since L^2 is complete.

Next suppose (A^n) is a Cauchy sequence in $\|\cdot\|_2$ where $\|A\|_p = \|\int_0^\infty |dA_s|\|_{L^p}$, $p \geq 1$. To show (A^n) converges it suffices to show a subsequence converges; therefore without loss of generality we can assume $\sum_n \|A^n\|_2 < \infty$.

Then $\sum A^n$ converges in $\|\cdot\|_1$ to a limit A. Moreover $\lim_{m\to\infty} \sum_{n\geq m} \int_0^\infty |dA_s^n|$ tends to 0 in L^1 and is dominated in L^2 by $\sum_n \int_0^\infty |dA_s^n|$. Therefore $\sum A^n$ converges to the limit A in $\|\cdot\|_2$ as well. To see that the limit A is natural, observe that for M a bounded martingale we have

$$\begin{aligned} E\{M_\infty A_\infty\} &= \lim_{n\to\infty} E\{M_\infty \sum_n A^n_\infty\} \\ &= \lim_{n\to\infty} \sum_n E\{\int_0^\infty M_{s-} dA^n_s\} \\ &= \lim_{n\to\infty} E\{\sum_n \int_0^\infty M_{s-} dA^n_s\} \\ &= \lim_{n\to\infty} E\{\int_0^\infty M_{s-} d(\sum A^n_s)\} \\ &= E\{\int_0^\infty M_{s-} dA_s\}, \end{aligned}$$

whence A is natural (Theorem 2 of Chap. III). This establishes completeness. □

For convenience we recall here the definition of $\mathbb{L}$:

Definition. $\mathbb{L}$ (respectively $\mathbf{b}\mathbb{L}$) denotes the space of adapted processes with càglàd[1] (resp. bounded, càglàd) paths.

We first establish a useful technical lemma.

Lemma. *Let A be a natural FV process, and let H be in $\mathbb{L}$ such that $E\{\int_0^\infty |H_s||dA_s|\} < \infty$. Then the FV process $(\int_0^t H_s dA_s)_{t\geq 0}$ is also natural.*

Proof. We need to show $E\{[M, H\cdot A]_\infty\} = 0$ for any bounded martingale M, where $H\cdot A_t = \int_0^t H_s dA_s$. We have $[M, H\cdot A]_\infty = \int_0^\infty H_s d[M,A]_s = [H\cdot M, A]_\infty$. The martingale $H\cdot M$ is locally bounded and by stopping it follows, since A is natural, that $E\{[H\cdot M, A]_\infty\} = 0$ which implies that $E\{[M, H\cdot A]_\infty\} = 0$ and therefore $H\cdot A$ is natural. □

Definition. The **predictable σ-algebra** $\mathcal{P}$ on $\mathbb{R}_+\times\Omega$ is the smallest σ-algebra making all processes in $\mathbb{L}$ measurable. That is, $\mathcal{P} = \sigma\{H : H \in \mathbb{L}\}$. We let $\mathbf{b}\mathcal{P}$ denote bounded processes that are $\mathcal{P}$-measurable.

The results that follow will enable us to extend the class of stochastic integrands from $\mathbf{b}\mathbb{L}$ to $\mathbf{b}\mathcal{P}$, with $X \in \mathcal{H}^2$ (and $X_0 = 0$). First we observe that if $H \in \mathbf{b}\mathbb{L}$ and $X \in \mathcal{H}^2$, then the stochastic integral $H\cdot X \in \mathcal{H}^2$. Also if $X = \overline{N} + \overline{A}$ is the canonical decomposition of X, then $H\cdot\overline{N} + H\cdot\overline{A}$ is

[1] càglàd denotes "left continuous with right limits".

the canonical decompositon of $H \cdot X$ by the preceding lemma. Moreover,

$$\|H \cdot X\|_{\mathcal{H}^2} = \|(\int_0^\infty H_s^2 d[\overline{N},\overline{N}]_s)^{\frac{1}{2}}\|_{L^2} + \|\int_0^\infty |H_s||d\overline{A}_s|\|_{L^2}.$$

The key idea in extending our integral is to notice that $[\overline{N},\overline{N}]$ and $\overline{A}$ are FV processes, and therefore ω-by-ω the integrals $\int_0^t H_s^2(\omega)d[\overline{N},\overline{N}]_s(\omega)$ and $\int_0^t |H_s||d\overline{A}_s|$ make sense for any $H \in \mathbf{b}\mathcal{P}$ and not just $H \in \mathsf{L}$.

Definition. Let $X \in \mathcal{H}^2$ with $X = \overline{N} + \overline{A}$ its canonical decomposition, and let H, $J \in \mathbf{b}\mathcal{P}$. We **define** $d_X(H,J)$ **by:**

$$d_X(H,J) \equiv \|(\int_0^\infty (H_s - J_s)^2 d[\overline{N},\overline{N}]_s)^{1/2}\|_{L^2} + \|\int_0^\infty |H_s - J_s||d\overline{A}_s|\|_{L^2}.$$

Theorem 2. *For $X \in \mathcal{H}^2$ the space $\mathbf{b}\mathsf{L}$ is dense in $\mathbf{b}\mathcal{P}$ under $d_X(\cdot,\cdot)$.*

Proof. We use the monotone class theorem. Define

$\mathcal{A} = \{H \in \mathbf{b}\mathcal{P} :$ for any $\epsilon > 0$, there exists $J \in \mathbf{b}\mathsf{L}$ such that $d_X(H,J) < \epsilon\}$.

Trivially $\mathcal{A}$ contains $\mathbf{b}\mathsf{L}$. If $H^n \in \mathcal{A}$ and H^n increases to H with H bounded, then $H \in \mathbf{b}\mathcal{P}$, and by the dominated convergence theorem if $\delta > 0$ then for some $N(\delta)$, $n > N(\delta)$ implies $d_X(H,H^n) < \delta$. Since each $H^n \in \mathcal{A}$, we choose $n_0 > N(\delta)$ and there exists $J \in \mathbf{b}\mathsf{L}$ such that $d_X(J,H^{n_0}) < \delta$. Therefore given $\epsilon > 0$, by taking $\delta = \epsilon/2$ we can find $J \in \mathbf{b}\mathsf{L}$ such that $d_X(J,H) < \epsilon$, and therefore $H \in \mathcal{A}$. An application of the monotone class theorem yields the result. □

Theorem 3. *Let $X \in \mathcal{H}^2$ and $H^n \in \mathbf{b}\mathsf{L}$ such that H^n is Cauchy under d_X. Then $H^n \cdot X$ is Cauchy in $\mathcal{H}^2$.*

Proof. Since $\|H^n \cdot X - H^m \cdot X\|_{\mathcal{H}^2} = d_X(H^n,H^m)$, the theorem is immediate. □

Theorem 4. *Let $X \in \mathcal{H}^2$ and $H \in \mathbf{b}\mathcal{P}$. Suppose $H^n \in \mathbf{b}\mathsf{L}$ and $J^m \in \mathbf{b}\mathsf{L}$ are two sequences such that $\lim_n d_X(H^n,H) = \lim_m d_X(J^m,H) = 0$. Then $H^n \cdot X$ and $J^m \cdot X$ tend to the same limit in $\mathcal{H}^2$.*

Proof. Let $Y = \lim_{n\to\infty} H^n \cdot X$ and $Z = \lim_{m\to\infty} J^m \cdot X$, where the limits are taken in $\mathcal{H}^2$. For $\epsilon > 0$, by taking n and m large enough we have:

$$\begin{aligned}\|Y - Z\|_{\mathcal{H}^2} &\leq \|Y - H^n \cdot X\|_{\mathcal{H}^2} + \|H^n \cdot X - J^m \cdot X\|_{\mathcal{H}^2} + \|J^m \cdot X - Z\|_{\mathcal{H}^2} \\ &\leq 2\epsilon + \|H^n \cdot X - J^m \cdot X\|_{\mathcal{H}^2} \\ &\leq 2\epsilon + d_X(H^n,J^m) \\ &\leq 2\epsilon + d_X(H^n,H) + d_X(H,J^m) \\ &\leq 4\epsilon,\end{aligned}$$

and the result follows. □

We are now in a position to define the stochastic integral for $H \in \mathbf{b}\mathcal{P}$ (and $X \in \mathcal{H}^2$).

Definition. Let X be a semimartingale in $\mathcal{H}^2$ and let $H \in \mathbf{b}\mathcal{P}$. Let $H^n \in \mathbf{bL}$ be such that $\lim_{n\to\infty} d_X(H^n, H) = 0$. The **stochastic integral** $H \cdot X$ is the (unique) semimartingale $Y \in \mathcal{H}^2$ such that $\lim_{n\to\infty} H^n \cdot X = Y$ in $\mathcal{H}^2$. We write $H \cdot X = (\int_0^t H_s dX_s)_{t\geq 0}$.

We have defined our stochastic integral for predictable integrands and semimartingales in $\mathcal{H}^2$ as limits of our (previously defined) stochastic integrals. In order to investigate the properties of this more general integral, we need to have approximations converging uniformly. The next theorem and its corollary give us this.

Theorem 5. *Let X be a semimartingale in $\mathcal{H}^2$. Then* $E\{(\sup_t|X_t|)^2\} \leq 8\|X\|^2_{\mathcal{H}^2}$.

Proof. For a process H, let $H^* = \sup_t|H_t|$. Let $X = \overline{N} + \overline{A}$ be the canonical decomposition of X. Then

$$X^* \leq \overline{N}^* + \int_0^\infty |d\overline{A}_s|.$$

Doob's maximal quadratic inequality (Theorem 20 of Chap. I) yields:

$$E\{(\overline{N}^*)^2\} \leq 4E\{\overline{N}^2_\infty\} = 4E\{[\overline{N}, \overline{N}]_\infty\},$$

and using $(a+b)^2 \leq 2a^2 + 2b^2$ we have

$$\begin{aligned} E\{(X^*)^2\} &\leq 2E\{(\overline{N}^*)^2\} + 2E\{(\int_0^\infty |d\overline{A}_s|)^2\} \\ &\leq 8E\{[\overline{N}, \overline{N}]_\infty\} + 2\|\int_0^\infty |d\overline{A}_s|\|^2_{L^2}, \\ &\leq 8\|X\|^2_{\mathcal{H}^2}. \end{aligned}$$

□

Corollary. *Let (X^n) be a sequence of semimartingales converging to X in $\mathcal{H}^2$. Then there exists a subsequence (n_k) such that* $\lim_{n_k\to\infty}(X^{n_k} - X)^* = 0$ *a.s.*

Proof. By Theorem 5 we know that $(X^n - X)^* = \sup_t|X^n_t - X_t|$ converges to 0 in L^2. Therefore there exists a subsequence converging a.s. □

We next investigate some of the properties of this generalized stochastic integral. Almost all of the properties established in Chap. II (Sect. 5) still hold.[2]

[2] Indeed, it is an open question whether or not Theorem 16 of Chap. II extends to integrands in $\mathbf{b}\mathcal{P}$. See the discussion at the end of this section.

Theorem 6. *Let* $X, Y \in \mathcal{H}^2$ *and* $H, K \in \mathbf{b}\mathcal{P}$. *Then*

$$(H+K)\cdot X = H\cdot X + K\cdot X,$$

and

$$H\cdot(X+Y) = H\cdot X + H\cdot Y.$$

Proof. One need only check that it is possible to take a sequence $H^n \in \mathbf{bL}$ that approximates H in both d_X and d_Y. □

Theorem 7. *Let* T *be a stopping time. Then* $(H\cdot X)^T = H1_{[0,T]}\cdot X = H\cdot(X^T)$.

Proof. Note that $1_{[0,T]} \in \mathbf{bL}$, so $H1_{[0,T]} \in \mathbf{b}\mathcal{P}$. Also, X^T is clearly still in $\mathcal{H}^2$. Since we know this result is true for $H \in \mathbf{bL}$ (Theorem 12 of Chap. II), the result follows by uniform approximation, using the Corollary of Theorem 5. □

Theorem 8. *The jump process* $(\Delta(H\cdot X)_s)_{s\geq 0}$ *is indistinguishable from* $(H_s(\Delta X_s))_{s\geq 0}$.

Proof. Recall that for a process J, $\Delta J_t = J_t - J_{t-}$, the jump of J at time t. (Note that $H\cdot X$ and X are càdlàg semimartingales, so Theorem 8 makes sense.) By Theorem 13 of Chap. II we know the result is true for $H \in \mathbf{bL}$. Let $H \in \mathbf{b}\mathcal{P}$, and let $H^n \in \mathbf{bL}$ such that $\lim_{n\to\infty} d_X(H^n, H) = 0$. By the Corollary of Theorem 5, there exists a subsequence (n_k) such that

$$\lim_{n_k\to\infty}(H^{n_k}\cdot X - H\cdot X)^* = 0 \quad a.s.$$

This implies that, considered as processes,

$$\lim_{n_k\to\infty}\Delta(H^{n_k}\cdot X) = \Delta(H\cdot X),$$

outside of an evanescent set.[3] Since each $H^{n_k} \in \mathbf{bL}$, we have $\Delta(H^{n_k}\cdot X) = H^{n_k}(\Delta X)$, outside of another evanescent set. Combining these, we have

$$\begin{aligned}\lim_{n_k\to\infty} H^{n_k}(\Delta X)1_{\{\Delta X\neq 0\}} &= \lim_{n_k\to\infty}\Delta(H^{n_k}\cdot X)1_{\{\Delta X\neq 0\}}\\ &= \Delta(H\cdot X)1_{\{\Delta X\neq 0\}},\end{aligned}$$

and therefore

$$\lim_{n_k\to\infty} H^{n_k}1_{\{\Delta X\neq 0\}} = \frac{\Delta(H\cdot X)}{\Delta X}1_{\{\Delta X\neq 0\}}.$$

[3] A set $\Lambda \subset \mathbf{R}_+\times\Omega$ is *evanescent* if 1_Λ is a process that is indistinguishable from the zero process.

In particular, the above implies that $\lim_{n_k\to\infty} H_t^{n_k}(\omega)$ exists for all (t,ω) in $\{\Delta X \neq 0\}$, a.s. We next form:

$$\Lambda = \{\omega\colon \text{there exists } t>0 \text{ such that } \lim_{n_k\to\infty} H_t^{n_k}(\omega) \neq H_t(\omega) \text{ and } \Delta X_t(\omega)\neq 0\}.$$

Suppose $P(\Lambda) > 0$. Then

$$(*)\qquad \begin{aligned} d_X(H^{n_k}, H) \geq \|1_\Lambda\{\int_0^\infty (H_s^{n_k} - H_s)^2 d(\sum_{0<u\leq s} (\Delta \overline{N}_u)^2)\}\|_{L^2} \\ + \|1_\Lambda\{\int_0^\infty |H_s^{n_k} - H_s| d(\sum_{0<u\leq s} |\Delta \overline{A}_u|)\}\|_{L^2}, \end{aligned}$$

and if $\Delta X_s \neq 0$, then $|\Delta \overline{N}_s| + |\Delta \overline{A}_s| > 0$. The left side of (*) tends to 0 as $n_k \to \infty$, and the right side of (*) does not. Therefore $P(\Lambda) = 0$, and we conclude $\Delta(H\cdot X) = \lim_{n_k\to\infty} H^{n_k}\Delta X = H\Delta X$. □

Corollary. *Let $X \in \mathcal{H}^2$, $H \in \mathbf{b}\mathcal{P}$, and T a finite stopping time. Then $H\cdot(X^{T-}) = (H\cdot X)^{T-}$.*

Proof. By Theorem 8, $(H\cdot X)^{T-} = (H\cdot X)^T - H_T\Delta X_T 1_{\{t\geq T\}}$. On the other hand, $X^{T-} = X^T - \Delta X_T 1_{\{t\geq T\}}$. Let $A_t = \Delta X_T 1_{\{t\geq T\}}$. By the bilinearity (Theorem 6), $H\cdot(X^{T-}) = H\cdot(X^T) - H\cdot A$. Since $H\cdot(X^T) = (H\cdot X)^T$ by Theorem 7, and $H\cdot A = H_T\Delta X_T 1_{\{t\geq T\}}$, the result follows. □

The next three theorems all involve the same simple proofs: the result is known to be true for processes in $\mathbf{bL}$; let $(H^n) \in \mathbf{bL}$ approximate $H \in \mathbf{b}\mathcal{P}$ in $d_X(\cdot,\cdot)$, and by the Corollary of Theorem 5 let n_k be a subsequence such that

$$\lim_{n_k\to\infty} (H^{n_k}\cdot X - H\cdot X)^* = 0 \text{ a.s.};$$

Then use the uniform convergence to obtain the desired result. We state these theorems, therefore, without proofs.

Theorem 9. *Let $X \in \mathcal{H}^2$ have paths of finite variation on compacts, and $H \in \mathbf{b}\mathcal{P}$. Then $H\cdot X$ agrees with a path-by-path Lebesgue-Stieltjes integral.*

Theorem 10 (Associativity). *Let $X \in \mathcal{H}^2$ and $H, K \in \mathbf{b}\mathcal{P}$. Then $K\cdot X \in \mathcal{H}^2$ and $H\cdot(K\cdot X) = (HK)\cdot X$.*

Theorem 11. *Let $X \in \mathcal{H}^2$ be a (square integrable) martingale, and $H \in \mathbf{b}\mathcal{P}$. Then $H\cdot X$ is a square integrable martingale.*

Theorem 12. *Let $X, Y \in \mathcal{H}^2$ and $H, K \in \mathbf{b}\mathcal{P}$. Then*

$$[H\cdot X, K\cdot Y]_t = \int_0^t H_s K_s d[X,Y]_s, \quad (t\geq 0),$$

and in particular

$$[H \cdot X, H \cdot X]_t = \int_0^t H_s^2 d[X,X]_s, \quad (t \geq 0).$$

Proof. As in the proof of Theorem 29 of Chap. II, it suffices to show

$$[H \cdot X, Y]_t = \int_0^t H_s d[X,Y]_s.$$

Let $(H^n) \in \mathbf{b}\mathbb{L}$ such that $\lim_{n\to\infty} d_X(H^n, H) = 0$. Let $T^m = \inf\{t > 0 : |Y_t| > m\}$. Then (T^m) are stopping times increasing to ∞ a.s. and $|Y_-^{T^m}| \leq m$.[4] Since it suffices to show the result holds on $[0, T^m)$, each m, we can assume without loss of generality that Y_- is in $\mathbf{b}\mathbb{L}$. Moreover the dominated convergence theorem gives $\lim_{n\to\infty} d_X(H^n Y_-, HY_-) = 0$. By Theorem 29 of Chap. II we have

$$[H^n \cdot X, Y]_t = \int_0^t H_s^n d[X,Y]_s \qquad (t \geq 0),$$

and again by dominated convergence

$$\lim_{n\to\infty} [H^n \cdot X, Y] = \int_0^t H_s d[X,Y]_s \qquad (t \geq 0).$$

It remains only to show $\lim_{n\to\infty}[H^n \cdot X, Y] = [H \cdot X, Y]$. Let $Z^n = H^n \cdot X$, and let n_k be a subsequence such that $\lim_{n_k\to\infty}(Z^{n_k} - Z)^* = 0$ a.s., where $Z = H \cdot X$ (by the Corollary to Theorem 5). Integration by parts yields

$$\begin{aligned}[Z^{n_k}, Y] &= Z^{n_k} Y - (Y_-) \cdot Z^{n_k} - (Z_-^{n_k}) \cdot Y \\ &= Z^{n_k} Y - (Y_- H^{n_k}) \cdot X - (Z_-^{n_k}) \cdot Y,\end{aligned}$$

where we have used associativity (Theorem 10). We take limits:

$$\begin{aligned}\lim_{n_k\to\infty} [Z^{n_k}, Y] &= ZY - Y_- \cdot (H \cdot X) - Z_- \cdot Y \\ &= ZY - Y_- \cdot (Z) - Z_- \cdot Y \\ &= [Z, Y] = [H \cdot X, Y].\end{aligned}$$

□

At this point the reader may wonder how to calculate in practice a canonical decomposition of a semimartingale X in order to verify that $X \in \mathcal{H}^2$. Fortunately Theorem 13 will show that $\mathcal{H}^2$ is merely a mathematical convenience.

Lemma. *Let A be an FV process with $A_0 = 0$ and $\int_0^\infty |dA_s| \in L^2$. Then $A \in \mathcal{H}^2$. Moreover $\|A\|_{\mathcal{H}^2} \leq 6 \| \int_0^\infty |dA_s| \|_{L^2}$.*

---

[4] Y_- denotes the left continuous version of Y.

Proof. If we can prove the result for A increasing then the general result will follow by decomposing $A = A^+ - A^-$. Therefore we assume without loss of generality that A is increasing. Hence as we noted at the end of Sect. 3 of Chap. III, the compensator $\widetilde{A}$ of A is also increasing and $E\{\widetilde{A}_\infty\} = E\{A_\infty\} < \infty$.

Let M be a martingale bounded by a constant k. Since $A - \widetilde{A}$ is a local martingale, Corollary 2 to Theorem 27 of Chap. II (together with Theorem 17 of Chap. III) shows that

$$L = M(A - \widetilde{A}) - [M, A - \widetilde{A}]$$

is a local martingale. Moreover:

$$\begin{aligned}\sup_t |L_t| &\leq k(A_\infty + \widetilde{A}_\infty) + 2k \sum_s |\Delta(A - \widetilde{A})_s| \\ &\leq 3k(A_\infty + \widetilde{A}_\infty) \in L^1.\end{aligned}$$

Therefore L is a uniformly integrable martingale (Theorem 47 of Chap. I) and $E\{L_\infty\} = E\{L_0\} = 0$. Hence

$$\begin{aligned}E\{M_\infty(A - \widetilde{A})_\infty\} &= E\{[M, A - \widetilde{A}]_\infty\} \\ &= E\{[M, A]_\infty\} - E\{[M, \widetilde{A}]_\infty\} \\ &= E\{[M, A]_\infty\},\end{aligned}$$

because $\widetilde{A}$ is natural. By the Kunita-Watanabe inequality (the Corollary to Theorem 25 of Chap. II)

$$\begin{aligned}E\{|[M, A]_\infty|\} &\leq (E\{[M, M]_\infty\} E\{[A, A]_\infty\})^{1/2} \\ &\leq \frac{1}{2} E\{[M, M]_\infty\} + \frac{1}{2} E\{[A, A]_\infty\},\end{aligned}$$

where the second inequality uses $2ab \leq a^2 + b^2$. However

$$E\{[M, M]_\infty\} = E\{M_\infty^2\}$$

(Corollary 4 of Theorem 27 of Chap. II) and also $[A, A]_\infty \leq A_\infty^2$ a.s. Therefore

$$E\{M_\infty(A - \widetilde{A})_\infty\} \leq \frac{1}{2} E\{M_\infty^2\} + \frac{1}{2} E\{A_\infty^2\}.$$

Since M is an arbitrary bounded martingale we are free to choose

$$M_\infty = (A - \widetilde{A})_\infty 1_{\{|(A - \widetilde{A})_\infty| \leq n\}},$$

and we obtain

$$\frac{1}{2} E\{(A - \widetilde{A})_\infty^2 1_{\{|(A - \widetilde{A})_\infty| \leq n\}}\} \leq \frac{1}{2} E\{A_\infty^2\},$$

and using the monotone convergence theorem we conclude

$$E\{(A-\widetilde{A})^2_\infty\} \le E\{A^2_\infty\}.$$

Consequently

$$E\{\widetilde{A}^2_\infty\} \le 2E\{A^2_\infty\} + 2E\{(A-\widetilde{A})^2_\infty\} \le 4E\{A^2_\infty\} < \infty,$$

and $A - \widetilde{A}$ is a square integrable martingale, and

$$\|A\|_{\mathcal{H}^2} = \|(A-\widetilde{A})_\infty\|_{L^2} + \|\widetilde{A}_\infty\|_{L^2} \le 3\|A_\infty\|_{L^2}$$

for A increasing. □

Remarks. The constant 6 can be improved to $1+\sqrt{8} \le 4$ by not decomposing A into A^+ and A^-. This lemma can also be proved using the Burkholder-Gundy inequalities (see Meyer [8], p. 347).

In Chap. V we use an alternative norm for semimartingales which we denote $\|\cdot\|_{\underline{\underline{H}}^p}$, $1 \le p < \infty$. The preceding lemma shows that the norms $\|\cdot\|_{\mathcal{H}^2}$ and $\|\cdot\|_{\underline{\underline{H}}^2}$ are equivalent.

The restrictions of integrands to $\mathbf{b}\mathcal{P}$ and semimartingales to $\mathcal{H}^2$ are mathematically convenient but not necessary. A standard method of relaxing such hypothesis is to consider cases where they hold *locally*. Recall that **a property π is said to hold locally** for a process X if there exists a sequence of stopping times $(T^n)_{n\ge 0}$ such that $0 = T^0 \le T^1 \le T^2 \le \cdots \le T^n \le \ldots$ and $\lim_{n\to\infty} T^n = \infty$ a.s., and such that $X^{T^n}1_{\{T^n>0\}}$ has property π for each n. Since we are assuming our semimartingales X satisfy $X_0 = 0$, we could as well require only that X^{T^n} has property π for each n. A related condition is that a property hold *prelocally*.

Definition. A property π is said to hold **prelocally** for a process X with $X_0 = 0$ if there exists a sequence of stopping times $(T^n)_{n\ge 1}$ increasing to ∞ a.s. such that X^{T^n-} has property π for each $n \ge 1$.

Recall that $X^{T-} = X_t 1_{(0\le t<T)} + X_{T-}1_{(t\ge T)}$. The next theorem shows that the restriction of semimartingales to $\mathcal{H}^2$ is not really a restriction at all.

Theorem 13. *Let X be a semimartingale, $X_0 = 0$. Then X is prelocally in $\mathcal{H}^2$. That is, there exists a nondecreasing sequence of stopping times (T^n), $\lim_{n\to\infty} T^n = \infty$ a.s., such that $X^{T^n-} \in \mathcal{H}^2$ for each n.*

Proof. Recall that $X^{T^n-} = X_t 1_{[0,T^n)} + X_{T^n-}1_{[T^n,\infty)}$. By the Bichteler-Dellacherie theorem (Theorem 22 of Chap. III) we can write $X = M + A$, where M is a local martingale and A is an FV process. By Theorem 13 of Chap. III we can further take M to have bounded jumps. Let β be the bound for the jumps of M. We define:

$$T^n = \inf\{t > 0 : [M,M]_t > n \text{ or } \int_0^t |dA_s| > n\}.$$

Then let $Y = X^{T^n-}$. Then Y has bounded jumps and hence it is a special semimartingale (Theorem 19 of Chap. III). Moreover

$$Y = X^{T^n-} = M^{T^n} + A^{T^n-} - (\Delta M_{T^n})1_{[T^n,\infty)},$$

or

$$Y = L + C,$$

where $L = M^{T^n}$ and $C = A^{T^n-} - (\Delta M_{T^n})1_{[T^n,\infty)}$. Then $[L,L] \leq n + \beta^2$, so L is a martingale in $\mathcal{H}^2$ (Corollary 4 to Theorem 27 of Chap. II), and also

$$\int_0^\infty |dC_s| \leq n + |\Delta M_{T^n}| \leq 2n + \beta,$$

hence $C \in \mathcal{H}^2$ by the Lemma. Therefore $X^{T^n-} = L + C \in \mathcal{H}^2$. □

We are now in a position to define the stochastic integral *for an arbitrary semimartingale*, as well as for predictable processes which need not be bounded.

Let X be a semimartingale in $\mathcal{H}^2$. To define a stochastic integral for predictable processes H which are not necessarily bounded (written: $H \in \mathcal{P}$), we approximate them with $H^n \in \mathbf{b}\mathcal{P}$.

Definition. Let $X \in \mathcal{H}^2$ with canonical decomposition $X = \overline{N} + \overline{A}$. We say $H \in \mathcal{P}$ is **$(\mathcal{H}^2, X)$-integrable** if

$$E\{\int_0^\infty H_s^2 d[\overline{N},\overline{N}]_s\} + E\{(\int_0^\infty |H_s||d\overline{A}_s|)^2\} < \infty.$$

Theorem 14. *Let X be a semimartingale and let $H \in \mathcal{P}$ be $(\mathcal{H}^2, X)$-integrable. Let $H^n = H1_{\{|H|\leq n\}} \in \mathbf{b}\mathcal{P}$. Then $H^n \cdot X$ is a Cauchy sequence in $\mathcal{H}^2$.*

Proof. Since $H^n \in \mathbf{b}\mathcal{P}$, each n, the stochastic integrals $H^n \cdot X$ are defined. Note also that $\lim_{n\to\infty} H^n = H$ and that $|H^n| \leq |H|$, each n. Then

$$\|H^n \cdot X - H^m \cdot X\|_{\mathcal{H}^2} = d_X(H^n, H^m)$$

$$\|(\int_0^\infty (H_s^n - H_s^m)^2 d[\overline{N},\overline{N}]_s)^{1/2}\|_{L^2} + \|\int_0^\infty |H_s^n - H_s^m||d\overline{A}_s|\|_{L^2},$$

and the result follows by two applications of the dominated convergence theorem of Lebesgue. □

Definition. Let X be a semimartingale in $\mathcal{H}^2$, and let $H \in \mathcal{P}$ be $(\mathcal{H}^2, X)$ integrable. The **stochastic integral** $H \cdot X$ is defined to be $\lim_{n\to\infty} H^n \cdot X$, with convergence in $\mathcal{H}^2$, where $H^n = H1_{\{|H|\leq n\}}$.

Note that $H \cdot X$ in the preceding definition exists by Theorem 14. We can "localize" the above theorem by allowing both more general $H \in \mathcal{P}$ and arbitrary semimartingales with the next definition.

Definition. Let X be a semimartingale and $H \in \mathcal{P}$. The stochastic integral $H \cdot X$ *is said to exist* if there exists a sequence of stopping times T^n increasing to ∞ a.s. such that $X^{T^n-} \in \mathcal{H}^2$, each $n \geq 1$, and such that H is $(\mathcal{H}^2, X^{T^n-})$-integrable for each n. In this case we say H is X**-integrable**, *written* $H \in L(X)$, and we define the **stochastic integral** by

$$H \cdot X = H \cdot (X^{T^n-}), \qquad \text{on } [0, T^n),$$

each n.

Note that if $m > n$ then

$$H^k \cdot (X^{T^m-})^{T^n-} = H^k \cdot (X^{T^m \wedge T^n-}) = H^k \cdot (X^{T^n-}),$$

where $H^k = H1_{\{|H|\leq k\}}$, by the Corollary of Theorem 8. Hence taking limits we have $H \cdot (X^{T^m-})^{T^n-} = H \cdot (X^{T^n-})$, and the stochastic integral is well-defined for $H \in L(X)$. Moreover let R^ℓ be another sequence of stopping times such that $X^{R^\ell-} \in \mathcal{H}^2$ and such that H is $(\mathcal{H}^2, X^{R^\ell-})$-integrable, for each ℓ. Again using the Corollary of Theorem 8 combined with taking limits we see that

$$H \cdot (X^{R^\ell-}) = H \cdot (X^{T^n-})$$

on $[0, R^\ell \wedge T^n)$, each $\ell \geq 1$ and $n \geq 1$. Thus in this sense the definition of the stochastic integral *does not depend on the particular sequence of stopping times*.

If $H \in \mathbf{b}\mathcal{P}$ (i.e., H is bounded), then $H \in L(X)$ for all semimartingales X, since every semimartingale is prelocally in $\mathcal{H}^2$ by Theorem 13.

Definition. A process H is said to be **locally bounded** if there exists a sequence of stopping times $(S^m)_{m\geq 1}$ increasing to ∞ a.s. such that $(H_{t\wedge S^m} 1_{\{S^m>0\}})_{t\geq 0}$ is bounded, each $m \geq 1$.

Note that any process in $\mathbb{L}$ is locally bounded. The next example is sufficiently important that we state it as a theorem.

Theorem 15. *Let X be a semimartingale and let $H \in \mathcal{P}$ be locally bounded. Then $H \in L(X)$. That is, the stochastic integral $H \cdot X$ exists.*

Proof. Let $(S^m)_{m\geq 1}$, $(T^n)_{n\geq 1}$ be two sequences of stopping times, each increasing to ∞ a.s., such that $H^{S^m}1_{\{S^m>0\}}$ is bounded for each m, and $X^{T^n-} \in \mathcal{H}^2$ for each n. Define $R^n = \min(S^n, T^n)$. Then $H = H^{R^n}1_{\{R^n>0\}}$ on $(0, R^n)$ and hence it is bounded there. Since X^{R^n-} charges only $(0, R^n)$, we have that H is $(\mathcal{H}^2, X^{R^n-})$-integrable for each $n \geq 1$. Therefore using the sequence R^n which increases to ∞ a.s., we are done. □

We now turn our attention to the properties of this more general integral. Many of the properties are simple extensions of earlier theorems and we omit their proofs. Note that trivially the stochastic integral $H \cdot X$, for $H \in L(X)$, is also a semimartingale.

Theorem 16. *Let X be a semimartingale and let $H, J \in L(X)$. Then $\alpha H + \beta J \in L(X)$ and $(\alpha H + \beta J) \cdot X = \alpha H \cdot X + \beta J \cdot X$. That is, $L(X)$ is a linear space.*

Proof. Let (R^m) and (T^n) be sequences of stopping times such that H is $(\mathcal{H}^2, X^{R^m-})$-integrable, each m, and J is $(\mathcal{H}^2, X^{T^n-})$-integrable, each n. Taking $S^n = R^n \wedge T^n$, it is easy to check that $\alpha H + \beta J$ is $(\mathcal{H}^2, X^{S^n-})$-integrable for each n. □

Theorem 17. *Let X, Y be semimartingales and suppose $H \in L(X)$ and $H \in L(Y)$. Then $H \in L(X+Y)$ and $H \cdot (X+Y) = H \cdot X + H \cdot Y$.*

Theorem 18. *Let X be a semimartingale and $H \in L(X)$. The jump process $(\Delta(H \cdot X)_s)_{s\geq 0}$ is indistinguishable from $(H_s(\Delta X_s))_{s\geq 0}$.*

Theorem 19. *Let T be a stopping time, X a semimartingale, and $H \in L(X)$. Then*

$$(H \cdot X)^T = H1_{[0,T]} \cdot X = H \cdot (X^T).$$

Letting $\infty-$ equal ∞, we have moreover

$$(H \cdot X)^{T-} = H \cdot (X^{T-}).$$

Theorem 20. *Let X be a semimartingale with paths of finite variation on compacts. Let $H \in L(X)$ be such that the Stieltjes integral $\int_0^t |H_s||dX_s|$ exists a.s., each $t \geq 0$. Then the stochastic integral $H \cdot X$ agrees with a path-by-path Stieltjes integral.*

Theorem 21. *Let X be a semimartingale with $K \in L(X)$. Then $H \in L(K \cdot X)$ if and only if $HK \in L(X)$, in which case $H \cdot (K \cdot X) = (HK) \cdot X$.*

Theorem 22. *Let X, Y be semimartingales and let $H \in L(X)$, $K \in L(Y)$. Then*

$$[H \cdot X, K \cdot Y]_t = \int_0^t H_s K_s d[X,Y]_s \quad (t \geq 0).$$

Note that in Theorem 22 since $H \cdot X$ and $H \cdot Y$ are semimartingales, the quadratic covariation exists and the content of the theorem is the formula. Indeed, Theorem 22 gives a necessary condition for H to be in $L(X)$: That $\int_0^t H_s^2 d[X,X]_s$ exists and is finite for all $t \geq 0$. The next theorem (Theorem 23) is a special case of Theorem 25, but we include it because of the simplicity of its proof.

Theorem 23. *Let X be a semimartingale, let $H \in L(X)$, and suppose Q is another probability with $Q \ll P$. If $H_Q \cdot X$ exists, it is Q-indistinguishable from $H_P \cdot X$.*

Proof. $H_Q \cdot X$ denotes the stochastic integral computed under Q. By Theorem 14 of Chap. II we know that $H_Q \cdot X = H_P \cdot X$ for $H \in \mathbb{L}$, and therefore if $X \in \mathcal{H}^2$ for both P and Q, they are equal for $H \in \mathbf{b}\mathcal{P}$ by the Corollary of Theorem 5. Let $(R^\ell)_{\ell \geq 1}$, $(T^n)_{n \geq 1}$ be two sequences of stopping times increasing to ∞ a.s. such that H is $(\mathcal{H}^2, X^{R^\ell -})$-integrable (Q), and H is $(\mathcal{H}^2, X^{T^n -})$-integrable (P), each ℓ and n. Let $S^m = R^m \wedge T^m$, so that H is $(\mathcal{H}^2, X^{S^m -})$-integrable under both P and Q. Then $H \cdot X = \lim_{n \to \infty} H^n \cdot X$ on $[0, S^m)$ in both $d_X(P)$ and $d_X(Q)$, where $H^n = H 1_{\{|H| \leq n\}} \in \mathbf{b}\mathcal{P}$. Since $H_P^n \cdot X = H_Q^n \cdot X$, each n, the limits are also equal. □

Much more than Theorem 23 is indeed true, as we will see in Theorem 25, which contains Theorem 23 as a special case. We need several preliminary results.

Lemma. *Let $X \in \mathcal{H}^2$ and $X = \overline{N} + \overline{A}$ be its canonical decomposition. Then*

$$E\{[\overline{N}, \overline{N}]_\infty\} \leq E\{[X,X]_\infty\}.$$

Proof. First observe that

$$[X,X] = [\overline{N}, \overline{N}] + 2[\overline{N}, \overline{A}] + [\overline{A}, \overline{A}].$$

It suffices to show $E\{[\overline{N}, \overline{A}]_\infty\} = 0$, since then

$$E\{[\overline{N}, \overline{N}]_\infty\} = E\{[X,X]_\infty - [\overline{A}, \overline{A}]_\infty\},$$

and the result follows since $[\overline{A}, \overline{A}]_\infty \geq 0$. Note that

$$E\{|[\overline{N}, \overline{A}]_\infty|\} \leq (E\{[\overline{N}, \overline{N}]_\infty\})^{1/2} (E\{[\overline{A}, \overline{A}]_\infty\})^{1/2} < \infty,$$

by the Kunita-Watanabe inequalities. Also $E\{[M,\overline{A}]_\infty\}=0$ for all bounded martingales because $\overline{A}$ is natural. Since bounded martingales are dense in the space of L^2 martingales, there exists a sequence $(M^n)_{n\geq 1}$ of bounded martingales such that $\lim_{n\to\infty}E\{[M^n-\overline{N},M^n-\overline{N}]_\infty\}=0$. Again using the Kunita-Watanabe inequalities we have

$$E\{|[\overline{N}-M^n,\overline{A}]_\infty|\}\leq (E\{[\overline{N}-M^n,\overline{N}-M^n]_\infty\})^{1/2}(E\{[\overline{A},\overline{A}]_\infty\})^{1/2}$$

and therefore $\lim_{n\to\infty}E\{[M^n,\overline{A}]_\infty\}=E\{[\overline{N},\overline{A}]_\infty\}$. Since $E\{[M^n,\overline{A}]_\infty\}=0$, each n, it follows that $E\{[\overline{N},\overline{A}]_\infty\}=0$. □

Note that in the preceding proof we established the useful equality $E\{[X,X]_t\}=E\{[\overline{N},\overline{N}]_t\}+E\{[\overline{A},\overline{A}]_t\}$ for a semimartingale $X\in\mathcal{H}^2$ with canonical decomposition $X=\overline{N}+\overline{A}$.

Theorem 24. *For X a semimartingale in $\mathcal{H}^2$,*

$$\|X\|_{\mathcal{H}^2}\leq \sup_{|H|\leq 1}\|(H\cdot X)^*_\infty\|_{L^2}+2\|[X,X]^{1/2}_\infty\|_{L^2}\leq 5\|X\|_{\mathcal{H}^2}.$$

Proof. By Theorem 5 for $|H|\leq 1$

$$\|(H\cdot X)^*_\infty\|_{L^2}\leq\sqrt{8}\|H\cdot X\|_{\mathcal{H}^2}\leq\sqrt{8}\|X\|_{\mathcal{H}^2}.$$

Since

$$\begin{aligned}2\|[X,X]^{1/2}_\infty\|_{L^2}&\leq 2\|[M,M]^{1/2}_\infty\|_{L^2}+2\|\int_0^\infty|dA_s|\|_{L^2}\\&=2\|X\|_{\mathcal{H}^2},\end{aligned}$$

where $X=M+A$ is the canonical decomposition of X, we have the right inequality.

For the left inequality we have $\|[M,M]^{1/2}_\infty\|_{L^2}\leq\|[X,X]^{1/2}_\infty\|_{L^2}$ by the lemma preceding this theorem. Moreover if $|H|\leq 1$, then

$$\begin{aligned}\|(H\cdot A)_t\|_{L^2}&\leq\|(H\cdot X)_t\|_{L^2}+\|(H\cdot M)\|_{L^2}\\&\leq\|(H\cdot X)^*_\infty\|_{L^2}+\|[M,M]^{1/2}_\infty\|_{L^2}.\end{aligned}$$

Next take $H=\frac{dA}{|dA|}$; this exists as a predictable process since A is natural if and only if it is predictable (Sect. 8 of Chap. III), and A predictable implies that $|A|$, the total variation process, is also predictable. Therefore we have

$$\begin{aligned}\|X\|_{\mathcal{H}^2}&=\|[M,M]^{1/2}_\infty\|_{L^2}+\|\int_0^\infty|dA_s|\|_{L^2}\\&=\|[M,M]^{1/2}_\infty\|_{L^2}+\|(H\cdot A)_\infty\|_{L^2}\\&\leq\|[M,M]^{1/2}_\infty\|_{L^2}+\|(H\cdot X)^*_\infty\|_{L^2}+\|[M,M]^{1/2}_\infty\|_{L^2}\\&\leq\|(H\cdot X)^*_\infty\|_{L^2}+2\|[X,X]^{1/2}_\infty\|_{L^2}.\end{aligned}$$

□

We present as a Corollary to Theorem 24 the equivalence of the two pseudonorms $\sup_{|H|\leq 1}\|(H\cdot X)^*_\infty\|_{L^2}$ and $\|X\|_{\mathcal{H}^2}$. We will not have need of this Corollary in this book,[5] but it is a pretty and useful result nevertheless. It is originally due to Yor [7], and it is actually true for all p, $1\leq p<\infty$. (See also Dellacherie-Meyer [2], pp. 303-305.)

Corollary. *For a semimartingale X (with $X_0=0$),*

$$\|X\|_{\mathcal{H}^2}\leq 3\sup_{|H|\leq 1}\|(H\cdot X)^*_\infty\|_{L^2}\leq 9\|X\|_{\mathcal{H}^2},$$

and in particular $\sup_{|H|\leq 1}\|(H\cdot X)^*_\infty\|_{L^2}<\infty$ *if and only if* $\|X\|_{\mathcal{H}^2}<\infty$.

Proof. By Theorem 5 if $\|X\|_{\mathcal{H}^2}<\infty$ and $|H|\leq 1$ we have

$$\|(H\cdot X)^*_\infty\|_{L^2}\leq\sqrt{8}\|H\cdot X\|_{\mathcal{H}^2}\leq\sqrt{8}\|X\|_{\mathcal{H}^2}.$$

Thus we need to show only the left inequality. By Theorem 24 it will suffice to show that

$$\|[X,X]^{1/2}_\infty\|_{L^2}\leq\sup_{|H|\leq 1}\|(H\cdot X)^*_\infty\|_{L^2},$$

for a semimartingale X with $X_0=0$.

To this end fix a $t>0$ and let $0=T_0\leq T_1\leq\cdots\leq T_n=t$ be a random partition of $[0,t]$. Choose $\epsilon_1,\dots,\epsilon_n$ non-random and each equal to 1 or -1 and let $H=\sum_{i=1}^n\epsilon_i 1_{(T_{i-1},T_i]}$. Then H is a simple predictable process and

$$(H\cdot X)_\infty=\sum_{i=1}^n\epsilon_i(X_{T_i}-X_{T_{i-1}}).$$

Let $\alpha=\sup_{|H|\leq 1}\|(H\cdot X)^*_\infty\|_{L^2}$. We then have:

$$\alpha^2\geq E\{(H\cdot X)^2_\infty\}=\sum_{i,j=1}^n\epsilon_i\epsilon_j E\{(X_{T_i}-X_{T_{i-1}})(X_{T_j}-X_{T_{j-1}})\}.$$

If we next average over all sequences $\epsilon_1,\dots,\epsilon_n$ taking values in the space $\{\pm 1\}^n$, we deduce

$$\alpha^2\geq\sum_{i=1}^n E\{(X_{T_i}-X_{T_{i-1}})^2\}=E\{\sum_{i=1}^n(X_{T_i}-X_{T_{i-1}})^2\}.$$

Next let $\sigma_m=\{T_i^m\}$ be a sequence of random partitions of $[0,t]$ tending to the identity. Then $\sum_i(X_{T_i^m}-X_{T_{i-1}^m})^2$ converges in probability to $[X,X]_t$.

[5] However the Corollary does give some insight into the relationship between Theorems 12 and 14 in Chap. V.

Let $\{m_k\}$ be a subsequence so that $\sum_i (X_{T_i^{m_k}} - X_{T_{i-1}^{m_k}})^2$ converges to $[X,X]_t$ a.s. Finally by Fatou's lemma we have

$$E\{[X,X]_t\} \leq \liminf_{m_k\to\infty} E\{\sum_i (X_{T_i^{m_k}} - X_{T_{i-1}^{m_k}})^2\} \leq \alpha^2.$$

Letting t tend to ∞ we conclude that

$$E\{[X,X]_\infty\} \leq \alpha^2 = \sup_{|H|\leq 1} \|(H\cdot X)^*_\infty\|^2_{L^2}.$$

It remains to show that if $\sup_{|H|\leq 1}\|(H\cdot X)^*_\infty\|_{L^2} < \infty$, then $X \in \mathcal{H}^2$. We will show the contrapositive: If $X \notin \mathcal{H}^2$, then $\sup_{|H|\leq 1}\|(H\cdot X)^*_\infty\|_{L^2} = \infty$. Indeed, let T_n be stopping times increasing to ∞ such that X^{T_n-} is in $\mathcal{H}^2$ for each n (cf. Theorem 13). Then

$$\|X^{T_n-}\|_{\mathcal{H}^2} \leq 3 \sup_{|H|\leq 1} \|(H\cdot X^{T_n-})^*_\infty\|_{L^2} \leq 9\|X^{T_n-}\|_{\mathcal{H}^2}.$$

Letting n tend to ∞ gives the result. □

Before proving Theorem 25 we need two technical lemmas.

Lemma 1. *Let A be a nonnegative increasing FV process and let Z be a positive uniformly integrable martingale. Let T be a stopping time such that $A = A^{T-}$ (that is, $A_\infty = A_{T-}$) and let k be a constant such that $Z \leq k$ on $[0,T)$. Then*

$$E\{A_\infty Z_\infty\} \leq kE\{A_\infty\}.$$

Proof. Since $A_{0-} = Z_{0-} = 0$, by integration by parts

$$\begin{aligned} A_tZ_t &= \int_0^t A_{s-}dZ_s + \int_0^t Z_{s-}dA_s + [A,Z]_t \\ &= \int_0^t A_{s-}dZ_s + \int_0^t Z_s dA_s \end{aligned}$$

where the second integral in the preceding is a path by path Stieltjes integral. Let R^n be stopping times increasing to ∞ a.s. that reduce the local martingale $(\int_0^t A_{s-}dZ_s)_{t\geq 0}$. Since dA_s charges only $[0,T)$ we have

$$\int_0^t Z_s dA_s \leq \int_0^t k dA_s \leq kA_\infty$$

for every $t \geq 0$. Therefore

$$\begin{aligned} E\{(AZ)_{R^n}\} &= E\{\int_0^{R^n} A_{s-}dZ_s\} + E\{\int_0^{R^n} Z_s dA_s\} \\ &\leq 0 + E\{kA_\infty\}. \end{aligned}$$

The result follows by Fatou's lemma. □

Lemma 2. *Let X be a semimartingale with $X_0 = 0$, let Q be another probability with $Q \ll P$, and let $Z_t = E_P\{\frac{dQ}{dP}|\mathcal{F}_t\}$. If T is a stopping time such that $Z_t \leq k$ on $[0,T)$ for a constant k, then*

$$\|X^{T-}\|_{\mathcal{H}^2(Q)} \leq 5\sqrt{k}\|X^{T-}\|_{\mathcal{H}^2(P)}.$$

Proof. By Theorem 24 we have

$$\begin{aligned}
\|X^{T-}\|_{\mathcal{H}^2(Q)} &\leq \sup_{|H|\leq 1} E_Q\{((H\cdot X^{T-})^*)^2\}^{1/2} + 2E_Q\{[X^{T-},X^{T-}]^{1/2}\} \\
&\leq \sup_{|H|\leq 1} E_P\left\{\frac{dQ}{dP}((H\cdot X^{T-})^*)^2\right\}^{1/2} + 2E_P\left\{\frac{dQ}{dP}[X^{T-},X^{T-}]^{1/2}\right\} \\
&\leq \sqrt{k}\sup_{|H|\leq 1} E_P\{((H\cdot X^{T-})^*)^2\}^{1/2} + 2\sqrt{k}E_P\{[X^{T-},X^{T-}]^{1/2}\}.
\end{aligned}$$

where we have used Lemma 1 on both terms to obtain the last inequality above. The result follows by the right inequality of Theorem 24. □

Note in particular that an important consequence of Lemma 2 is that if $Q \ll P$ with $\frac{dQ}{dP}$ bounded, then $X \in \mathcal{H}^2(P)$ implies that $X \in \mathcal{H}^2(Q)$ as well, with the estimate $\|X\|_{\mathcal{H}^2(Q)} \leq 5\sqrt{k}\|X\|_{\mathcal{H}^2(P)}$, where k is the bound for $\frac{dQ}{dP}$. Note further that this result (without the estimate) is obvious if one uses the equivalent pseudonorm given by the Corollary to Theorem 24, since

$$\begin{aligned}
\sup_{|H|\leq 1} E_Q\{((H\cdot X)^*_\infty)^2\} &= \sup_{|H|\leq 1} E_P\{\frac{dQ}{dP}((H\cdot X)^*_\infty)^2\} \\
&\leq k \sup_{|H|\leq 1} E_P\{((H\cdot X)^*_\infty)^2\},
\end{aligned}$$

where again k is the bound for $\frac{dQ}{dP}$.

Theorem 25. *Let X be a semimartingale and $H \in L(X)$. If $Q \ll P$ then $H \in L(X)$ under Q as well and $H_Q \cdot X = H_P \cdot X$, Q a.s.*

Proof. Let T^n be a sequence of stopping times increasing to ∞ a.s. such that H is $(X^{T^n-},\mathcal{H}^2)$-integrable (P), each $n \geq 1$. Let $Z_t = E_P\{\frac{dQ}{dP}|\mathcal{F}_t\}$, the càdlàg version. Define $S^n = \inf\{t > 0 : |Z_t| > n\}$ and set $R^n = S^n \wedge T^n$. Then $X^{R^n-} \in \mathcal{H}^2(P)\cap\mathcal{H}^2(Q)$ by Lemma 2, and H is $(\mathcal{H}^2, X^{R^n-})$ integrable (P). We need to show H is $(\mathcal{H}^2, X^{R^n-})$ integrable (Q), which will in turn imply that $H \in L(X)$ under Q. Let $X^{R^n-} = N + C$ be the canonical decomposition under Q. Let $H^m = H1_{\{|H|\leq m\}}$. Then

$$\begin{aligned}
&(E_Q\{\int (H^m_s)^2 d[N,N]_s\})^{1/2} + \|\int |H^m_s||dC_s|\|_{L^2(Q)} \\
&\qquad = \|H^m\cdot X^{R^n-}\|_{\mathcal{H}^2(Q)} \leq 5\sqrt{n}\|H^m\cdot X^{R^n-}\|_{\mathcal{H}^2(P)} \\
&\qquad \leq 5\sqrt{n}\|H\cdot X^{R^n-}\|_{\mathcal{H}^2(P)} < \infty,
\end{aligned}$$

and then by monotone convergence we see that H is $(\mathcal{H}^2, X^{R^n-})$ integrable (Q). Thus $H \in L(X)$ under Q, and it follows that $H_Q \cdot X = H_P \cdot X$, Q a.s. □

Theorem 25 can be used to extend Theorem 20 in a way analogous to the extension of Theorem 17 by Theorem 18 in Chap. II:

Theorem 26. *Let X, $\overline{X}$ be two semimartingales, and let $H \in L(X)$, $\overline{H} \in L(\overline{X})$. Let $A = \{\omega : H_\bullet(\omega) = \overline{H}_\bullet(\omega)$ and $X_\bullet(\omega) = \overline{X}_\bullet(\omega)\}$, and let $B = \{\omega : t \to X_t(\omega)$ is of finite variation on compacts$\}$. Then $H \cdot X = \overline{H} \cdot \overline{X}$ on A, and $H \cdot X$ is equal to a path by path Lebesgue-Stieltjes integral on B.*

Proof. Without loss of generality assume $P(A) > 0$. Define Q by $Q(\Lambda) = P(\Lambda|A)$. Then $Q \ll P$ and therefore $H \in L(X)$, $\overline{H} \in L(\overline{X})$ under Q as well as under P by Theorem 25. However under Q the processes H and $\overline{H}$ as well as X and $\overline{X}$ are indistinguishable. Thus $H_Q \cdot X = \overline{H}_Q \cdot \overline{X}$ and hence $H \cdot X = \overline{H} \cdot \overline{X}$ $P-$ a.s. on A by Theorem 25, since $Q \ll P$.

The second assertion has an analogous proof (see the proof of Theorem 18 of Chap. II). □

Note that one can use stopping times to localize the result of Theorem 26. the proof of the following Corollary is analogous to the proof of the Corollary of Theorem 18 of Chap. II.

Corollary. *With the notation of Theorem 26, let S, T be two stopping times with $S < T$. Define*

$$C = \{\omega : H_t(\omega) = \overline{H}_t(\omega); X_t(\omega) = \overline{X}_t(\omega); S(\omega) < t \leq T(\omega)\}$$

$$D = \{\omega : t \to X_t(\omega) \text{ is of finite variation on } S(\omega) < t < T(\omega)\}$$

Then $H \cdot X^T - H \cdot X^S = \overline{H} \cdot \overline{X}^T - \overline{H} \cdot \overline{X}^S$ on C and $H \cdot X^T - H \cdot X^S$ equals a path-by-path Lebesgue-Stieltjes integral on D.

Theorem 27. *Let P_k be a sequence of probabilities such that X is a P_k-semimartingale for each k. Let $R = \sum_{k=1}^\infty \lambda_k P_k$ where $\lambda_k \geq 0$, each k, and $\sum_{k=1}^\infty \lambda_k = 1$. Let $H \in L(X)$ under R. Then $H \in L(X)$ under P_k and $H_R \cdot X = H_{P_k} \cdot X$, P_k a.s., for all k such that $\lambda_k > 0$.*

Proof. If $\lambda_k > 0$ then $P_k \ll R$. Moreover since $P_k(\Lambda) \leq \frac{1}{\lambda_k} R(\Lambda)$, it follows that $H \in L(X)$ under P_k. The result then follows by Theorem 25. □

We now turn to the relationship of stochastic integration to martingales and local martingales. In Theorem 11 we saw that if M is a square integrable martingale and $H \in \mathbf{b}\mathcal{P}$, then $H \cdot M$ is also a square integrable martingale.

When M is *locally* square integrable we have a simple sufficient condition for H to be in $L(M)$.

Lemma. *Let M be a square integrable martingale and let $H \in \mathcal{P}$ be such that $E\{\int_0^\infty H_s^2 d[M,M]_s\} < \infty$. Then $H \cdot M$ is a square integrable martingale.*

Proof. If $H^k \in \mathbf{b}\mathcal{P}$, then $H^k \cdot M$ is a square integrable martingale by Theorem 11. Taking $H^k = H1_{\{|H| \leq k\}}$, and since H is $(\mathcal{H}^2, M)$ integrable, by Theorem 14 $H^k \cdot M$ converges in $\mathcal{H}^2$ to $H \cdot M$ which is hence a square integrable martingale. □

Theorem 28. *Let M be a locally square integrable local martingale, and let $H \in \mathcal{P}$. The stochastic integral $H \cdot M$ exists (i.e., $H \in L(M)$) and is a locally square integrable local martingale if there exists a sequence of stopping times $(T^n)_{n \geq 1}$ increasing to ∞ a.s. such that $E\{\int_0^{T^n} H_s^2 d[M,M]_s\} < \infty$.*

Proof. We assume that M is a square integrable martingale stopped at the time T^n. The result follows by applying the lemma. □

Theorem 29. *Let M be a local martingale, and let $H \in \mathcal{P}$ be locally bounded. Then the stochastic integral $H \cdot M$ is a local martingale.*

Proof. By stopping we may, as in the proof of Theorem 17 of Chap. III, assume that H is bounded, M is uniformly integrable, and that $M = N + A$ where N is a bounded martingale and A is of integrable variation. We know that there exists R^k increasing to ∞ a.s. such that $M^{R^k-} \in \mathcal{H}^2$, and since $H \in \mathbf{b}\mathcal{P}$ there exist processes $H^\ell \in \mathbf{b}\mathbb{L}$ such that $\|H^\ell \cdot M^{R^k-} - H \cdot M^{R^k-}\|_{\mathcal{H}^2}$ tends to zero. In particular, $H^\ell \cdot M^{R^k-}$ tends to $H \cdot M^{R^k-}$ in ucp. Therefore we can take H^ℓ such that $H^\ell \cdot M^{R^k-}$ tends to $H \cdot M$ in ucp, with $H^\ell \in \mathbf{b}\mathbb{L}$; finally without loss of generality we assume $H^\ell \cdot M$ converges to $H \cdot M$ in ucp. Since $H^\ell \cdot M = H^\ell \cdot N + H^\ell \cdot A$ and $H^\ell \cdot N$ converges to $H \cdot N$ in ucp, we deduce $H^\ell \cdot A$ converges to $H \cdot A$ in ucp as well. Let $0 \leq s < t$ and assume $Y \in \mathbf{b}\mathcal{F}_s$. Therefore, since A is of integrable total variation, and since $H^\ell \cdot A$ is a martingale for $H^\ell \in \mathbf{b}\mathbb{L}$ (see Theorem 17 of Chap. III), we have

$$\begin{aligned} E\{Y \int_{s+}^t H_u dA_u\} &= E\{Y \int_{s+}^t \lim_{\ell \to \infty} H_u^\ell dA_u\} \\ &= E\{Y \lim_{\ell \to \infty} \int_{s+}^t H_u^\ell dA_u\} \\ &= \lim_{\ell \to \infty} E\{Y \int_{s+}^t H_u^\ell dA_u\} \\ &= 0, \end{aligned}$$

where we have used Lebesgue's dominated convergence theorem both for the Stieltjes integral ω-by-ω (taking a subsequence if necessary to have a.s.

convergence) and for the expectation. We conclude that $(\int_0^t H_s dA_s)_{t\geq 0}$ is a martingale, hence $H \cdot M = H \cdot N + H \cdot A$ is also a martingale under the assumptions made; therefore it is a local martingale. □

In the proof of Theorem 29 the hypothesis that $H \in \mathcal{P}$ was locally bounded was used to imply that if $M = N + A$ with N having locally bounded jumps and A an FV local martingale, then the two proceses

$$\int_0^t H_s^2 d[N,N]_s \quad \text{and} \quad \int_0^t |H_s||dA_s|$$

are locally integrable. Thus one could weaken the hypothesis that H is locally bounded, but it would lead to an awkward statement. *The general result, that M a local martingale and $H \in L(M)$ implies that $H \cdot M$ is a local martingale, is not true*! See Emery [4, pp. 152, 153] for a surprisingly simple counterexample.

Corollary. *Let M be a local martingale, $M_0 = 0$, and let T be a predictable stopping time. Then M^{T-} is a local martingale.*

Proof. The notation M^{T-} means:

$$M_t^{T-} = M_t 1_{(t<T)} + M_{T-} 1_{(t\geq T)},$$

where $M_{0-} = M_0 = 0$. Let $(S^n)_{n\geq 1}$ be a sequence of stopping times announcing T. On $\{T > 0\}$,

$$\lim_{n\to\infty} 1_{[0,S^n]} = 1_{[0,T)},$$

and since $1_{[0,S^n]}$ is a left continuous, adapted process it is predictable; hence $1_{[0,T)}$ is predictable. But $\int_0^t 1_{[0,T)}(s)dM_s = M_t^{T-}$ by Theorem 18, and hence it is a local martingale by Theorem 29. □

Note that if M is a local martingale and T is an arbitrary stopping time, it is not true in general that M^{T-} is still a local martingale.

Since a continuous local martingale is locally square integrable, the theory is particularly nice.

Theorem 30. *Let M be a continuous local martingale and let $H \in \mathcal{P}$ be such that $\int_0^t H_s^2 d[M,M]_s < \infty$ a.s., each $t \geq 0$. Then the stochastic integral $H \cdot M$ exists (i.e., $H \in L(M)$) and it is a continuous local martingale.*

Proof. Since M is continuous, we can take

$$R^k = \inf\{t > 0 : |M_t| > k\}.$$

Then $|M_{t\wedge R^k}| \leq k$ and therefore M is locally bounded, hence locally square integrable. Also M continuous implies $[M,M]$ is continuous, whence if

$$T^k = \inf\{t > 0 : \int_0^t H_s^2 d[M,M]_s > k\},$$

we see that $(\int_0^t H_s^2 d[M,M]_s)_{t\geq 0}$ is also locally bounded. Then $H \cdot M$ is a locally square integrable local martingale by Theorem 28. The stochastic integral $H \cdot M$ is continuous because $\Delta(H \cdot M) = H(\Delta M)$ and $\Delta M = 0$ by hypothesis. □

In the classical case where the continuous local martingale M equals B, a standard Brownian motion, Theorem 30 yields that if $H \in \mathcal{P}$ and $\int_0^t H_s^2 ds < \infty$ a.s., each $t \geq 0$, then the stochastic integral $(H \cdot B_t)_{t\geq 0} = (\int_0^t H_s dB_s)_{t\geq 0}$ exists, since $[B,B]_t = t$.

Corollary. *Let X be a continuous semimartingale with (unique) decomposition $X = M + A$. Let $H \in \mathcal{P}$ be such that*

$$\int_0^t H_s^2 d[M,M]_s + \int_0^t |H_s||dA_s| < \infty \quad a.s.$$

each $t \geq 0$. Then the stochastic integral $(H \cdot X)_t = \int_0^t H_s dX_s$ exists and it is continuous.

Proof. By the Corollary of Theorem 19 of Chap. III, we know that M and A have continuous paths. The integral $H \cdot M$ exists by Theorem 30. Since $H \cdot A$ exists as a Stieltjes integral, it is easy to check that $H \in L(A)$, since A is continuous, and the result follows from Theorem 20. □

In the preceding corollary the semimartingale X is continuous, hence $[X,X] = [M,M]$ and the hypothesis can be written equivalently as

$$\int_0^t H_s^2 d[X,X]_s + \int_0^t |H_s||dA_s| < \infty \quad \text{a.s.}$$

each $t \geq 0$.

We end our treatment of martingales with a special case that yields a particularly simple condition for H to be in $L(M)$.

Theorem 31. *Let M be a local martingale with jumps bounded by a constant β. Let $H \in \mathcal{P}$ be such that $\int_0^t H_s^2 d[M,M]_s < \infty$ a.s., $t \geq 0$, and $E\{H_T^2\} < \infty$ for any bounded stopping time T. Then the stochastic integral $(\int_0^t H_s dM_s)_{t\geq 0}$ exists and it is a local martingale.*

Proof. Let $R^n = \inf\{t > 0 : \int_0^t H_s^2 d[M,M]_s > n\}$, and let $T^n = \min(R^n, n)$. Then T^n are bounded stopping times increasing to ∞ a.s. Note that

$$\begin{aligned} E\{\int_0^{T^n} H_s^2 d[M,M]_s\} &\leq n + E\{H_{T^n}^2 (\Delta M_{T^n})^2\} \\ &\leq n + \beta^2 E\{H_{T^n}^2\} < \infty, \end{aligned}$$

and the result follows from Theorem 28. □

The next theorem is, of course, an especially important theorem: the dominated convergence theorem for stochastic integrals.

Theorem 32 (Dominated Convergence). *Let X be a semimartingale, and let $H^m \in \mathcal{P}$ be a sequence converging a.s. to a limit H. If there exists a process $G \in L(X)$ such that $|H^m| \leq G$, all m, then H^m, H are in $L(X)$ and $H^m \cdot X$ converges to $H \cdot X$ in ucp.*

Proof. First note that if $|J| \leq G$ with $J \in \mathcal{P}$, then $J \in L(X)$: Indeed, let $(T^n)_{n\geq 1}$ increase to ∞ a.s. such that G is $(\mathcal{H}^2, X^{T^n-})$-integrable for each n. Then clearly

$$E\{\int_0^\infty J_s^2 d[\overline{N},\overline{N}]_s\} + E\{(\int_0^\infty |J_s||d\overline{A}_s|)^2\}$$
$$\leq E\{\int_0^\infty G_s^2 d[\overline{N},\overline{N}]_s\} + E\{(\int_0^\infty |G_s||d\overline{A}_s|)^2\} < \infty,$$

and thus J is $(\mathcal{H}^2, X^{T^n-})$ integrable for each n. (Here $\overline{N}+\overline{A}$ is the canonical decomposition of X^{T^n-}).

To show convergence in ucp, it suffices to show uniform convergence in probability on intervals of the form $[0,t_0]$ for t_0 fixed. Let $\epsilon > 0$ be given, and choose n such that $P(T^n < t_0) < \epsilon$, where $X^{T^n-} \in \mathcal{H}^2$ and G is $(\mathcal{H}^2, X^{T^n-})$-integrable. Let $X^{T^n-} = \overline{N} + \overline{A}$, the canonical decomposition. Then

$$E\{\sup_{t\leq t_0} |H^m \cdot X^{T^n-} - H \cdot X^{T^n-}|^2\}$$
$$\leq 2E\{\sup_{t\leq t_0} |(H^m - H)\cdot \overline{N}|^2\} + 2E\{(\int_0^{t_0} |H_s^m - H_s||d\overline{A}_s|)^2\}.$$

The second term tends to zero by the dominated convergence theorem of Lebesgue. Since $|H^m - H| \leq 2G$, the integral $(H^m - H)\cdot\overline{N}$ is a square-integrable martingale (Corollary of Theorem 28). Therefore using Doob's maximal quadratic inequality:

$$E\{\sup_{t\leq t_0}|(H^m - H)\cdot \overline{N}|^2\} \leq 4E\{|(H^m - H)\cdot \overline{N}_{t_0}|^2\}$$
$$= 4E\{\int_0^{t_0} (H_s^m - H_s)^2 d[\overline{N},\overline{N}]_s\},$$

and again this tends to zero by the dominated convergence theorem. Since convergence in L^2 implies convergence in probability, we conclude for $\delta > 0$:

$$\limsup_{m\to\infty} P\{\sup_{t\le t_0} |H^m \cdot X_t - H \cdot X_t| > \delta\}$$
$$\le \limsup_{m\to\infty} P\{\sup_{t\le t_0} |((H^m - H)\cdot X^{T^n -})_t| > \delta\} + P(T^n < t_0)$$
$$\le \epsilon,$$

and since ϵ is arbitrary, the limit is zero. □

We use the Dominated Convergence Theorem to prove a seemingly innocuous result. Generalizations however are delicate as we indicate following the proof.

Theorem 33. *Let $(\mathcal{F}_t)_{t\ge0}$ and $(\mathcal{G}_t)_{t\ge0}$ be two filtrations satisfying the usual hypotheses and suppose $\mathcal{F}_t \subset \mathcal{G}_t$, each $t \ge 0$, and that X is a semimartingale for both $(\mathcal{F}_t)_{t\ge0}$ and $(\mathcal{G}_t)_{t\ge0}$. Let H be locally bounded and predictable for $(\mathcal{F}_t)_{t\ge0}$. Then the stochastic integrals $H_{\mathcal{F}}\cdot X$ and $H_{\mathcal{G}}\cdot X$ both exist, and they are equal.*[6]

Proof. It is trivial that H is locally bounded and predictable for $(\mathcal{G}_t)_{t\ge0}$ as well. By stopping, we can assume without loss of generality that H is bounded. Let

$$\mathcal{H} = \{\text{all bounded, } \mathcal{F}\text{-predictable } H \text{ such that } H_{\mathcal{F}}\cdot X = H_{\mathcal{G}}\cdot X\}.$$

Then $\mathcal{H}$ is clearly a monotone vector space, and $\mathcal{H}$ contains the multiplicative class **bL** by Theorem 16 of Chap. II. Thus using Theorem 32 and the Monotone Class Theorem we are done. □

It is surprising that the assumption that H be locally bounded is important. Indeed, Jeulin [**1**, pp.46,47] has exhibited an example which shows that Theorem 33 is false in general.

Theorem 33 is not an exact generalization of Theorem 16 of Chap. II. Indeed, suppose $(\mathcal{F}_t)_{t\ge0}$ and $(\mathcal{G}_t)_{t\ge0}$ are two arbitrary filtrations such that X is a semimartingale for both $(\mathcal{F}_t)_{t\ge0}$ and $(\mathcal{G}_t)_{t\ge0}$, and H is bounded and predictable for both of them. If $\mathcal{I}_t = \mathcal{F}_t \cap \mathcal{G}_t$, then X is still an $(\mathcal{I}_t)_{t\ge0}$ semimartingale by Stricker's theorem, but it is not true in general that H is $(\mathcal{I}_t)_{t\ge0}$ predictable. It is an open question as to whether or not $H_{\mathcal{F}}\cdot X = H_{\mathcal{G}}\cdot X$ in this situation. For a partial result, see Zheng [**1**].

[6] $H_{\mathcal{F}}\cdot X$ and $H_{\mathcal{G}}\cdot X$ denote the stochastic integrals computed with the filtrations $(\mathcal{F}_t)_{t\ge0}$ and $(\mathcal{G}_t)_{t\ge0}$ respectively.

3. Martingale Representation

In this section we will be concerned with martingales, rather than semimartingales. The question of martingale representation is the following: given a collection $\mathcal{A}$ of martingales (or local martingales), when can *all* martingales (or all local martingales) be represented as stochastic integrals with respect to processes in $\mathcal{A}$? This question is surprisingly important in applications, and it is particularly interesting (in Finance Theory, for example) when $\mathcal{A}$ consists of just one element.

Throughout this section we assume as given an underlying complete, filtered probability space $(\Omega, \mathcal{F}, (\mathcal{F}_t)_{t\geq 0}, P)$ satisfying the usual hypotheses. We begin by considering only square integrable martingales; later we indicate how to extend these results to locally square integrable local martingales.

Definition. The space $\mathbf{M}^2$ of square integrable martingales is all martingales M such that $\sup_t E\{M_t^2\} < \infty$, and $M_0 = 0$ a.s. Notice that if $M \in \mathbf{M}^2$, then $\lim_{t\to\infty} E\{M_t^2\} = E\{M_\infty^2\} < \infty$, and $M_t = E\{M_\infty|\mathcal{F}_t\}$. Thus each $M \in \mathbf{M}^2$ can be identified with its terminal value M_∞. We can endow $\mathbf{M}^2$ with a **norm**: $\|M\| = E\{M_\infty^2\}^{1/2} = E\{[M,M]_\infty\}^{1/2}$, and also with an **inner product**: $(M,N) = E\{M_\infty N_\infty\}$, for $M, N \in \mathbf{M}^2$. It is evident that $\mathbf{M}^2$ is a Hilbert space and that its dual space is also $\mathbf{M}^2$.

The next definition is a key idea in the theory of martingale representation. It differs slightly from the customary definition because we are assuming all martingales are zero at time $t = 0$. If we did not have this hypothesis, we would have to add the condition that for any event $\Lambda \in \mathcal{F}_0$, any martingale M in the subspace F, then $M1_\Lambda \in F$.

Definition. A closed subspace F of $\mathbf{M}^2$ is called a **stable subspace** if it is **stable under stopping** (that is, if $M \in F$ and if T is a stopping time, then $M^T \in F$).[7]

Theorem 34. *Let F be a closed subspace of $\mathbf{M}^2$. Then the following are equivalent:*

a) F is closed under the operation: For $M \in F$, $(M - M^t)1_\Lambda \in F$ for $\Lambda \in \mathcal{F}_t$, any $t \geq 0$;
b) F is a stable subspace;
c) if $M \in F$ and H is bounded, predictable, then $(\int_0^t H_s dM_s)_{t\geq 0} = H \cdot M \in F$;
d) If $M \in F$ and H is predictable with $E\{\int_0^\infty H_s^2 d[M,M]_s\} < \infty$, then $H \cdot M \in F$.

[7] Recall that $M_t^T = M_{t\wedge T}$, and $M_0 = 0$.

Proof. Property (d) obviously implies (c), and it is simple that (c) implies (b). To get (b) implies (a), let $T = t_\Lambda$, where

$$t_\Lambda = \begin{cases} t & \text{if } \omega \in \Lambda \\ \infty & \text{if } \omega \notin \Lambda. \end{cases}$$

Then $T = t_\Lambda$ is a stopping time when $\Lambda \in \mathcal{F}_t$, and $(M - M^t)1_\Lambda = M - M^T$; since F is assumed stable, both M and M^T are in F. It remains to show only that (a) implies (d). Note that if H is simple predictable of the special form:

$$H_t = 1_{\Lambda_0}1_{\{0\}} + \sum_{i=1}^{n} 1_{\Lambda_i}1_{(t_i,t_{i+1}]}$$

with $\Lambda_i \in \mathcal{F}_{t_i}$, $0 \leq i \leq n$, $0 = t_0 \leq t_1 \leq \cdots \leq t_{n+1} < \infty$, then $H \cdot M \in F$ whenever $M \in F$. Linear combinations of such processes are dense in $\mathbf{bL}$ which in turn is dense in $\mathbf{b}\mathcal{P}$ under $d_M(\cdot,\cdot)$ by Theorem 2. But then $\mathbf{b}\mathcal{P}$ is dense in the space of predictable processes $\mathcal{H}$ such that $E\{\int_0^\infty H_s^2 d[M,M]_s\} < \infty$, as is easily seen (cf. Theorem 14). Therefore (a) implies (d) and the theorem is proved. □

Since the arbitrary intersection of closed, stable subspaces is still closed and stable, we can make the following definition.

Definition. Let $\mathcal{A}$ be a subset of $\mathbf{M}^2$. The **stable subspace generated by** $\mathcal{A}$, denoted $\mathcal{S}(\mathcal{A})$, is the intersection of all closed, stable subspaces containing $\mathcal{A}$.

We can identify a martingale $M \in \mathbf{M}^2$ with its terminal value $M_\infty \in L^2$. Therefore another martingale $N \in \mathbf{M}^2$ is **orthogonal** to M if $E\{N_\infty M_\infty\} = 0$. There is however another, stronger notion of orthogonality for martingales in $\mathbf{M}^2$.

Definition. Two martingales $N, M \in \mathbf{M}^2$ are said to be **strongly orthogonal** if their product $L = NM$ is a (uniformly integrable) martingale.

Note that if $N, M \in \mathbf{M}^2$ are strongly orthogonal, then NM a (uniformly integrable) martingale implies that $[N, M]$ is also a local martingale by Corollary 2 of Theorem 27 of Chap. II; it is a uniformly integrable martingale by the Kunita-Watanabe inequality (Theorem 25 of Chap. II). Thus $M, N \in \mathbf{M}^2$ *are strongly orthogonal if and only if* $[M, N]$ *is a uniformly integrable martingale.* If N and M are strongly orthogonal then $E\{N_\infty M_\infty\} = E\{L_\infty\} = E\{L_0\} = 0$, so strong orthogonality implies orthogonality. The converse is not true however. For example let $M \in \mathbf{M}^2$, and let $Y \in \mathcal{F}_0$, independent of M, with $P(Y = 1) = P(Y = -1) = \frac{1}{2}$. Let $N_t = YM_t$, $t \geq 0$. Then $N \in \mathbf{M}^2$ and

$$E\{N_\infty M_\infty\} = E\{YM_\infty^2\} = E\{Y\}E\{M_\infty^2\} = 0,$$

so M and N are orthogonal. However $MN = YM^2$ is not a martingale (unless $M = 0$) because $E\{YM_t^2|\mathcal{F}_0\} = YE\{M_t^2|\mathcal{F}_0\} \neq 0 = YM_0^2$.

Definition. For a subset $\mathcal{A}$ of $\mathbf{M}^2$ we let $\mathcal{A}^\perp$ (respectively $\mathcal{A}^\times$) denote the set of all elements of $\mathbf{M}^2$ orthogonal (resp. strongly orthogonal) to each element of $\mathcal{A}$.

Lemma 1. *If $\mathcal{A}$ is any subset of $\mathbf{M}^2$, then $\mathcal{A}^\times$ is (closed and) stable.*

Proof. Let M^n be a sequence of $\mathcal{A}^\times$ converging to M, and let $N \in \mathcal{A}$. Then $M^n N$ is a martingale for each n and $\mathcal{A}^\times$ will be shown to be closed if MN is also one, or equivalently that $[M,N]$ is a martingale. However

$$E\{|[M^n,N]-[M,N]_t|\} = E\{|[M^n-M,N]_t|\}$$
$$\leq (E\{[M^n-M,M^n-M]_t\})^{1/2}(E\{[N,N]_t\})^{1/2}$$

by the Kunita-Watanabe inequalities. It follows that $[M^n,N]_t$ converges to $[M,N]_t$ in L^1, and therefore $[M,N]$ is a martingale, and $\mathcal{A}^\times$ is closed. Also $\mathcal{A}^\times$ is stable because $M \in \mathcal{A}^\times$, $N \in \mathcal{A}$ implies $[M^T,N]=[M,N]^T$ is a martingale and thus M^T is strongly orthogonal to N. □

Lemma 2. *Let N, M be in $\mathbf{M}^2$. Then the following are equivalent:*

a) M and N are strongly orthogonal;
b) $\mathcal{S}(M)$ and N are strongly orthogonal;
c) $\mathcal{S}(M)$ and $\mathcal{S}(N)$ are strongly orthogonal;
d) $\mathcal{S}(M)$ and N are weakly orthogonal;
e) $\mathcal{S}(M)$ and $\mathcal{S}(N)$ are weakly orthogonal.

Proof. If M and N are strongly orthogonal, let $\mathcal{A}=\{N\}$ and then $M \in \mathcal{A}^\times$. Since $\mathcal{A}^\times$ is a closed stable subspace by Lemma 1, $\mathcal{S}(M) \subset \{N\}^\times$. Therefore (b) holds and hence (a) implies (b). The same argument yields that (b) implies (c). That (c) implies (e) which implies (d) is obvious. It remains to show that (d) implies (a).

Suppose N is weakly orthogonal to $\mathcal{S}(M)$. It suffices to show that $[N,M]$ is a martingale. By Theorem 21 of Chap. I it suffices to show $E\{[N,M]_T\}=0$ for any stopping T. However $E\{[N,M]_T\} = E\{[N,M^T]_\infty\}=0$, since N is orthogonal to M^T which is in $\mathcal{S}(M)$. □

Theorem 35. *Let $M^1,\dots,M^n \in \mathbf{M}^2$, and suppose M^i, M^j are strongly orthogonal for $i \neq j$. Then $\mathcal{S}(M^1,\dots,M^n)$ consists of the set of stochastic integrals*

$$H^1\cdot M^1+\dots+H^n\cdot M^n=\sum_{i=1}^n H^i\cdot M^i,$$

where H^i is predictable and

$$E\{\int_0^\infty (H_s^i)^2 d[M^i,M^i]_s\}<\infty, \qquad 1\leq i\leq n.$$

Proof. Let $\mathcal{I}$ denote the space of processes $\sum_{i=1}^n H^i\cdot M^i$, where H^i satisfy the hypotheses of the theorem. By Theorem 34 any closed, stable subspace must contain $\mathcal{I}$. It is simple to check that $\mathcal{I}$ is stable, so we need to show only that $\mathcal{I}$ is closed. Let

$$L_M^2 = \{H \in \mathcal{P} : E\{\int_0^\infty H_s^2 d[M,M]_s\} < \infty\}.$$

Then the mapping $(H^1, H^2, \dots H^n) \to \sum_{i=1}^n H^i \cdot M^i$ is an isometry from $L_{M^1}^2 \oplus \cdots \oplus L_{M^n}^2$ into $\mathbf{M}^2$; since it is a Hilbert space isometry its image $\mathcal{I}$ is complete, and therefore closed. □

Theorem 36. *Let $\mathcal{A}$ be a subset of $\mathbf{M}^2$ which is stable. Then $\mathcal{A}^\perp$ is a stable subspace, and if $M \in \mathcal{A}^\perp$ then M is strongly orthogonal to $\mathcal{A}$. That is, $\mathcal{A}^\perp = \mathcal{A}^\times$, and $\mathcal{S}(\mathcal{A}) = \mathcal{A}^{\perp\perp} = \mathcal{A}^{\times\perp} = \mathcal{A}^{\times\times}$.*

Proof. We first show that $\mathcal{A}^\perp = \mathcal{A}^\times$. Let $M \in \mathcal{A}$ and $N \in \mathcal{A}^\perp$. Since N is orthogonal to $\mathcal{S}(M)$, by Lemma 2 N and M are strongly orthogonal. Therefore $\mathcal{A}^\perp \subset \mathcal{A}^\times$. However clearly $\mathcal{A}^\times \subset \mathcal{A}^\perp$, whence $\mathcal{A}^\times = \mathcal{A}^\perp$, and thus $\mathcal{A}^\perp = \mathcal{A}^\times$ is a stable subspace by Lemma 1.

By the above applied to $\mathcal{A}^\perp$, we have that $(\mathcal{A}^\perp)^\perp = (\mathcal{A}^\perp)^\times$.

It remains to show that $\mathcal{S}(\mathcal{A}) = \mathcal{A}^{\perp\perp}$. Since $\mathcal{A}^{\perp\perp} = \overline{\mathcal{A}}$, the closure of $\mathcal{A}$ in $\mathbf{M}^2$, it suffices to show that $\overline{\mathcal{A}}$ is stable. However it is simple to check that condition (a) of Theorem 34 is satisfied for $\overline{\mathcal{A}}$, since it already is satisfied for $\mathcal{A}$, and we conclude $\overline{\mathcal{A}}$ is a stable subspace. □

Corollary 1. *Let $\mathcal{A}$ be a stable subspace of $\mathbf{M}^2$. Then each $M \in \mathbf{M}^2$ has a decomposition $M = A + B$, with $A \in \mathcal{A}$ and $B \in \mathcal{A}^\times$.*

Proof. $\mathcal{A}$ is a closed subspace of $\mathbf{M}^2$, so each $M \in \mathbf{M}^2$ has a decomposition into $M = A + B$ with $A \in \mathcal{A}$ and $B \in \mathcal{A}^\perp$. However $\mathcal{A}^\perp = \mathcal{A}^\times$ by Theorem 36. □

Corollary 2. *Let $M, N \in \mathbf{M}^2$, and let L be the projection of N onto $\mathcal{S}(M)$, the stable subspace generated by M. Then there exists a predictable process H such that $L = H \cdot M$.*

Proof. We know that such an L exists by Corollary 1. Since $\{M\}$ consists of just one element we can apply Theorem 35 to obtain the result. □

Definition. Let $\mathcal{A}$ be finite set of martingales in $\mathbf{M}^2$. We say that $\mathcal{A}$ **has the (predictable) representation property** if $\mathcal{I} = \mathbf{M}^2$, where

$$\mathcal{I} = \{X : X = \sum_{i=1}^n H^i \cdot M^i, M^i \in \mathcal{A}\},$$

each H^i predictable such that

$$E\{\int_0^\infty (H_s^i)^2 d[M^i, M^i]_s\} < \infty, \quad 1 \le i \le n.$$

Corollary 3. *Let $\mathcal{A} = \{M^1, \ldots, M^n\} \subset \mathbf{M}^2$, and suppose M^i, M^j are strongly orthogonal for $i \neq j$. Suppose further that if $N \in \mathbf{M}^2$, $N \perp \mathcal{A}$ in the strong sense implies that $N = 0$. Then $\mathcal{A}$ has the predictable representation property.*

Proof. By Theorem 35 we have $\mathcal{S}(\mathcal{A}) = \mathcal{I}$. The hypotheses imply that $\mathcal{S}(\mathcal{A})^{\perp} = \{0\}$, hence $\mathcal{S}(\mathcal{A}) = \mathbf{M}^2$. □

Stable subspaces and predictable representation can be viewed from an alternative standpoint. Up to this point we have assumed as given and fixed an underlying space $(\Omega, \mathcal{F}, (\mathcal{F}_t)_{t \geq 0}, P)$, and a set of martingales $\mathcal{A}$ in $\mathbf{M}^2$. We will see that the property that $\mathcal{S}(\mathcal{A}) = \mathbf{M}^2$, intimately related to predictable representation (cf. Theorem 35), is actually a property of the probability measure P, considered as one element among the collection of probability measures that make all elements of $\mathcal{A}$ square-integrable $(\mathcal{F}_t)_{t \geq 0}$ martingales.

Our first observation is that since the filtration $(\mathcal{F}_t)_{t \geq 0}$ is assumed to be P-complete, it is reasonable to consider only probability measures that are absolutely continuous with respect to P.

Definition. Let $\mathcal{A} \subset \mathbf{M}^2$. **The set of $\mathbf{M}^2$ martingale measures for** $\mathcal{A}$, denoted $\mathcal{M}^2(\mathcal{A})$, is the set of all probability measures Q defined on $\bigvee_{0 \leq t < \infty} \mathcal{F}_t$ such that

(i) $Q \ll P$,
(ii) $Q = P$ on $\mathcal{F}_0$,
(iii) if $X \in \mathcal{A}$ then $X \in \mathbf{M}^2(Q)$,

where $\mathbf{M}^2(Q)$ denotes all $(Q, (\mathcal{F}_t)_{t \geq 0})$ square integrable martingales.

Lemma. *The set $\mathcal{M}^2$ is a convex set.*

Proof. Let Q and $R \in \mathcal{M}^2(\mathcal{A})$, and let $S = \lambda Q + (1 - \lambda)R$, $0 < \lambda < 1$. Then for $X \in \mathcal{M}^2(\mathcal{A})$,

$$\sup_t E_S\{M_t^2\} = \sup_t [\lambda E_Q\{M_t^2\} + (1 - \lambda)E_R\{M_t^2\}] < \infty,$$

since Q, $R \in \mathcal{M}^2(\mathcal{A})$. Also if $H \in b\mathcal{F}_s$, $s < t$, then

$$\begin{aligned} E_S\{M_t H\} &= \lambda E_Q\{M_t H\} + (1 - \lambda)E_R\{M_t H\} \\ &= \lambda E_Q\{M_s H\} + (1 - \lambda)E_R\{M_s H\} \\ &= E_S\{M_s H\}, \end{aligned}$$

and $S \in \mathcal{M}^2(\mathcal{A})$. □

Definition. A measure $Q \in \mathcal{M}^2(\mathcal{A})$ is an **extremal point** of $\mathcal{M}^2(\mathcal{A})$ if whenever $Q = \lambda R + (1 - \lambda)S$ with $R, S \in \mathcal{M}^2(\mathcal{A})$, $R \neq S$, $0 \leq \lambda \leq 1$, then $\lambda = 0$ or 1.

Theorem 37. *Let $\mathcal{A} \subset \mathbf{M}^2$. If $\mathcal{S}(\mathcal{A}) = \mathbf{M}^2$ then P is an extremal point of $\mathcal{M}^2(\mathcal{A})$.*

Proof. Suppose P is not extremal. We will show that $\mathcal{S}(\mathcal{A}) \neq \mathbf{M}^2$. Since P is not extremal, there exist $Q, R \in \mathcal{M}^2(\mathcal{A})$, $Q \neq R$, such that $P = \lambda Q + (1-\lambda)R$, $0 < \lambda < 1$. Let

$$L_\infty = \frac{dQ}{dP},$$

and let $L_t = E\{\frac{dQ}{dP}|\mathcal{F}_t\}$. Then $1 = \frac{dP}{dP} = \lambda L_\infty + (1-\lambda)\frac{dR}{dP} \geq \lambda L_\infty$ a.s., hence $L_\infty \leq \frac{1}{\lambda}$ a.s. Therefore L is a bounded martingale with $L_0 = 1$ (since $Q = P$ on $\mathcal{F}_0$), and thus $L - L_0$ is a nonconstant martingale in $\mathbf{M}^2(P)$. However if $X \in \mathcal{A}$ and $H \in b\mathcal{F}_s$, then X is a Q-martingale and for $s < t$:

$$\begin{aligned} E_P\{X_t L_t H\} &= E_P\{X_t L_\infty H\} \\ &= E_Q\{X_t H\} \\ &= E_Q\{X_s H\} \\ &= E_P\{X_s L_\infty H\} \\ &= E_P\{X_s L_s H\}, \end{aligned}$$

and XL is a P-martingale. Therefore $X(L - L_0)$ is a P-martingale, and $L - L_0 \in \mathbf{M}^2$ and it is strongly orthogonal to $\mathcal{A}$. By Theorem 36 we cannot have $\mathcal{S}(\mathcal{A}) = \mathbf{M}^2$. □

Theorem 38. *Let $\mathcal{A} \subset \mathbf{M}^2$. If P is an extremal point of $\mathcal{M}^2(\mathcal{A})$, then every bounded P martingale strongly orthogonal to $\mathcal{A}$ is null.*

Proof. Let L be a bounded nonconstant martingale strongly orthogonal to $\mathcal{A}$. Let c be a bound for $|L|$, and set:

$$dQ = (1 - \frac{L_\infty}{2c})dP; \; dR = (1 + \frac{L_\infty}{2c})dP.$$

We have Q, $R \in \mathcal{M}^2(\mathcal{A})$, and $P = \frac{1}{2}Q + \frac{1}{2}R$ is a decomposition that shows that P is not extremal, a contradiction. □

Theorem 39. *Let $\mathcal{A} = \{M^1, \ldots, M^n\} \subset \mathbf{M}^2$, with M^i continuous and M^i, M^j strongly orthogonal for $i \neq j$. Suppose P is an extremal point of $\mathcal{M}^2(\mathcal{A})$. Then*

a) every stopping time is accessible;
b) every bounded martingale is continuous;
c) every uniformly integrable martingale is continuous;
d) $\mathcal{A}$ has the predictable representation property.

Proof. (a) Suppose T is a totally inaccessible stopping time and $P(T<\infty)>0$. By Theorem 11 of Chap. III there exists a martingale M with $\Delta M_T =$

$1_{\{T<\infty\}}$ and M continuous elsewhere. Moreover the martingale M can be taken of finite variation with $M_0 = 0$. Therefore, since each M^i is continuous, $[M, M^i] = 0$. Moreover taking (for example)

$$R^n = \inf\{t > 0 : |M_t| > n\},$$

$|M^{R^n}| \leq n + 1$ and thus M is locally bounded. Indeed, we then have M^{R^n} are bounded martingales strongly orthogonal to M^i. By Theorem 38 we have $M^{R^n} = 0$ for each n; since $\lim_{n\to\infty} R^n = \infty$ a.s., we conclude $M = 0$, a contradiction.

(b) Let M be a bounded martingale which is not continuous, and assume $M_0 = 0$. Let $T^\epsilon = \inf\{t > 0 : |\Delta M_t| > \epsilon\}$. Then there exists $\epsilon > 0$ such that for $T = T^\epsilon$, $P\{|\Delta M_T| > 0\} > 0$. By part (a) the stopping time T is accessible, hence without loss we may assume that T is predictable. Therefore M^{T-} is a bounded martingale by the Corollary to Theorem 29, whence $N = M^T - M^{T-} = \Delta M_T 1_{\{t\geq T\}}$ is also a bounded martingale. However N is a finite variation bounded martingale, hence $[N, M^i] = 0$ each i; that is, N is a bounded martingale strongly orthogonal to $\mathcal{A}$, hence $N = M^T - M^{T-} = 0$ by Theorem 38, and we conclude that M is continuous.

(c) Let M be a uniformly integrable martingale closed by M_∞. Define

$$M_t^n = E\{M_\infty 1_{\{|M_\infty|\leq n\}} | \mathcal{F}_t\}.$$

Then M^n are bounded martingales and therefore continuous by part (b). However

$$P\{\sup_t |M_t^n - M_t| > \epsilon\} \leq \frac{1}{\epsilon} E\{|M_\infty^n - M_\infty|\}$$

by an inequality of Doob[8], and the right side tends to 0 as n tends to ∞. Therefore there exists a subsequence (n_k) such that $\lim_{k\to\infty} M_t^{n_k} = M_t$ a.s., uniformly in t. Therefore M is continuous.

(d) By Corollary 3 of Theorem 36 it suffices to show that if $N \in \mathcal{A}^\times$ then $N = 0$. Suppose $N \in \mathcal{A}^\times$. Then N is continuous by (c). Therefore N is locally bounded; hence by stopping N must be 0 by Theorem 38. □

The next theorem allows us to consider subspaces generated by countably infinite collections of martingales.

Theorem 40. *Let $M \in \mathbf{M}^2$, $Y^n \in \mathbf{M}^2$, $n \geq 1$, and suppose Y_∞^n converges to Y_∞ in L^2 and that there exists a sequence $H^n \in L(M)$ such that $Y_t^n = \int_0^t H_s^n dM_s$, $n \geq 1$. Then there exists a predictable process H in $L(M)$ such that $Y_t = \int_0^t H_s dM_s$.*

Proof. If Y_∞^n converges to Y_∞ in L^2, then Y^n converges to Y in $\mathbf{M}^2$. By Theorem 35 we have that $\mathcal{S}(M) = \mathcal{I}(M)$, the stochastic integrals with respect to M. Moreover $Y^n \in \mathcal{S}(M)$, each n. Therefore Y is in the closure

[8] See, for example, Breiman [1], page 88.

of $\mathcal{S}(M)$; but $\mathcal{S}(M)$ is closed, so $Y \in \mathcal{S}(M) = \mathcal{I}(M)$, and the theorem is proved. □

Theorem 41. *Let $\mathcal{A} = \{M^1, M^2, \dots, M^n, \dots\}$, with $M^i \in \mathbf{M}^2$, and suppose there exist disjoint predictable sets Λ^i such that $1_{\Lambda^i} d[M^i, M^i] = d[M^i, M^i]$, $i \geq 1$. Let $A_t = \sum_{i=1}^{\infty} \int_0^t 1_{\Lambda_i}(s) d[M^i, M^i]_s$. Suppose that (a): $E\{A_\infty\} < \infty$ and (b): For $\mathcal{F}^i \subset \mathcal{F}_\infty$ such that for any $X^i \in b\mathcal{F}^i$, we have $X_t^i = E\{X^i|\mathcal{F}_t\} = \int_0^t H_s^i dM_s^i$, $t \geq 0$, for some predictable process H^i.*

Then $M = \sum_{i=1}^{\infty} M^i$ exists and is in $\mathbf{M}^2$, and for any $Y \in b\bigvee_i \mathcal{F}^i$, if $Y_t = E\{Y|\mathcal{F}_t\}$, we have that $Y_t = \int_0^t H_s dM_s$, for the martingale $M = \sum_{i=1}^{\infty} M^i$ and for some $H \in L(M)$.

Proof. Let $N^n = \sum_{i=1}^n M^i$. Then $[N^n, N^n]_t = \sum_{i=1}^n \int_0^t 1_{\Lambda_i}(s) d[M^i, M^i]_s$, hence $E\{(N_\infty^n)^2\} = E\{[N^n, N^n]_\infty\} \leq E\{A_\infty\}$, and N^n is Cauchy in $\mathbf{M}^2$ with limit equal to M. By hypothesis we have that if $X^i \in b\mathcal{F}^i$ then

$$\begin{aligned} E\{X^i|\mathcal{F}_t\} = X_t^i &= \int_0^t H_s^i dM_s^i \\ &= \int_0^t 1_{\Lambda^i} H_s^i dM_s^i \\ &= \int_0^t 1_{\Lambda^i} H_s^i d(M_s^i + \sum_{i \neq j} M_s^j) \\ &= \int_0^t 1_{\Lambda^i} H_s^i dM_s. \end{aligned}$$

Therefore if $i \neq j$ we have that

$$\begin{aligned} [X^i, X^j] &= \int_0^t 1_{\Lambda^i} H_s^i 1_{\Lambda^j} H_s^j d[M, M]_s \\ &= \int_0^t 1_{\Lambda^i \cap \Lambda^j} H_s^i H_s^j d[M, M]_s \\ &= 0, \end{aligned}$$

since $\Lambda^i \cap \Lambda^j = \emptyset$, by hypothesis. However using integration by parts we have

$$\begin{aligned} X_t^i X_t^j &= \int_0^t X_{s-}^i dX_s^j + \int_0^t X_{s-}^j dX_s^i + [X^i, X^j]_t \\ &= \int_0^t X_{s-}^i dX_s^j + \int_0^t X_{s-}^j dX_s^i \\ &= \int_0^t X_{s-}^i 1_{\Lambda^j} H_s^j dM_s + \int_0^t X_{s-}^j 1_{\Lambda^i} H_s^i dM_s \\ &= \int_0^t H_s^{i,j} dM_s, \end{aligned}$$

where $H^{i,j}$ is defined in the obvious way. By iteration we have predictable representation for all finite products $\prod_{i\le n} X^i$. The monotone class theorem together with Theorem 40 then yields the result. □

We conclude this section by applying these results to a very important special case: n-dimensional Brownian motion. This example shows how these results can be easily extended to locally square integrable local martingales.

Theorem 42. *Let $X = (X^1, \dots, X^n)$ be an n-dimensional Brownian motion and let $(\mathcal{F}_t)_{0\le t\le\infty}$ denote its completed natural filtration. Then every locally square integrable local martingale M for $(\mathcal{F}_t)_{t\ge 0}$ has a representation*

$$M_t = M_0 + \sum_{i=1}^{n} \int_0^t H_s^i dX_s^i,$$

where H^i is (predictable, and) in $L(X^i)$.

Proof. Fix t_0, $0 < t < t_0$, and assume X is stopped at t_0. Then letting $\mathcal{A} = \{X^1, \dots, X^n\}$, we have that $\mathcal{A} \subset \mathbf{M}^2$. Let $\mathcal{M}^2(\mathcal{A})$ be all probability measures Q such that $Q \ll P$, and under Q all of $\mathcal{A} \subset \mathbf{M}^2(Q)$.[9] Since $[X^i, X^j]$ is the same process under Q as under P we have by Lévy's Theorem (Theorem 39 of Chap. II) that $X = (X^1, \dots, X^n)$ is a Q Brownian motion as well. Therefore for bounded Borel functions $f_1, \dots, f_n$ we have that $f_1(X_{t_1}) f_2(X_{t_2} - X_{t_1}) \dots f_n(X_{t_n} - X_{t_{n-1}})$, for $t_1 < t_2 \cdots < t_n \le t_0$ have the same expectation for Q as they do for P. Thus $P = Q$ on $\mathcal{F}_{t_0}$, and we conclude that $\mathcal{M}^2(\mathcal{A})$ is the singleton $\{P\}$. The probability law P is then trivially extremal for $\mathcal{M}^2(\mathcal{A})$, and therefore by Theorem 39 we know that $\mathcal{A}$ has the predictable representation property.

Suppose M is locally square integrable and $M_t = M_{t\wedge t_0}$. Let $(T^k)_{k\ge 0}$ be a sequence of stopping times increasing to ∞ a.s., $T_0 = 0$, and such that $M^{T^k} \in \mathbf{M}^2$, each $k \ge 0$. Then by the preceding there exist processes $H^{i,k}$ such that

$$M^{T^k} = \sum_{i=1}^{n} H^{i,k} \cdot X^i, \quad \text{each } k \ge 0.$$

By defining $H^i = \sum_{k=1}^{\infty} H^{i,k} 1_{(T^{k-1}, T^k]}$, we have that $H^i \in L(X^i)$ and also

$$M = \sum_{i=1}^{n} H^i \cdot X^i.$$

Next let $(t_\ell)_{\ell\ge 0}$ be fixed times increasing to ∞ with $t_0 = 0$. For each t_ℓ we know there exist $H^{i,\ell}$ such that for a given locally square integrable local martingale N

$$N_{t\wedge t_\ell} = \sum_{i=1}^{n} (H^{i,\ell} \cdot X^i)_{t\wedge t_\ell}.$$

[9] $\mathbf{M}^2(Q)$ denotes all square integrable Q-martingales.

By defining $H^i = \sum_{\ell=1}^{\infty} H^{i,\ell} 1_{(t_{\ell-1},t_\ell]}$ one easily checks that $H^i \in L(X^i)$ and that

$$N_t = \sum_{i=1}^{n} H^i \cdot X^i. \qquad \square$$

Corollary 1. *Let $(\mathcal{F}_t)_{t\geq 0}$ be the completed natural filtration of an n-dimensional Brownian motion. Then every local martingale M for $(\mathcal{F}_t)_{t\geq 0}$ is continuous.*

Proof. In the proof of Theorem 42 we saw that the underlying probability law P is extremal for $\mathcal{A} = \{X^1, \dots, X^n\}$. Therefore by Theorem 39(c), every uniformly integrable martingale is continuous. The Corollary follows by stopping. $\square$

Corollary 2. *Let $X = (X^1, \dots, X^n)$ be an n-dimensional Brownian motion and let $(\mathcal{F}_t)_{0\leq t\leq \infty}$ be its completed natural filtration. Then every local martingale M for $(\mathcal{F}_t)_{t\geq 0}$ has a representation*

$$M_t = M_0 + \sum_{i=1}^{n} \int_0^t H_s^i dX_s^i$$

where H^i are predictable.

Proof. By Corollary 1 any local martingale M is continuous, hence it is locally square integrable. It remains only to apply Theorem 42. $\square$

Corollary 3. *Let $X = (X^1, \dots, X^n)$ be an n-dimensional Brownian motion and let $(\mathcal{F}_t)_{0\leq t\leq \infty}$ be its completed natural filtration. Let $Z \in \mathcal{F}_\infty$ be in L^1. Then there exist H^i predictable in $L(X^i)$ with $\int_0^\infty (H_s^i)^2 ds < \infty$ a.s. such that*

$$Z = E\{Z\} + \sum_{i=1}^{n} \int_0^\infty H_s^i dX_s^i.$$

Proof. Let $Z_t = E\{Z|\mathcal{F}_t\}$, taking the càdlàg (and hence continuous) version. By Corollary 2 we have

$$Z_t = Z_0 + \sum_{i=1}^{n} \int_0^t H_s^i dX_s^i.$$

By Theorem 41 of Chap. II we have that $Z_t = B_{[Z,Z]_t}$ for some Brownian motion B, $0 < t < \infty$. Letting t tend to ∞ shows that $Z_\infty = B_{\lim_{t\to\infty}[Z,Z]_t}$, which shows that $[Z,Z]_\infty < \infty$ a.s. However

$$\sum_{i=1}^{n} \int_0^\infty (H_s^i)^2 ds = \lim_{t\to\infty} \sum_{i=1}^{n} \int_0^t (H_s^i)^2 ds = \lim_{t\to\infty} [Z,Z]_t = [Z,Z]_\infty < \infty$$

a.s. Finally, take $t=\infty$, observe that $Z_\infty = Z$ and that $\mathcal{F}_0$ is a.s. trivial and hence Z_0 is constant and therefore $Z_0 = E\{Z\}$. □

Corollary 4. *Let $X=(X^1,\dots,X^n)$ be an n-dimensional Brownian motion and let $(\mathcal{F}_t)_{0\le t\le\infty}$ be its completed natural filtration. Let $Z \in L^1(\mathcal{F}_\infty)$ and $Z>0$ a.s. Then there exist J^i predictable with $\int_0^\infty (J_s^i)^2 ds < \infty$ a.s. such that*

$$Z = E\{Z\}\exp\left(\sum_{i=1}^n \int_0^\infty J_s^i dX_s^i - \frac{1}{2}\int_0^\infty (J_s^i)^2 ds\right).$$

Proof. By Corollary 3 there exist predictable H^i such that if $Z_t = E\{Z|\mathcal{F}_t\}$, then

$$Z_t = E\{Z\} + \sum_{i=1}^n \int_0^t H_s^i dX_s^i.$$

Therefore

$$\ell n(Z_t) = \ell n(Z_0) + \sum_{i=1}^n \int_0^t \frac{1}{Z_s} H_s^i dX_s^i - \frac{1}{2}\sum_{i=1}^n \int_0^t \frac{1}{Z_s^2}(H_s^i)^2 ds.$$

(Note that since $(Z_t)_{t\ge 0}$ is continuous and never 0, $\frac{1}{Z}$ is locally bounded.) The proof is completed by setting $J_s^i = \frac{1}{Z_s}H_s^i$ and taking exponentials of both sides. □

Corollary 5. *Let $(\mathcal{F}_t)_{t\ge 0}$ be the completed natural filtration of an n-dimensional Brownian motion. If T is a totally inaccessible stopping time, then $T=\infty$ a.s..*

Proof. This is merely Theorem 39(a). □

4. Stochastic Integration Depending on a Parameter

The results of this section are of a technical nature, but they are needed for our subsequent investigation of semimartingale local times. Nevertheless they have intrinsic interest. For example Theorems 45 and 46 are a type of Fubini theorems for stochastic integration. A more comprehensive treatment of stochastic integration depending on a parameter can be found in Stricker-Yor [1] and Jacod [1]. Throughout this section $(A,\mathcal{A})$ denotes a measurable space.

Theorem 43. *Let $Y^n(a,t,\omega)$ be a sequence of processes that are (i) $\mathcal{A}\otimes\mathcal{B}(\mathbb{R}_+)\otimes\mathcal{F}$ measurable, and (ii) for each fixed a the process $Y^n(a,t,\omega)$ is* *càdlàg.*[10] *Suppose $Y^n(a,t,\cdot)$ converges ucp for each $a \in A$. Then there exists an $\mathcal{A} \otimes \mathcal{B}(\mathbf{R}_+) \otimes \mathcal{F}$ measurable process $Y = Y(a,t,\omega)$ such that*

(a) *$Y(a,t,\cdot) = \lim_{n\to\infty} Y^n(a,t,\cdot)$ with convergence in ucp;*
(b) *for each $a \in A$, Y is a.s. càdlàg.*

Moreover there exists a subsequence $n_k(a)$ depending measurably on a such that $\lim_{n_k(a)\to\infty} Y_t^{n_k(a)} = Y_t$ uniformly in t on compacts, a.s.

Proof. Let $S^a_{u,i,j} = \sup_{t\leq u}|Y^i(a,t,\cdot) - Y^j(a,t,\cdot)|$. Since Y^i is càdlàg in t the function $(a,\omega) \to S^a_{u,i,j}$ is $\mathcal{A} \otimes \mathcal{F}$ measurable. By hypothesis we have $\lim_{i,j\to\infty} S^a_{u,i,j} = 0$ in probability. Let $n_0(a) = 1$, and define inductively:

$$n_k(a) = \inf\{m > \max(k, n_{k-1}(a)) : \sup_{i,j\geq m} P(S^a_{k,i,j} > 2^{-k}) \leq 2^{-k}\}.$$

We then define

$$Z^k(a,t,\omega) = Y^{n_k(a)}(a,t,\omega).$$

Since each $a \to n_k(a)$ is measurable, so also is Z^k. Define

$$T^a_{u,i,j} = \sup_{t\leq u}|Z^i(a,t,\omega) - Z^j(a,t,\omega)|;$$

then also $(a,\omega) \to T^a_{u,i,j}(\omega)$ is jointly measurable, since Z^i have càdlàg paths (in t). Moreover by our construction $P(T^a_{k,k,k+m} > 2^{-k}) \leq 2^{-k}$ for any $m \geq 1$. The Borel-Cantelli lemma then implies that $\lim_{i,j\to\infty} T^a_{u,i,j} = 0$ *almost surely*, which in turn implies that

$$\lim_{i\to\infty} Z^i(a,t,\cdot) \quad \text{exists a.s. },$$

with convergence uniform in t. Let Λ^a be the set where Z^i converges uniformly (note that $\Lambda^a \in \mathcal{A} \otimes \mathcal{F}$ and $P(\Lambda^a) = 1$, each fixed a), and define

$$Y(a,t,\omega) = \begin{cases} \lim_{i\to\infty} Z^i(a,t,\omega), & \omega \in \Lambda^a \\ 0 & \omega \notin \Lambda^a. \end{cases}$$

Then Y is càdlàg thanks to the uniform convergence, and it is jointly measurable. □

Theorem 44. *Let X be a semimartingale with $X_0 = 0$ a.s. and let $H(a,t,\omega) = H^a_t(\omega)$ be $\mathcal{A} \otimes \mathcal{P}$*[11] *measurable and bounded. Then there is a function $Z(a,t,\omega) \in \mathcal{A} \otimes \mathcal{B}(\mathbf{R}_+) \otimes \mathcal{F}$ such that for each $a \in A$, $Z(a,t,\omega)$ is a càdlàg, adapted version of the stochastic integral $\int_0^t H^a_s dX_s$.*

Proof. Let $\mathcal{H} = \{H \in b\mathcal{A}\otimes\mathcal{P}$ such that the conclusion of the theorem holds$\}$. If $K = K(t,\omega) \in b\mathcal{P}$ and $f = f(a) \in b\mathcal{A}$ and if $H(a,t,\omega) = f(a)K(t,\omega)$,

[10] "càdlàg" is the French acronym for "right continuous with left limits".
[11] Recall that $\mathcal{P}$ denotes the predictable σ-algebra.

then

$$\int_0^t H(a,s,\cdot)dX_s = \int_0^t f(a)K(s,\cdot)dX_s$$
$$= f(a)\int_0^t K(s,\cdot)dX_s,$$

and thus clearly $H = fK$ is in $\mathcal{H}$. Also note that $\mathcal{H}$ is trivially a vector space, and that H of the form $H = fK$ generate $\mathbf{b}\mathcal{A}\otimes\mathcal{P}$.

Next let $H^n \in \mathcal{H}$ and suppose that H^n converges boundedly to a process $H \in \mathbf{b}\mathcal{A}\otimes\mathcal{P}$. By Theorem 32 (for example) we have that $H^n \cdot X$ converges uniformly in t in probability on compacts, for each a. Therefore $H \in \mathcal{H}$, and an application of the monotone class theorem yields the result. □

Corollary. *Let X be a semimartingale ($X_0 = 0$ a.s.), and let $H(a,t,\omega) = H_t^a(\omega) \in \mathcal{A}\otimes\mathcal{P}$ be such that for each a the process $H^a \in L(X)$. Then there exists a function $Z(a,t,\omega) = Z_t^a \in \mathcal{A}\otimes\mathcal{B}(\mathbf{R}_+)\otimes\mathcal{F}$ such that for each a Z_t^a is an a.s. càdlàg version of $\int_0^t H_s^a dX_s$.*

Proof. By Theorem 32 the bounded processes $Z^{a,k} = H^a 1_{\{|H^a|\le k\}} \cdot X$ converge to $H^a \cdot X$ in ucp, each a. But $Z^{a,k}$ can be chosen càdlàg and jointly measurable by Theorem 44. The result now follows by Theorem 43. □

Theorem 45 (Fubini's Theorem). *Let X be a semimartingale, $H_t^a = H(a,t,\omega)$ be a bounded $\mathcal{A}\otimes\mathcal{P}$ measurable function, and let μ be a finite measure on $\mathcal{A}$. Let $Z_t^a = \int_0^t H_s^a dX_s$ be $\mathcal{A}\otimes\mathcal{B}(\mathbf{R}_+)\otimes\mathcal{F}$ measurable such that for each a, Z^a is a càdlàg version of $H^a \cdot X$. Then $Y_t = \int_A Z_t^a \mu(da)$ is a càdlàg version of $H \cdot X$, where $H_t = \int_A H_t^a \mu(da)$.*

Proof. By pre-stopping we may assume without loss of generality that $X \in \mathcal{H}^2$, and because the result holds for the finite variation part of the canonical decomposition of X by the ordinary Stieltjes Fubini theorem, we may further assume that X is a martingale with $E\{[X,X]_\infty\} < \infty$. Next suppose H_t^a is of the form $H(a,t,\omega) = K(t,\omega)f(a)$ where $K \in \mathbf{b}\mathcal{P}$ and f is bounded, measurable. Then $K \in L(X)$ and $\int |f(a)|\mu(da) < \infty$. In this case we have $Z_t^a = f(a)K \cdot X$, and moreover:

$$\int Z_t^a \mu(da) = \int f(a)K \cdot X\mu(da) = K \cdot X \int f(a)\mu(da)$$
$$= (\int f(a)\mu(da)K) \cdot X$$
$$= H \cdot X.$$

By linearity the same result holds for the vector space $\mathbf{V}$ generated by processes of the form $K(t,\omega)f(a)$ with $K \in \mathbf{b}\mathcal{P}$ and f bounded, measurable.

By the monotone class theorem it now suffices to show that if $H_n \in \mathbf{V}$ and $\lim_{n\to\infty} H_n = H$, then the result holds for H. Let $Z^a_{n,t} = H^a_n \cdot X$, the càdlàg version. Then by Jensen's and the Cauchy-Schwarz inequalities,

$$\begin{aligned}\frac{1}{\|\mu\|}(E\{\int_A \sup_t |Z^a_{n,t} - Z^a_t|\mu(da)\})^2 &\le E\{\int_A \sup_t |Z^a_{n,t} - Z^a_t|^2\mu(da)\}\\ &= \int_A E\{\sup_t |Z^a_{n,t} - Z^a_t|^2\}\mu(da)\\ &\le 4\int_A E\{(Z^a_{n,\infty} - Z^a_\infty)^2\}\mu(da)\\ &= 4\int_A E\{[Z^a_n - Z^a, Z^a_n - Z^a]_\infty\}\mu(da)\end{aligned}$$

by Doob's quadratic inequality for the martingales Z^a_n and Z^a, and by Corollary 3 of Theorem 27 of Chap. II. Continuing, the preceding equals:

$$= 4\int_A E\{\int_0^\infty (H^a_{n,s} - H^a_s)^2 d[X,X]_s\}\mu(da),$$

and the above tends to 0 by three applications of the dominated convergence theorem.

We conclude from the preceding that

$$\int_A \sup_t |Z^a_{n,t} - Z^a_t|\mu(da) < \infty \ a.s.$$

and therefore $\int_A |Z^a_t|\mu(da) < \infty$ for all t, a.s. Moreover

$$E\{\sup_t |\int_A Z^a_{n,t}\mu(da) - \int_A Z^a_t\mu(da)|\} \le E\{\int_A \sup_t |Z^a_{n,t} - Z^a_t|\mu(da)\}$$

which tends to 0. Therefore taking $H_{n,t} = \int H^a_{n,t}\mu(da)$ we have $H_n \cdot X_t = \int_A Z^a_{n,t}\mu(da)$ converges in ucp to $\int Z^a_t\mu(da)$. Since $H_n \cdot X$ converges to $H \cdot X$ by Theorem 32, we conclude $H \cdot X = \int Z^a_t\mu(da)$. □

The version of Fubini's theorem given in Theorem 45 suffices for the applications of it used in this book. Nevertheless it is interesting to determine under what more general conditions a Fubini-type theorem holds.

Theorem 46 (Fubini's Theorem; Second Version). *Let X be a semimartingale, let $H^a_t = H(a,t,\omega)$ be $\mathcal{A}\otimes\mathcal{P}$ measurable, let μ be a finite positive measure on A, and assume*

$$(\int_A (H^a_t)^2\mu(da))^{1/2} \in L(X).$$

Letting $Z^a_t = \int_0^t H^a_s dX_s$ be $\mathcal{A}\otimes\mathcal{B}(\mathbf{R}_+)\otimes\mathcal{F}$ measurable and Z^a càdlàg for each a, then $Y_t = \int_A Z^a_t\mu(da)$ exists and is a càdlàg version of $H\cdot X$, where $H_t = \int_A H^a_t\mu(da)$.

Proof. By pre-stopping we may assume without loss of generality that $X \in \mathcal{H}^2$ and that $\|H^a\|_{L^2(d\mu)}$ is $(\mathcal{H}^2, X)$ integrable. Let $X = \overline{N} + \overline{A}$ be the canonical decomposition of X. Then

$$E\{\int_0^\infty \int_A (H_s^a)^2 \mu(da) d[\overline{N}, \overline{N}]_s\} + E\{(\int_0^\infty \|H_s^a\|_{L^2(d\mu)} |d\overline{A}_s|)^2\} < \infty.$$

Next observe that

$$\begin{aligned} E\{\int_0^\infty \|H_s^a\|_{L^2(d\mu)} |d\overline{A}_s|\} &\geq E\{\| \int_0^\infty |H_s^a| |d\overline{A}_s| \|_{L^2(d\mu)}\} \\ &\geq cE\{\| \int_0^\infty |H_s^a| |d\overline{A}_s| \|_{L^1(d\mu)}\} \\ &= c \int_A E\{\int_0^\infty |H_s^a| |d\overline{A}_s|\} \mu(da). \end{aligned}$$

Also

$$E\{\int_0^\infty \int_A (H_s^a)^2 \mu(da) d[\overline{N}, \overline{N}]_s\} = \int_A E\{\int_0^\infty (H_s^a)^2 d[\overline{N}, \overline{N}]_s\} \mu(da),$$

and therefore $E\{\int_0^\infty |H_s^a| |d\overline{A}_s|\} < \infty$ and $E\{\int_0^\infty (H_s^a)^2 d[\overline{N}, \overline{N}]_s\} < \infty$ for μ almost all $a \in A$, whence $H^a \in L(X)$ for μ almost all $a \in A$.

Next define $H_n = H 1_{\{|H| \leq n\}}$, and the proof of Theorem 45 works as well here. □

The hypotheses of Theorem 46 are slightly unnatural, since they are not invariant under the transformation

$$H \to \frac{1}{\varphi(a)} H^a \qquad \mu \to \varphi(a) \mu(da)$$

where φ is any positive function such that $\int \varphi(a) \mu(da) < \infty$. This can be alleviated by replacing the assumption $(\int_A (H^a)^2 \mu(da))^{1/2} \in L(X)$ with $(\int \frac{(H^a)^2}{\varphi(a)} \mu(da))^{1/2} \in L(X)$ for some positive $\varphi \in L^1(d\mu)$. One can also relax the assumption on μ to be σ-finite rather than finite.

Example. The hypotheses of Theorem 46 are a bit strange, however they are in some sense best possible. We give here an example of a parameterized process $(H^a)_{a \in A}$, a positive, finite measure μ on A and a semimartingale X such that

(i) $(a, t) \to H_t^a$ is $\mathcal{A} \otimes \mathcal{P}$ measurable;
(ii) $H^a \in L(X)$, each $a \in A$;
(iii) $\int_A |H_t^a| \mu(da) \in L(X)$

but such that if $Z_t^a = \int_0^t H_s^a dX_s$ then $\int_A Z_t^a \mu(da)$ *does not exist as a Lebesgue integral.* Thus a straightforward extension of the classical Fubini theorem for Lebesgue integration does not hold.

Indeed, let $A = \mathbb{N} = \{1, 2, 3, \dots\}$ and let μ be any finite positive measure on A such that $\mu(\{a\}) > 0$ for all $a \in A$. Let X be standard Brownian motion, and let $t_0 < t_1 < t_2 < \dots$ be an increasing sequence in $[0, 1]$ and define

$$H_t^a = \frac{1}{a}\mu(\{a\})^{-1}(t_a - t_{a-1})^{-1/2}1_{\{t_{a-1}<t\leq t_a\}}.$$

Then

$$H_t = \sum_{a=1}^{\infty} \frac{1}{a}(t_a - t_{a-1})^{-1/2}1_{\{t_{a-1}<t\leq t_a\}}$$

is in $L^2(dt)$, whence $H_t = \int_A |H_t^a|\mu(da) \in L(X)$, and moreover if $t \geq 1$,

$$H \cdot X_t = \sum_{a=1}^{\infty} \frac{1}{a}(t_a - t_{a-1})^{-1/2}(X_{t_a} - X_{t_{a-1}}),$$

where the sum converges in L^2. However if $t \geq 1$ then

$$Z_t^a = H^a \cdot X_t = \frac{1}{a}\mu(\{a\})^{-1}(t_a - t_{a-1})^{-1/2}(X_{t_a} - X_{t_{a-1}}),$$

and

$$\int_A |Z_t^a|\mu(da) = \sum_{a=1}^{\infty} \frac{1}{a}(t_a - t_{a-1})^{-1/2}|X_{t_a} - X_{t_{a-1}}| = \infty \quad a.s.$$

because $(t_a - t_{a-1})^{-1/2}(X_{t_a} - X_{t_{a-1}})$ is an i.i.d. sequence and $\sum_{a=1}^{\infty} \frac{1}{a} = \infty$.

Note that this example can be modified to show that *we also cannot replace the assumption that*

$$\left(\int_A (H_t^a)^2\mu(da)\right)^{1/2} \in L(X)$$

with the weaker assumption that $(\int_A (H_t^a)^p\mu(da))^{1/p} \in L(X)$ for some $p < 2$.

5. Local Times

In Chap. II we established Itô's formula (Theorem 32 of Chap. II) which showed that if $f : \mathbb{R} \to \mathbb{R}$ is C^2 and X is a semimartingale, then $f(X)$ is again a semimartingale. That is, semimartingales are preserved under C^2 transformations. This property extends slightly: semimartingales are preserved under convex transformations, as Theorem 47 below shows. (Indeed, this is the best one can do in general: if $B = (B_t)_{t\geq 0}$ is standard Brownian motion and $Y_t = f(B_t)$ is a semimartingale, then f *must* be the difference of convex functions (See Çinlar-Jacod-Protter-Sharpe [1]). We establish a related result in Theorem 52, later in this section.) *Local times* for semimartingales appear in the extension of Itô's formula from $\mathcal{C}^2$ functions to convex functions (Theorem 51).

Theorem 47. *Let $f : \mathbb{R} \to \mathbb{R}$ be convex and let X be a semimartingale. Then $f(X)$ is a semimartingale and one has*

$$f(X_t) - f(X_0) = \int_{0+}^{t} f'(X_{s-})dX_s + A_t$$

where f' is the left derivative of f and A is an adapted, right continuous, increasing process. Moreover $\Delta A_t = f(X_t) - f(X_{t-}) - f'(X_{t-})\Delta X_t$.

Proof. First suppose $|X|$ is bounded by n, and in $\mathcal{H}^2$, and that $X_0 = 0$. Let g be a positive $\mathcal{C}^\infty$ function with compact support in $(-\infty, 0]$ such that $\int_{-\infty}^{\infty} g(s)ds = 1$. Let $f_n(t) = n\int_{-\infty}^{\infty} f(t+s)g(ns)ds$. Then f_n is convex and $\mathcal{C}^2$ and moreover f_n' increases to f' as n tends to ∞. By Itô's formula

$$f_n(X_t) - f_n(X_0) = \int_0^t f_n'(X_{s-})dX_s + A_t^n$$

where

$$A_t^n = \sum_{0<s\leq t}\{f_n(X_s) - f_n(X_{s-}) - f_n'(X_{s-})\Delta X_s\} + \frac{1}{2}\int_0^t f_n''(X_{s-})d[X,X]_s^c.$$

The convexity of f implies that A^n is an increasing process. Letting n tend to ∞, we obtain

$$(*) \qquad f(X_t) - f(X_0) = \int_0^t f'(X_{s-})dX_s + A_t$$

where $\lim_{n\to\infty} A_t^n = A_t$ in L^2, and where the convergence of the stochastic integral terms is in $\mathcal{H}^2$ on $[0,t]$.

We now compare the jumps on both sides of the equation (*). Since $\int_0^0 f'(X_{s-})dX_s = 0$ we have that $A_0 = 0$. When $t > 0$, the jump of the left side of (*) is $f(X_t) - f(X_{t-})$, while the jump of the right side equals $f'(X_{t-})\Delta X_t + \Delta A_t$; therefore $\Delta A_t = f(X_t) - f(X_{t-}) - f'(X_{t-})\Delta X_t$, and the theorem is established for $|X|$ bounded by n and in $\mathcal{H}^2$.

Now let X be an arbitrary semimartingale with $X_0 = 0$. By Theorem 13 we know there exists a sequence of stopping times $(T^n)_{n\geq 1}$, increasing to ∞ a.s. such that $X^{T^n-} \in \mathcal{H}^2$ for each n. An examination of the proof of Theorem 13 shows that there is no loss of generality in further assuming that $|X^{T^n-}| \leq n$, also. Then let $Y^n = X1_{[0,T^n)}$ and we have

$$f(Y_t^n) - f(Y_0^n) = \int_0^t f'(Y_{s-}^n)dY_s^n + A_t^n,$$

which is equivalent to saying

$$f(X_t) - f(X_0) = \int_0^t f'(X_{s-})dX_s + A_t^n$$

on $[0, T^n)$. One easily checks that $(A^{n+1})^{T^n-} = (A^n)^{T^n-}$, and we can define $A = A^n$ on $[0, T^n)$, each n.

The above extends without difficulty to functions $g : \mathbf{R}^2 \to \mathbf{R}$ of the form $g(X_t, H)$ where H is an $\mathcal{F}_0$ measurable random variable and $x \to g(x, y)$ is convex for every y. For general X we take $\hat{X}_t = X_t - X_0$, and then $f(X_t) = f(\hat{X}_t + X_0) = g(\hat{X}_t, X_0)$, where $g(x, y) = f(x + y)$. This completes the proof. □

Notation. For x a real variable let x^+, x^- be the functions $x^+ = \max(x, 0)$ and $x^- = -\min(x, 0)$. For x, y real variables, let $x \vee y \equiv \text{maximum}(x, y)$ and $x \wedge y \equiv \text{minimum}(x, y)$.

Corollary 1. *Let X be a semimartingale. Then $|X|$, X^+, X^- are all semimartingales.*

Proof. The functions $f(x) = |x|$, $g(x) = x^+$ and $h(x) = x^-$ are all convex; the result then follows by Theorem 47. □

Corollary 2. *Let X, Y be semimartingales. Then $X \vee Y$ and $X \wedge Y$ are semimartingales.*

Proof. Since semimartingales form a vector space and $x \vee y = \frac{1}{2}(|x - y| + x + y)$ and $x \wedge y = \frac{1}{2}(x + y - |x - y|)$, the result is an immediate consequence of Corollary 1. □

We can summarize the surprisingly broad stability properties of semimartingales.

Theorem 48. *The space of semimartingales is a vector space, an algebra, a lattice, and is stable under C^2, and more generally under convex transformations.*

Proof. In Chap. II we saw that semimartingales form a vector space (Theorem 1), an algebra (Corollary 2 of Theorem 22: Integration by Parts), and that they are stable under C^2 transformations (Theorem 32: Itô's Formula). That they form a lattice is by Corollary 2 above, and that they are stable under convex transformations is Theorem 47. □

Definition. The **sign function** is defined to be

$$\text{sign}(x) = \begin{cases} 1 & \text{if } x > 0 \\ -1 & \text{if } x \le 0. \end{cases}$$

Note that our definition of sign is not symmetric. We further define

$$(*) \qquad \begin{cases} h_0(x) = |x| \\ h_a(x) = |x - a|. \end{cases}$$

Then $\operatorname{sign}(x)$ is the *left derivative* of $h_0(x)$, and $\operatorname{sign}(x-a)$ is the left derivative of $h_a(x)$. Since $h_a(x)$ is convex by Theorem 47 we have for a semimartingale X:

$$(**)\qquad h_a(X_t)=|X_t-a|=|X_0-a|+\int_{0+}^{t}\operatorname{sign}(X_{s-}-a)dX_s+A_t^a,$$

where A_t^a is the increasing process of Theorem 47. Using (*) and (**) as defined above we can define the local time of an arbitrary semimartingale.

Definition. Let X be a semimartingale, and let h_a and A^a be as defined in (*) and (**) above. **The local time at a of X, denoted $L_t^a = L^a(X)_t$,** is defined to be the process given by

$$L_t^a = A_t^a - \sum_{0<s\le t}\{h_a(X_s)-h_a(X_{s-})-h_a'(X_{s-})\Delta X_s\}.$$

Notice that by the Corollary of Theorem 44 the integral $\int_{0+}^{t}\operatorname{sign}(X_{s-}-a)dX_s$ in (**) has a version which is jointly measurable in (a,t,ω) and càdlàg in t. Therefore so also does $(A_t^a)_{t\ge0}$, and finally so also does the local time L_t^a. *We always choose this jointly measurable, càdlàg version of the local time*, without any special mention. We further observe that the jumps of the process A^a defined in (**) are precisely $\sum_{s\le t}\{h_a(X_s)-h_a(X_{s-})-h_a(X_{s-})\Delta X_s\}$ (by Theorem 47), and therefore *the local time $(L_t^a)_{t\ge0}$ is continuous in t*. Indeed, the local time L^a is the continuous part of the increasing process A^a.

The next theorem is quite simple yet crucial to proving the properties of L^a that justify its name.

Theorem 49. *Let X be a semimartingale and let L^a be its local time at a. Then*

$$\begin{aligned}(X_t-a)^+-(X_0-a)^+ &= \int_{0+}^{t}1_{(X_{s-}>a)}dX_s+\sum_{0<s\le t}1_{(X_{s-}>a)}(X_s-a)^-\\ &\quad+\sum_{0<s\le t}1_{(X_{s-}\le a)}(X_s-a)^+ +\frac{1}{2}L_t^a;\end{aligned}$$

$$\begin{aligned}(X_t-a)^--(X_0-a)^- &= -\int_{0+}^{t}1_{(X_{s-}\le a)}dX_s+\sum_{0<s\le t}1_{(X_{s-}>a)}(X_s-a)^-\\ &\quad+\sum_{0<s\le t}1_{(X_{s-}\le a)}(X_s-a)^+ +\frac{1}{2}L_t^a.\end{aligned}$$

Proof. Applying Theorem 47 to the convex functions $f(x) = (x-a)^+$ and $g(x) = (x-a)^-$ we get

$$f(X_t) = f(X_0) + \int_{0+}^{t} f'(X_{s-})dX_s + C_t^+$$

$$g(X_t) = g(X_0) + \int_{0+}^{t} g'(X_{s-})dX_s + C_t^-$$

Next let

$$D_t^+ = C_t^+ - \sum_{0<s\leq t} \{f(X_s) - f(X_{s-}) - f'(X_{s-})\Delta X_s\},$$

$$D_t^- = C_t^- - \sum_{0<s\leq t} \{g(X_s) - g(X_{s-}) - g'(X_{s-})\Delta X_s\}$$

and subtracting the formulas we get $C_t^+ - C_t^- = 0$ and hence $D_t^+ = D_t^-$; also $D_t^+ + D_t^- = L_t^a$; thus $D_t^+ = D_t^- = \frac{1}{2}L_t^a$, and the proof is complete. □

The next theorem, together with the "occupation time density" formula (Corollary 1 of Theorem 51), are the traditional justifications for the terminology "local time".

Theorem 50. *Let X be a semimartingale, and let L_t^a be its local time at the level a, each $a \in \mathbf{R}$. For a.a. ω, the measure in t, $dL_t^a(\omega)$, is carried by the set $\{s : X_{s-}(\omega) = X_s(\omega) = a\}$.*

Proof. Since L_t^a has continuous paths, the measure $dL_t^a(\omega)$ is diffuse, and since $\{s : X_{s-}(\omega) = a\}$ and $\{s : X_{s-}(\omega) = X_s(\omega) = a\}$ differ by at most a countable set, it will suffice to show that $dL_t^a(\omega)$ is carried by the set $\{s : X_{s-}(\omega) = a\}$.

Suppose S, T are stopping times and that $0 < S \leq T$ such that $[S,T) \subset \{(s,\omega) : X_{s-}(\omega) < a\} \equiv \{X_- < a\}$. Then $X \leq a$ on $[S,T)$ as well, hence by the first equation in Theorem 49 we have:

$$(X-a)_T^+ - (X-a)_S^+ = \int_S^T 1_{(X_{s-}>a)}dX_s + \sum_{S<s\leq T} 1_{(X_{s-}>a)}(X_s - a)^-$$
$$+ \sum_{S<s\leq T} 1_{(X_{s-}<a)}(X_s - a)^+ + \frac{1}{2}(L_T^a - L_S^a).$$

However the left side of the above equation equals zero, and all terms on the right side except possibly $\frac{1}{2}(L_T^a - L_S^a)$ also equal zero. Therefore $\frac{1}{2}(L_T^a - L_S^a) = 0$, whence $L_T^a = L_S^a$.

Next for $r \in \mathbf{Q}$, the rationals, define the stopping times $S_r(\omega)$, $r > 0$, by

$$S_r(\omega) = \begin{cases} r & \text{if } X_{r-}(\omega) < a \\ \infty & \text{if } X_{r-}(\omega) \geq a. \end{cases}$$

Then define

$$T_r(\omega) = \inf\{t > S_r(\omega) : X_{t-}(\omega) \geq a\}.$$

Then $[S_r, T_r) \subset \{X_- < a\}$, and moreover the interior of the set $\{X_- < a\}$ equals $\bigcup_{r\in\mathbb{Q},\, r>0}(S_r, T_r)$. As we have now seen, dL^a does not charge the interior of the set $\{X_- < a\}$. This set is open on the left, hence it differs from its interior by an at most countable set, and since dL^a is diffuse it doesn't charge countable sets. Thus dL^a does not charge the set $\{X_- < a\}$ itself.

Analogously one can show dL^a does not charge $\{X_- > a\}$. Hence its support is contained in the set $\{X_- = a\}$, and we are done. □

The next theorem gives a very satisfying generalization of Itô's formula (Theorem 32 of Chap. II).

Theorem 51 (Meyer-Itô Formula). *Let f be the difference of two convex functions, let f' be its left derivative, and let μ be the signed measure (when restricted to compacts) which is the second derivative of f in the generalized function sense. Then the following equation holds:*

(*)

$$f(X_t) - f(X_0) = \int_{0+}^{t} f'(X_{s-})dX_s + \sum_{0<s\leq t} \{f(X_s) - f(X_{s-}) - f'(X_{s-})\Delta X_s\} + \frac{1}{2}\int_{-\infty}^{\infty} \mu(da)L_t^a,$$

where X is a semimartingale and $L_t^a = L_t^a(X)$ is its local time at a.

Proof. Notice that equation (*) is trivially true if f is an affine function (i.e., $f(x) = ax + b$). Next let us first assume that μ is a signed measure with finite total mass on a compact interval. Define a function g by

$$g(x) = \frac{1}{2}\int |x - y|\mu(dy).$$

It is well known that f and g differ by at most an affine function $h(x) = ax + b$. Therefore without loss of generality we can assume that the function f in (*) is of the form $f(x) = \frac{1}{2}\int |x - y|\mu(dy)$. Then $f'(x) = \frac{1}{2}\int \operatorname{sign}(x - y)\mu(dy)$ and $f''(x) = \mu(dx)$. Moreover if:

$$J_t^y = \sum_{0<s\leq t} |X_s - y| - |X_{s-} - y| - \operatorname{sign}(X_{s-} - y)\Delta X_s,$$

then

$$\frac{1}{2}\int J_t^y \mu(dy) = \sum_{0<s\leq t} \{f(X_s) - f(X_{s-}) - f'(X_{s-})\Delta X_s\}. \tag{1}$$

Also, letting $H_t^y = |X_t - y| - |X_0 - y|$, one has

$$\frac{1}{2}\int H_t^y \mu(dy) = f(X_t) - f(X_0). \tag{2}$$

Next consider $Z_t^y = \int_{0+}^t \operatorname{sign}(X_{s-} - y)dX_s$, and let Z^y be the $\mathcal{B}(\mathbf{R}) \otimes \mathcal{B}(\mathbf{R}_+) \otimes \mathcal{F}$ measurable version, which we know exists by Theorem 44. By Fubini's Theorem (Theorem 45) we have

$$\frac{1}{2}\int Z_t^y \mu(dy) = \int_{0+}^t f'(X_{s-})dX_s. \tag{3}$$

Since $L_t^a = H_t^a - J_t^a - Z_t^a$, we have

$$\frac{1}{2}\int L_t^a \mu(da) = \frac{1}{2}\int H_t^a \mu(da) - \frac{1}{2}\int J_t^a \mu(da) - \frac{1}{2}\int Z_t^a \mu(da) \tag{4}$$

and combining (4) with (1), (2), and (3), yields the formula (*).

It remains to reduce the general case to that of μ supported on a compact interval with finite total mass. By linearity it further suffices to treat the case where f is convex. Define

$$f_n(x) = \begin{cases} f(n) + f'(n)(x - n) & \text{if } x \geq n \\ f(x) & \text{if } -n \leq x \leq n \\ f(-n) + f'(-n)(x + n) & \text{if } x \leq -n \end{cases}$$

Then f_n is also convex and its second (generalized) derivative $\mu_n(dx)$ agrees with μ on $[-n, n)$ and is of finite total mass. Let $Y^n = f_n(X)$, and $T^n = \inf\{t > 0 : |X_t| \geq n\}$. Note that $Y^n = f(X)$ on $[0, T^n)$, and that $L_{T^n}^a = 0$ for all a, $|a| \geq n$, by Theorem 50. Therefore $\int L_t^a \mu_n(da) = \int L_t^a \mu(da)$ for $t \leq T^n$, and by the preceding (*) holds for $Y = f_n(X) = f(X)$ on $[0, T^n)$. Therefore (*) holds on $[0, T^n)$. Since the stopping times $(T^n)_{n \geq 1}$ increase to ∞ a.s., we have (*) holds on all of $\mathbf{R}_+ \times \Omega$, and the theorem is proved. □

The next formula gives an interpretation of semimartingale local time as an occupation density relative to the random "clock" $d[X, X]_s^c$.[12]

Corollary 1. *Let X be a semimartingale with local time $(L^a)_{a \in \mathbf{R}}$. Let g be a bounded Borel measurable function. Then a.s.*

$$\int_{-\infty}^{\infty} L_t^a g(a)da = \int_0^t g(X_{s-})d[X, X]_s^c.$$

Proof. Let f be convex and C^2. Comparing (*) of Theorem 51 with Itô's formula (Theorem 32 of Chap. II) shows that:

$$\int_{-\infty}^{\infty} L_t^a f''(a)da = \int_0^t f''(X_{s-})d[X, X]_s^c$$

[12] Recall that $[X, X]^c$ denotes the path-by-path continuous part of $t \to [X, X]_t$, with $[X, X]_0^c = 0$.

where $\mu(da)$ is of course $f''(a)da$. Since the above holds for any continuous and positive function f'', a monotone class argument shows that it must hold, up to a P-null set, for any bounded, Borel measurable function g. □

We record here an important special case of Corollary 1.

Corollary 2. *Let X be a semimartingale with local time $(L^a)_{a\in\mathbf{R}}$. Then*

$$[X,X]^c_t = \int_{-\infty}^{\infty} L^a_t da.$$

Corollary 3 (Meyer-Tanaka Formula). *Let X be a semimartingale with continuous paths. Then*

$$|X_t| = |X_0| + \int_{0+}^{t} sign(X_s)dX_s + L^0_t.$$

Proof. This is merely Theorem 51 with $f(x) = |x|$, which implies $\mu(da) = 2\epsilon_0(da)$, point mass at 0. The formula also follows trivially from the definition of L^0. □

When $X_t = B_t$, standard Brownian motion, then the term $M_t = \int_0^t \operatorname{sign}(B_s)dB_s$ is a continuous local martingale, and $[M,M]_t = \int_0^t \operatorname{sign}(B_s)^2 d[B,B]_s = [B,B]_t = t$. Therefore by Lévy's theorem (Theorem 38 of Chap. II) we know that M_t is another Brownian motion. We therefore have, where $B_0 = x$ for standard Brownian motion,

$$(**) \qquad |B_t| = |B_0| + \beta_t + L_t,$$

where $\beta_t = \int_0^t \operatorname{sign}(B_s)dB_s$ is a Brownian motion, and L_t is the local time of B at zero. Formula (**) is known as **Tanaka's Formula**.

Observe that if $f(x) = |x|$, then $f'' = 2\delta(x)$, where δ is the "delta function at 0", which of course is a generalized function, or "distribution". Thus Corollaries 1 and 3 give the intuitive interpretation of local time as $L^0_t = \int_0^t \delta(X_s)d[X,X]_s$, and $L^a_t = \int_0^t \delta(X_s - a)d[X,X]_s$, for continuous semimartingales.

Local times also allow us to give simple examples of functions f that are only slightly more general than convex functions and are such that $f(X)$ need not be a semimartingale when X is one. Let $f(x) = |x|^\alpha$, $0 < \alpha < 1$. Then f cannot be expressed as the difference of two convex functions because the slope of f becomes vertical as x decreases to zero.

Theorem 52. *Let X be a continuous local martingale with $X_0 = 0$, and let $0 < \alpha < \frac{1}{2}$. Then $Y_t = |X_t|^\alpha$ is not a semimartingale unless X is identically zero.*

Proof. Let us suppose that $Y = |X|^\alpha$ is a semimartingale. Let $\beta = \frac{1}{\alpha} > 2$. We then have

$$|X_t| = Y_t^\beta = \beta \int_0^t Y_s^{\beta-1} dY_s + \frac{\beta(\beta-1)}{2} \int_0^t Y_s^{\beta-2} d[Y,Y]_s,$$

since Y has continuous paths. Using Theorem 50 and the Meyer-Tanaka formula (Corollary 3 of Theorem 51) we have (where $L_t = L_t^0$, the local time of X at 0):

$$\begin{aligned} L_t &= \int_0^t 1_{\{X_s=0\}} dL_s \\ &= -\int_0^t 1_{\{X_s=0\}} \operatorname{sign}(X_s) dX_s + \int_0^t 1_{\{X_s=0\}} d|X|_s \end{aligned}$$

The integral $\int_0^t 1_{\{X_s=0\}} \operatorname{sign}(X_s) dX_s$ is identically zero: Let T^n be a sequence of stopping times increasing to ∞ a.s. such that X^{T^n} is a bounded martingale. Then

$$\begin{aligned} E\{(\int_0^{t\wedge T^n} 1_{(X_s=0)} \operatorname{sign}(X_s) dX_s)^2\} &= E\{\int_0^{t\wedge T^n} 1_{(X_s=0)} d[X,X]_s\} \\ &= E\{\int_{-\infty}^{\infty} L_{t\wedge T^n}^a 1_{\{0\}}(a) da\} = 0. \end{aligned}$$

Therefore $L_t = L_t^0 = \int_0^t 1_{\{X_s=0\}} d|X|_s$. Since $\{X_s = 0\}$ equals $\{Y_s = 0\}$, this becomes

$$\begin{aligned} L_t &= \int_0^t 1_{\{Y_s=0\}} d|X|_s \\ &= \beta \int_0^t Y_s^{\beta-1} 1_{\{Y_s=0\}} dY_s + \frac{\beta(\beta-1)}{2} \int_0^t Y_s^{\beta-2} 1_{\{Y_s=0\}} d[Y,Y]_s \\ &= \beta \int_0^t 0 dY_s + \frac{\beta(\beta-1)}{2} \int_0^t 0 d[Y,Y]_s \\ &= 0. \end{aligned}$$

Using the Meyer-Tanaka formula again we conclude that

$$|X_t| = \int_0^t \operatorname{sign}(X_s) dX_s.$$

Since X is a continuous local martingale, so also is the stochastic integral on the right side above (Theorem 30); it is also nonnegative, and equal to zero at 0. Such a local martingale must be identically zero. This completes the proof. □

It is worth noting that if X is a bounded, continuous martingale with $X_0 = 0$, then $Y_t = |X_t|^\alpha$, $0 < \alpha < \frac{1}{2}$ is an example of an *asymptotic martingale*, or AMART, which is not a semimartingale.[13]

We next wish to determine when there exists a version of the local time L_t^a which is jointly continuous in $(a,t) \to L_t^a$ a.s., or jointly right continuous in $(a,t) \to L_t^a$. We begin with the classical theorem of Kolmogorov which gives a sufficient condition for joint continuity. There are several versions of Kolmogorov's Theorem. We give here a quite general one because we will use it often in Chap. V. In this section we use only its Corollary, which can also be proved directly in a fairly simple way.

Before the statement of the theorem we establish some notation. The letter Δ will denote the dyadic rational points of the unit cube $[0,1]^n$ in $\mathbf{R}^n$, while Δ_m will denote all $x \in \Delta$ whose coordinates are of the form $k/2^m$, $0 \leq k \leq 2^m$.

Theorem 53 (Kolmogorov's Lemma). *Let (E,d) be a complete metric space, and let U^x be an E-valued random variable for all x dyadic rationals in $\mathbf{R}^n$. Suppose that for all x, y, we have $d(U^x, U^y)$ is a random variable and that there exist strictly positive constants ϵ, C, β such that*

$$E\{d(U^x, U^y)^\epsilon\} \leq C\|x-y\|^{n+\beta}.$$

Then for almost all ω the function $x \to U^x$ can be extended uniquely to a continuous function from $\mathbf{R}^n$ to E.

Proof. We prove the theorem for the unit cube $[0,1]^n$ and leave the extension to $\mathbf{R}^n$ to the reader. Two points x and y in Δ_m are *neighbors* if $\sup_i |x^i - y^i| = 2^{-m}$. We use Chebyshev's inequality on the inequality hypothesized to get

$$P\{d(U^x, U^y) \geq 2^{-\alpha m}\} \leq C 2^{\alpha\epsilon m} 2^{-m(n+\beta)}.$$

Let

$$\Lambda_m = \{\omega : \text{there exist neighbors } x, y \in \Delta_m \text{ with } d(U^x(\omega), U^y(\omega)) \geq 2^{-\alpha m}\}.$$

Since each $x \in \Delta_m$ has at most 3^n neighbors, and the cardinality of Δ_m is 2^{mn}, we have

$$P(\Lambda_m) \leq c 2^{m(\alpha\epsilon - \beta)}$$

where the constant $c = 3^n C$. Take α sufficiently small so that $\alpha\epsilon < \beta$. Then

$$P(\Lambda_m) \leq c 2^{-m\delta}$$

where $\delta = \beta - \alpha\epsilon > 0$. The Borel-Cantelli lemma then implies $P(\Lambda_m$ infinitely often$) = 0$. That is, there exists an m_0 such that for $m \geq m_0$ and every pair (u,v) of points of Δ_m that are neighbors,

$$d(U^u, U^v) \leq 2^{-\alpha m}.$$

[13] For a more elementary example of an asymptotic martingale that is not a semimartingale, see Gut [1, p. 7].

We now use the preceding to show that $x \to U^x$ is uniformly continuous on Δ and hence extendable uniquely to a continuous function on $[0,1]^n$. To this end, let $x, y \in \Delta$ be such that $\|x-y\| \leq 2^{-k-1}$. We will show that $d(U^x, U^y) \leq c2^{-\alpha k}$ for a constant c, and this will complete the proof.

Without loss of generality assume $k \geq m_0$. Then $x = (x^1, \dots, x^n)$ and $y = (y^1, \dots, y^n)$ in Δ with $\|x-y\| \leq 2^{-k-1}$ have dyadic expansions of the form

$$x^i = u^i + \sum_{j>k} a^i_j 2^{-j}$$

$$y^i = v^i + \sum_{j>k} b^i_j 2^{-j}$$

where a^i_j, b^i_j are each 0 or 1, and u, v are points of Δ_k which are either equal or neighbors.

Next set $u_0 = u$, $u_1 = u_0 + a_{k+1}2^{-k-1}$, $u_2 = u_1 + a_{k+2}2^{-k-2}$, etcetera. We also make analogous definitions for v_0, v_1, v_2, etcetera. Then u_{i-1} and u_i are equal or neighbors in Δ_{k+i}, each i, and analogously for v_{i-1} and v_i. Hence

$$d(U^x(\omega), U^u(\omega)) \leq \sum_{j=k}^{\infty} 2^{-\alpha j}$$

$$d(U^y(\omega), U^v(\omega)) \leq \sum_{j=k}^{\infty} 2^{-\alpha j}$$

and moreover

$$d(U^u(\omega), U^v(\omega)) \leq 2^{-\alpha k}.$$

The result now follows by the triangle inequality. □

Comment. If the complete metric space (E, d) in Theorem 53 is *separable*, then the hypothesis that $d(U^x, U^y)$ be measurable is satisfied. Often the metric spaces chosen are one of $\mathbf{R}$, $\mathbf{R}^d$, or the function space $\mathcal{C}$ with the sup norm and these are separable.

A complete metric space that arises often in Chap. V is the space $E = \mathcal{D}^n$ of càdlàg functions mapping $[0,\infty)$ into $\mathbf{R}^n$, topologized by uniform convergence on compacts. While this is a complete metric space, *it is not separable*. Indeed, a compatible metric is

$$d(f,g) = \sum_{n=1}^{\infty} \frac{1}{2^n}(1 \wedge \sup_{0 \leq s \leq n} |f(s) - g(s)|).$$

However if $f_\alpha(t) = 1_{[\alpha,\infty)}(t)$, then $d(f_\alpha, f_\beta) = \frac{1}{2}$ for all α, β with $0 \leq \alpha < \beta \leq 1$, and since there are uncountably many such α, β, the space is not separable. Fortunately, however, the condition that $d(U^x, U^y)$ be measurable is nevertheless satisfied in this case, due to the path regularity of the functions in the function space $\mathcal{D}^n$. (Note that in many other contexts the space $\mathcal{D}^n$ is endowed with the Skorokhod topology, and with this topology $\mathcal{D}^n$ is a complete metric space which is also separable; see for example Ethier-Kurtz [1] or Jacod-Shiryaev [1].)

We state as a corollary the form of Kolmogorov's Lemma that we will use in our study of local times.

Corollary. *Let $(X_t^a)_{t\geq 0, a\in \mathbb{R}^n}$ be a parameterized family of stochastic processes such that $t \to X_t^a$ is càdlàg a.s., each $a \in \mathbb{R}^n$. Suppose that*

$$E\{\sup_{s\leq t}|X_s^a - X_s^b|^\epsilon\} \leq C(t)\|a-b\|^{n+\beta}$$

for some $\epsilon, \beta > 0$, $C(t) > 0$. Then there exists a version $\hat{X}_t^a$ of X_t^a which is $\mathcal{B}(\mathbb{R}_+) \otimes \mathcal{B}(\mathbb{R}^n) \otimes \mathcal{F}$ measurable and which is càdlàg in t and uniformly continuous in a on compacts and is such that for all $a \in \mathbb{R}^n$, $t \geq 0$,

$$\hat{X}_t^a = X_t^a \quad a.s.$$

(The null set $\{\hat{X}_t^a \neq X_t^a\}$ can be chosen independently of t). In this section we will use the above corollary for parameterized processes X^a which are *continuous* in t. In this case the process obtained from the Corollary of Kolmogorov's Lemma, $\hat{X}^a$, *will be jointly continuous in* (a,t) almost surely.

Hypothesis A. For the remainder of this section we let X denote a semimartingale with the restriction that $\sum_{0<s\leq t}|\Delta X_s| < \infty$ a.s., each $t > 0$.

Observe that if $(\Omega, \mathcal{F}, (\mathcal{F}_t)_{t\geq 0}, P)$ is a probability space where $(\mathcal{F}_t)_{t\geq 0}$ is the completed minimal filtration of a Brownian motion $B = (B_t)_{t\geq 0}$, *then all semimartingales on this Brownian space verify Hypothesis A*: Indeed, by Corollary 1 of Theorem 42 all the local martingales are continuous. Thus if X is a semimartingale, let $X = M + A$ be a decomposition with M a local martingale and A a FV process. Then the jump processes ΔX and ΔA are equal, hence

$$\sum_{0<s\leq t}|\Delta X_s| = \sum_{0<s\leq t}|\Delta A_s| < \infty,$$

since A is of finite variation on compacts.

Let X be a semimartingale satisfying Hypothesis A, and let

$$J_t = \sum_{0\leq s\leq t}\Delta X_s;$$

that is, J is the process which is the sum of the jumps. Because of our hypothesis J is an FV process and thus also a semimartingale. Moreover $Y_t = X_t - J_t$ is a continuous semimartingale with $Y_0 = 0$, and we let

$$Y = M + A$$

be its (unique) decomposition, where M is a continuous local martingale and A is a continuous FV process with $M_0 = A_0 = 0$. Such a process M is then uniquely determined and we can write

$$M = X^c,$$

the *continuous local martingale part of* X.

Notation. We assume given a semimartingale X satisfying Hypothesis A. If Z is any other semimartingale we write

$$\hat{Z}_t^a = \int_{0+}^t 1_{\{X_{s-}>a\}} dZ_s \qquad (a \in \mathbb{R}).$$

This notation should not be confused with that of Kolmogorov's Lemma (Theorem 53). It is always assumed that we take the $\mathcal{B}(\mathbb{R}) \otimes \mathcal{B}(\mathbb{R}_+) \otimes \mathcal{F}$ measurable, càdlàg version of Z^a (cf. Theorem 44).

Before we can prove our primary regularity property (Theorem 56), we need two preliminary results. The first is a very special case of a family of martingale inequalities known as the Burkholder-Davis-Gundy inequalities.

Theorem 54 (Burkholder's Inequality). *Let X be a continuous local martingale with $X_0 = 0$, $2 \le p < \infty$, and T a finite stopping time. Then*

$$E\{(X_T^*)^p\} \le C_p E\{[X,X]_T^{p/2}\}$$

where $C_p = \{q^p(\frac{p(p-1)}{2})\}^{p/2}$, where $\frac{1}{p} + \frac{1}{q} = 1$.

Proof. By stopping, it suffices to consider the case where X and $[X,X]$ are bounded. By Itô's formula we have

$$|X_T|^p = p\int_0^T \operatorname{sign}(X_s)|X_s|^{p-1} dX_s + \frac{p(p-1)}{2}\int_0^T |X_s|^{p-2} d[X,X]_s.$$

By Doob's inequalities (Theorem 20 of Chap. I) we have (with $\frac{1}{p} + \frac{1}{q} = 1$):

$$\begin{aligned} E\{(X_T^*)^p\} &\le q^p E\{|X_T|^p\} \\ &= q^p E\{\frac{p(p-1)}{2}\int_0^T |X_s|^{p-2} d[X,X]_s\} \\ &\le q^p\left(\frac{p(p-1)}{2}\right) E\{(X_T^*)^{p-2}[X,X]_T\} \\ &\le q^p\left(\frac{p(p-1)}{2}\right) E\{(X_T^*)^p\}^{1-\frac{2}{p}} E\{[X,X]_T^{p/2}\}^{2/p}, \end{aligned}$$

with the last inequality by Hölder's inequality. Since $E\{(X_T^*)^p\}^{1-\frac{2}{p}} < \infty$, we divide both sides by it to obtain:

$$E\{(X_T^*)^p\}^{2/p} \leq q^p \left(\frac{p(p-1)}{2}\right) E\{[X,X]_T^{p/2}\}^{2/p},$$

and raising both sides to the power $p/2$ gives the result. □

Actually much more is true. Indeed for any local martingale (continuous or not) it is known that there exist constants c_p, C_p such that for a finite stopping time T

$$E\{(X_T^*)^p\}^{1/p} \leq c_p E\{[X,X]_T^{p/2}\}^{1/p} \leq C_p E\{(X_T^*)^p\}^{1/p}$$

for $1 \leq p < \infty$. See Sect. 3 of Chap. VII of Dellacherie-Meyer [**2**] for these and related results. When the local martingales are continuous some results even hold for $0 < p < 1$ (see, e.g., Barlow-Jacka-Yor [**1**], Table 4.1 on p. 162).

Theorem 55. *Let X be a semimartingale satisfying Hypothesis A. There exists a version of $(\hat{X}^c)_t^a$ such that $(a,t,\omega) \to (\hat{X}^c)_t^a(\omega)$ is $\mathcal{B}(\mathbb{R}) \otimes \mathcal{P}$ measurable, and everywhere jointly continuous in (a,t).*

Proof. Without loss of generality we may assume $X - X_0 \in \mathcal{H}^2$: If it is not, we can stop $X - X_0$ at T^n-; the continuous local martingale part of X^{T^n-} is just $(X^c)^{T^n}$. Suppose $-\infty < a < b < \infty$, and let

$$\alpha_t(a,b) = E\{(\int_0^t 1_{(b \geq X_{s-} > a)} d[X,X]_s^c)^2\}.$$

By Corollary 1 of Theorem 51 we have

$$\begin{aligned}
\alpha_t(a,b) &= E\{(\int_a^b L_t^u du)^2\} \\
&= (b-a)^2 E\{(\frac{1}{b-a}\int_a^b L_t^u du)^2\} \\
&\leq (b-a)^2 E\{\frac{1}{b-a}\int_a^b (L_t^u)^2 du\},
\end{aligned}$$

by the Cauchy-Schwarz inequality. The above implies

$$\alpha_t(a,b) \leq (b-a)^2 \sup_{u \in (a,b)} E\{(L_t^u)^2\}.$$

By the definition,

$$L_t^u \leq A_t^u \leq |X_t - X_0| - \int_{0+}^t \operatorname{sign}(X_{s-} - u) dX_s$$

and therefore

$$\begin{aligned} E\{(L_t^u)^2\} &\le 2E\{|X_t - X_0|^2\} + 2E\{(\int_{0+}^t \operatorname{sign}(X_{s-} - u)dX_s)^2\} \\ &\le 4\|X - X_0\|_{\mathcal{H}^2}^2 + 4\|\operatorname{sign}(X_{s-} - u)\cdot(X - X_0)\|_{\mathcal{H}^2}^2 \\ &\le 8\|X - X_0\|_{\mathcal{H}^2}^2 < \infty, \end{aligned}$$

and the bound is independent of u. Therefore

$$\alpha_t(a,b) \le (b-a)^2\Gamma,$$

for a constant $\Gamma < \infty$, and *independent of* t. Next using Burkholder's inequality (Theorem 54) we have

$$\begin{aligned} &E\{\sup_s |(\hat{X}^c)_s^b - (\hat{X}^c)_s^a|^4\} \\ &\qquad \le C_4 E\{(\int_0^\infty 1_{(b \ge X_{s-} > a)} d[X,X]_s^c)^2\} \\ &\qquad \le C_4 \sup_t \alpha_t(a,b) \\ &\qquad \le C_4 \Gamma (b-a)^2. \end{aligned}$$

The result now follows by applying Kolmogorov's Lemma (the Corollary of Theorem 53). □

We can now establish our primary result.

Theorem 56. *Let X be a semimartingale satisfying Hypothesis A. Then there exists a $\mathcal{B}(\mathbf{R}) \otimes \mathcal{P}$ measurable version of $(a,t,\omega) \to L_t^a(\omega)$ which is everywhere jointly right continuous in a and continuous in t. Moreover a.s. the limits $L_t^{a-} = \lim_{b\to a,\, b<a} L_t^b$ exist.*

Proof. Since X satisfies Hypothesis A, the process $J_t = \sum_{0\le s\le t} \Delta X_s$ is an FV semimartingale, and $Y = X - J$ is a continuous semimartingale. We let $Y = M + A$ be the (unique) decomposition of Y, with $M_0 = A_0 = 0$.

Then $X = M + A + J$. Further define

$$S_t^a = \sum_{0<s\le t} 1_{(X_{s-} > a)}(X_s - a)^- + \sum_{0<s\le t} 1_{(X_{s-} \le a)}(X_s - a)^+.$$

Observe that $|S_t^a| \le \sum_{0<s\le t} |\Delta X_s| < \infty$. By Theorem 49 we have

$$(X_t - a)^+ - (X_0 - a)^+ = (\hat{M})_t^a + (\hat{A})_t^a + (\hat{J})_t^a + S_t^a + \frac{1}{2} L_t^a.$$

Consider

$$(\hat{A})_t^a = \int_{0+}^t 1_{(X_{s-} > a)} dA_s.$$

We have

$$\lim_{\substack{a \to b \\ a > b}} (\hat{A})_t^a = \int_{0+}^t 1_{(X_{s-} > b)} dA_s$$

and

$$\lim_{\substack{a \to b \\ a < b}} (\hat{A})_t^a = \int_{0+}^t 1_{(X_{s-} \geq b)} dA_s$$

where the convergence is uniform in t on $[0, \tau]$, and $\tau > 0$. We have analogous results for $(\hat{J})_t^a$ and also for S_t^a, because it is dominated by $\sum_{0<s\leq t} |\Delta X_s| < \infty$. Since we already know that $(\hat{M})_t^a$ is continuous, the proof is complete. □

The next three corollaries are simple consequences of Theorem 56. Recall that for a semimartingale X satisfying Hypothesis A, we let $J_t = \sum_{0 \leq s \leq t} \Delta X_s$, $Y = X - J$, and $Y = M + A$ is the (unique) decomposition of Y, with $A_0 = 0$.

Corollary 1. *Let X be a semimartingale satisfying Hypothesis A. Let $X = M + A + J$ be a decomposition with M and A continuous and J the jump process of X, and let $(L_t^a)_{t \geq 0}$ be its local time at the level a. Then*

$$\begin{aligned} L_t^a - L_t^{a-} &= 2 \int_0^t 1_{\{X_{s-} = a\}} dA_s \\ &= 2 \int_0^t 1_{\{X_s = a\}} dA_s. \end{aligned}$$

An example of a semimartingale satisfying Hypothesis A but having a discontinuous local time is $X_t = |B_t|$ where B is standard Brownian motion with $B_0 = 0$. Here $L^a(X) = L^a(B) + L^{-a}(B)$ for $a > 0$, $L^0(X) = L^0(B)$, and $L^a(X) = 0$ for $a < 0$. John Walsh [1] has given a more interesting (but more complicated) example.

Corollary 2. *Let X be a semimartingale satisfying Hypothesis A. Let A be as in Corollary 1. The local time (L_t^a) is continuous in t and is continuous at $a = a_0$ if and only if*

$$\int_0^\infty 1_{(X_s = a_0)} |dA_s| = 0.$$

Observe that if $X = B$, a Brownian motion (or any continuous local martingale), then $A = 0$ and the local time of X can be taken everywhere jointly continuous.

Corollary 3. *Let X be a semimartingale satisfying Hypothesis A. Then for every (a,t) we have:*

$$L_t^a = \lim_{\epsilon \to 0} \frac{1}{\epsilon} \int_0^t 1_{(a \le X_s \le a+\epsilon)} d[X,X]_s^c, \quad a.s.$$

and

$$L_t^{a-} = \lim_{\epsilon \to 0} \frac{1}{\epsilon} \int_0^t 1_{(a-\epsilon \le X_s \le a)} d[X,X]_s^c, \quad a.s$$

The lack of symmetry in the above formula stems from the definition of local time, where we defined $\operatorname{sign}(x)$ in an asymmetric way:

$$\operatorname{sign}(x) = \begin{cases} 1 & x > 0 \\ -1 & x \le 0. \end{cases}$$

A symmetrized result follows trivially:

$$\frac{L_t^a + L_t^{a-}}{2} = \lim_{\epsilon \to 0} \frac{1}{2\epsilon} \int_0^t 1_{(|X_s - a| \le \epsilon)} d[X,X]_s^c, \quad \text{a.s.}$$

Corollary 3 is intuitively very appealing, and justifies thinking of local time as an *occupation time density*. Also if $X = B$, a Brownian motion, then Corollary 3 becomes the classical result:

$$L_t^a = \lim_{\epsilon \to 0} \frac{1}{2\epsilon} \int_0^t 1_{(a-\epsilon, a+\epsilon)}(B_s) ds, \quad \text{a.s.}$$

Local times have interesting properties when viewed as processes with "time" fixed, and the space variable "a" as the parameter. The Ray-Knight theorem giving the distribution of $(L_T^{1-a})_{0 \le a < 1}$, the Brownian local time sampled at the hitting time of 1, as the square of a two dimensional Bessel diffusion, is one example.

The Meyer-Itô formula (Theorem 51) can be extended in a different direction, which gives rise to the Bouleau-Yor formula, allowing non-convex functions of semimartingales. The key idea is that the function $a \to L_U^a$ induces a measure on $\mathbf{R}$ if L is the local time of a semimartingale X satisfying Hypothesis A and U is a positive random variable.

Theorem 57. *Let X be a semimartingale satisfying Hypothesis A, U a positive random variable, L^a the local times of X. Then the operation*

$$f \to \sum_{i=1}^n f_i (L_U^{a_{i+1}} - L_U^{a_i}),$$

where $f(x) = \sum f_i 1_{(a_i, a_{i+1}]}(x)$, can be extended uniquely to a vector measure on $\mathcal{B}(\mathbf{R})$ with values in L^0.

Proof. Recall that L^0 denotes finite valued random variables. By Theorem 49 we know that

$$\frac{1}{2}L_t^a = (X_t - a)^- - (X_0 - a)^- + \int_{0+}^t 1_{(X_{s-}\le a)}dX_s$$
$$- \sum_{0<s\le t} 1_{(X_{s-}>a)}(X_s - a)^- - \sum_{0<s\le t} 1_{(X_{s-}\le a)}(X_s - a)^+.$$

We write S_t^a for the last two sums on the right side of the equation above. Note first that $(X_U - a)^- - (X_0 - a)^-$ is Lipschitz continuous in a and hence absolutely continuous in a, therefore it induces a measure in a. Also the function $a \to S_U^a$ is càdlàg and of finite variation in a, and moreover $\int |d_a S_U^a| \le 2\sum_{0<s\le U} |\Delta X_s|$. Finally consider the stochastic integral: let $X_t^a = \int_{0+}^t 1_{(X_{s-}\le a)}dX_s$. Then

$$\sum_i f_i(X_U^{a_{i+1}} - X_U^{a_i}) = \int_{0+}^U \sum_i f_i 1_{(a_i<X_{s-}\le a_{i+1})}dX_s$$
$$= \int_{0+}^U f(X_{s-})dX_s.$$

Therefore by the dominated convergence theorem for stochastic integrals (Theorem 32) we have that $\sum_i f_i(X_U^{a_{i+1}} - X_U^{a_i})$ extends to a measure with values in L^0 and moreover $\int f(a)d_aX_U^a = \int_{0+}^U f(X_{s-})dX_s$, for f bounded, Borel measurable on $\mathbf{R}$. Thus $d_aL_U^a$ can be defined as the sum of three L^0-valued measures on $(\mathbf{R}, \mathcal{B}(\mathbf{R}))$, and the theorem is proved. □

In the proof of Theorem 57 we proved the following result, which is important enough to state as a corollary.

Corollary. *Let X be a semimartingale satisfying Hypothesis A, U a positive random variable, and $X_t^a = \int_0^t 1_{(X_{s-}\le a)}dX_s$. Then $d_aX_U^a$ can be defined as an L^0-valued measure, and for $f \in b\mathcal{B}(\mathbf{R})$ we have*

$$\int_{\mathbf{R}} f(a)d_aX_U^a = \int_{0+}^U f(X_{s-})dX_s.$$

Combining Theorem 57 and its Corollary yields the Bouleau-Yor formula. We leave its proof to the reader.

Theorem 58 (Bouleau-Yor Formula). *Let X be a semimartingale satisfying Hypothesis A, U a positive random variable, f a bounded, Borel function, and $F(x) = \int_0^x f(u)du$. Then*

$$F(X_U) - F(X_0) = \int_{0+}^U f(X_{s-})dX_s - \frac{1}{2}\int f(a)d_aL_U^a$$
$$+ \sum_{0<s\le U} \{F(X_s) - F(X_{s-}) - f(X_{s-})\Delta X_s\}.$$

Combining Theorem 58 with the Meyer-Itô formula (Theorem 51) we obtain the following relationship.

Corollary. *Let X be a semimartingale satisfying Hypothesis A, and let T be a finite-valued random variable. Let L^a be the local times of X and let f be a C^1 function. Then*

$$\int_{-\infty}^{\infty} f'(a)L_T^a da = -\int_{-\infty}^{\infty} f(a)d_a L_T^a.$$

6. Azema's Martingale

In order to prove the regularity properties of local times in Sect. 5, we needed *Hypothesis A*: a semimartingale X satisfies Hypothesis A if X is such that $\sum_{0<s\leq t}|\Delta X_s| < \infty$ a.s., each $t > 0$. This hypothesis was needed to prove Theorems 55 through 58 and their corollaries. We present here a counterexample, known as Azema's martingale, that shows that these results do not hold for general semimartingales. In particular, Theorems 62 and 63 show that Azema's martingale has a local time that is zero at every level a except $a = 0$, and therefore there is no regular version in the space variable.

Let $(\Omega, \mathcal{F}, (\mathcal{F}_t)_{t\geq 0}, (B_t)_{t\geq 0}, P)$ be a standard Brownian motion, $B_0 = 0$, with the filtration $(\mathcal{F}_t)_{t\geq 0}$ completed (and hence right continuous). We define

$$\operatorname{sign}(x) = \begin{cases} 1 & \text{if } x > 0 \\ -1 & \text{if } x \leq 0. \end{cases}$$

Set $\mathcal{G}_t^0 = \sigma\{\operatorname{sign}(B_s); s \leq t\}$, and let $(\mathcal{G}_t)_{t\geq 0}$ denote the filtration $(\mathcal{G}_t^0)_{t\geq 0}$ completed. (It is then right continuous as a consequence of the strong Markov property of Brownian motion.)

Let $M_t = E\{B_t|\mathcal{G}_t\}, t \geq 0$. Then for $s < t$,

$$E\{M_t|\mathcal{G}_s\} = E\{B_t|\mathcal{G}_s\} = E\{E\{B_t|\mathcal{F}_s\}|\mathcal{G}_s\} = E\{B_s|\mathcal{G}_s\} = M_s,$$

and M is a $\mathcal{G}$-martingale. We always take the càdlàg version of M.

Definition. The process $M_t = E\{B_t|\mathcal{G}_t\}$ is called **Azema's martingale.**

Fundamentally related to Azema's martingale is the process which gives *the last exit from zero* before time t. We define:

$$g_t = \sup\{s \leq t : B_s = 0\}.$$

To study the process g we need the *reflection principle for Brownian motion*, which we proved in Chap. I (Theorem 33 and its Corollary) and which we recall here for convenience.

Theorem 59. *Let B be standard Brownian motion, $B_0 = 0$, and $c > 0$. Then*

$$P(\sup_{s\leq t} B_s > c) = 2P(B_t > c).$$

An elementary conditioning argument provides a useful Corollary.

Corollary. *Let B be standard Brownian motion, $B_0 = 0$, and $0 < s < t$. Then*

$$P(\sup_{s<u\leq t} B_u > 0, B_s < 0) = 2P(B_t > 0, B_s < 0).$$

Theorem 60 is known as *Lévy's arcsine law* for the last exit from zero.

Theorem 60. *The process g_t is $(\mathcal{G}_t)_{t\geq 0}$ adapted, and $P(g_t \leq s) = \frac{2}{\pi}\arcsin\sqrt{\frac{s}{t}}$, $0 \leq s \leq t$.*

Proof. For $s \leq t$ we have almost surely:

$$\{g_t \leq s\} = (\bigcap_{\substack{s<u<t\\ u\in\mathbb{Q}}} \{B_u > 0\}) \bigcup (\bigcap_{\substack{s<u<t\\ u\in\mathbb{Q}}} \{B_u < 0\})$$

from which it follows that g_t is $\mathcal{G}_t$ adapted.

To calculate the distribution of g_t we observe for $s < t$ (using the symmetry of B):

$$\begin{aligned} P(g_t > s) &= 2P(g_t > s, B_s < 0) \\ &= 2P(\sup_{s<u\leq t} B_u > 0, B_s < 0) \\ &= 4P(B_t > 0, B_s < 0), \end{aligned}$$

by the Corollary of Theorem 59. Since (B_s, B_t) are jointly Gaussian, there exist X, Y that are independent $N(0,1)$ random variables with

$$B_s = \sqrt{s}X \qquad \text{and} \qquad B_t = \sqrt{s}X + \sqrt{t-s}\,Y.$$

Then $\{B_s < 0\} = \{X < 0\}$, and $\{B_t > 0\} = \{Y > -\frac{\sqrt{s}}{\sqrt{t-s}}X\}$. To calculate $P(\{Y > -\frac{\sqrt{s}}{\sqrt{t-s}}X, X < 0\})$, use polar coordinates:

$$\begin{aligned} P(B_t > 0, B_s < 0) &= \frac{1}{2\pi}\int_{\pi/2}^{\pi-\arcsin\sqrt{\frac{s}{t}}}\int_0^\infty e^{-r^2/2} r\,dr\,d\theta \\ &= \frac{1}{2\pi}(\frac{\pi}{2} - \arcsin\sqrt{\frac{s}{t}}) \\ &= \frac{1}{4} - \frac{1}{2\pi}\arcsin\sqrt{\frac{s}{t}}. \end{aligned}$$

Therefore $4P(B_t > 0, B_s < 0) = 1 - \frac{2}{\pi}\arcsin\sqrt{\frac{s}{t}}$ and $P(g_t \leq s) = \frac{2}{\pi}\arcsin\sqrt{\frac{s}{t}}$. □

Theorem 61. *Azema's martingale M is given by $M_t = \text{sign}(B_t)\sqrt{\frac{\pi}{2}}\sqrt{t - g_t}$, $t \geq 0$.*

Proof. By definition $M_t = E\{B_t|\mathcal{G}_t\} = E\{\text{sign}(B_t)|B_t||\mathcal{G}_t\} = \text{sign}(B_t)E\{|B_t||\mathcal{G}_t\}$. However $E\{|B_t||\mathcal{G}_t\} = E\{|B_t||g_t;\ \text{sign}(B_s), s \leq t\}$. But the process $\text{sign}(B_s)$ is independent of $|B_t|$ for $s < g_t$, and $\text{sign}(B_s) = \text{sign}(B_t)$ for $g_t < s \leq t$, hence $\text{sign}(B_s)$ is constant after g_t. It follows that

$$E\{|B_t||\mathcal{G}_t\} = E\{|B_t||g_t\}.$$

Therefore if we can show $E\{|B_t||g_t = s\} = \sqrt{\frac{\pi}{2}}\sqrt{t-s}$, the proof will be complete.

Given $0 < s < t$ we define:

$$T = t \wedge \inf\{u > s : B_u = 0\}.$$

Note that T is a bounded stopping time. Then

$$\begin{aligned} E\{|B_t|; g_t \leq s\} &= 2E\{B_t; g_t \leq s \text{ and } B_s > 0\} \\ &= 2E\{B_t; T = t \text{ and } B_s > 0\} \\ &= 2E\{B_T; B_s > 0\} \\ &= 2E\{B_s; B_s > 0\}, \end{aligned}$$

since $B_T = 0$ on $\{T \neq t\}$, and the last step uses that B is a martingale, and $s \leq T$. The last term above equals $E\{|B_s|\} = \sqrt{\frac{2}{\pi}}\sqrt{s}$. Therefore

$$\begin{aligned} E\{|B_t||g_t = s\} &= \frac{\frac{d}{ds}E\{|B_t|; g_t \leq s\}}{\frac{d}{ds}P(g_t \leq s)} \\ &= \sqrt{\frac{\pi}{2}}\sqrt{t-s}, \end{aligned}$$

since $\frac{d}{ds}P(g_t \leq s) = \frac{1}{\pi\sqrt{s(t-s)}}$, by Theorem 60. □

Let $\mathcal{H}_t^0 = \sigma\{M_s; s \leq t\}$, the minimal filtration of Azema's martingale M, and let $(\mathcal{H}_t)_{t\geq 0}$ be the completed filtration.

Corollary. *The two filtrations $(\mathcal{H}_t)_{t\geq 0}$ and $(\mathcal{G}_t)_{t\geq 0}$ are equal.*

Proof. By definition $M_t \in \mathcal{G}_t$, each $t \geq 0$, hence $\mathcal{H}_t \subset \mathcal{G}_t$. However by Theorem 61 $\text{sign}(M_t) = \text{sign}(B_t)$, thus $\mathcal{G}_t \subset \mathcal{H}_t$. □

Next we turn our attention to the local times of Azema's martingale. For a semimartingale $(X_t)_{t\geq 0}$, let $(L_t^a(X))_{t\geq 0} = L^a(X)$ denote its local time at the level a.

Theorem 62. *The local time at zero of Azema's martingale is not identically zero.*

Proof. By the Meyer-Itô formula (Theorem 51) for $f(x) = |x|$ we have:

$$|M_t| = \int_0^t \operatorname{sign}(M_{s-})dM_s + L_t^0(M) + \sum_{0<s\leq t} \{|M_s| - |M_{s-}| - \operatorname{sign}(M_{s-})\Delta M_s\}.$$

Set $Y_t = |M_t|$. Since $\operatorname{sign}(M_{s-})\Delta M_s = \Delta Y_s$, we have that $|M_t| - L_t^0(M) =$ martingale. If $L^0(M)$ were the zero process, then $Y = |M|$ would be a nonnegative martingale with $Y_0 = 0$, hence identically zero. Since Y is not identically zero, the theorem is proved. □

Theorem 63. *The local times at all levels except zero of Azema's martingale are* 0. *That is,* $L_t^a(M) = 0$, $t \geq 0$, *if* $a \neq 0$.

Proof. By Theorem 50 we know that the measure $dL_s^a(\omega)$ on $\mathbf{R}_+$ is carried by the set $\{s : M_{s-}(\omega) = M_s(\omega) = a\}$, a.s., for each a. However if $a \neq 0$, this set is countable. Since $s \to L_s^a$ is a.s. continuous, it cannot charge a countable set. Therefore dL_s^a is the zero measure for $a \neq 0$. Since $L_0^a = 0$, we have $L_t^a \equiv 0$ for $a \neq 0$. □

Theorems 62 and 63 together show that Azema's martingale provides a counterexample to a general version of Theorem 56, for example. Therefore a hypothesis such as Hypothesis A is necessary, and also we see that Azema's martingale does not satisfy Hypothesis A. Since all semimartingales for the Brownian motion filtration $(\mathcal{F}_t)_{t\geq 0}$ do satisfy Hypothesis A, as noted in Sect. 5, we conclude:

Theorem 64. *Azema's martingale M is not a semimartingale for the Brownian filtration* $(\mathcal{F}_t)_{t\geq 0}$.

While the local time at zero, $L^0(M)$, is non-trivial by Theorem 62, more information is available: It is actually the same as the Brownian local time, as Theorem 66 below shows. First we need a preliminary result.

Theorem 65. *The local time $L^0(B)$ of Brownian motion is $(\mathcal{G}_t)_{t\geq 0}$ adapted.*

Theorem 65 will be proved if one can express $L^0(B)$ as the a.s. limit of measurable functionals of the zero set of Brownian motion. Such a result is classical: See, e.g., Kingman [**1**, p. 730].

Theorem 66. *The local times at zero of Brownian motion and Azema's martingale are the same. That is, $L^0(B) = L^0(M)$.*

Proof. As we saw in the proof of Theorem 62, $|M_t| - L_t^0(M) =$ martingale. Since the process $L^0(M)$ is nondecreasing, continuous and adapted, it is natural. If A is another natural process such that $|M| - A =$ martingale, then $L^0(M) = A$ by (for example) Theorem 18 of Chap. III. Thus it suffices to show that $|M| - L^0(B) =$ martingale, for the filtration $(\mathcal{G}_t)_{t\geq 0}$.

By Tanaka's formula, $|B_t| - L_t^0(B) = N_t$, where N is an $(\mathcal{F}_t)_{t\geq 0}$-martingale. Then:

$$E\{|B_t| - L_t^0(B)|\mathcal{G}_t\} = E\{N_t|\mathcal{G}_t\}$$

and by Theorem 65:

$$E\{|B_t||\mathcal{G}_t\} - L_t^0(B) = \text{ martingale.}$$

Since

$$\begin{aligned} E\{|B_t||\mathcal{G}_t\} &= E\{\text{sign}(B_t)B_t|\mathcal{G}_t\} \\ &= \text{sign}(B_t)E\{B_t|\mathcal{G}_t\} \\ &= \text{sign}(B_t)M_t \\ &= |M_t|, \end{aligned}$$

we have $|M_t| - L_t^0(B) =$ martingale, and the theorem is proved. □

Corollary 1. *The process $|M| - L^0(B)$ is a $(\mathcal{G}_t)_{t\geq 0}$ martingale.*

Proof. This corollary is proved at the end of the proof of Theorem 66. □

Corollary 2. *The process $1_{(B_t>0)}\sqrt{\frac{\pi}{2}}\sqrt{t-g_t} - \frac{1}{2}L_t^0(B)$ is a $(\mathcal{G}_t)_{t\geq 0}$ martingale.*

Proof. Since $1_{(M_{s-}>0)}M_s^- = 1_{(M_{s-}\leq 0)}M_s^+ = 0$, it follows from Theorem 49 that $M_t^+ - \frac{1}{2}L_t^0(M) =$ martingale. However $M_t^+ = 1_{(B_t>0)}\sqrt{\frac{\pi}{2}}\sqrt{t-g_t}$ by Theorem 61, and $L_t^0(M) = L_t^0(B)$ by Theorem 66, and the theorem follows. □

The next theorem shows that Azema's martingale is quadratic pure jump.

Theorem 67. *For Azema's martingale M, $[M,M]^c \equiv 0$, and $[M,M]_t = \frac{\pi}{2}g_t$.*

Proof. By Corollary 2 of Theorem 51,

$$[M,M]_t^c = \int_{-\infty}^{\infty} L_t^a\,da.$$

By Theorem 63 then, $[M,M]^c \equiv 0$. We conclude that $[M,M]_t = \sum_{0<s\leq t}(\Delta M_s)^2$, and it follows from Theorem 61 that $[M,M]_t = \frac{\pi}{2}g_t$. □

Theorem 67 allows us to compute the compensation of $[M,M]$; that is, the unique natural, increasing process A such that $[M,M]-A =$ martingale. This process A is denoted $\langle M,M\rangle$ in the literature, and it was defined in Sect. 3 of Chap. III.

Theorem 68. *Let M be Azema's martingale. Then $[M,M]_t - \frac{\pi}{4}t =$ martingale. That is, $\langle M,M\rangle_t = \frac{\pi}{4}t$.*

Proof. Recall that $M_t^2 - [M,M]_t =$ martingale (cf. Theorem 27 of Chap. II). By Theorems 61 and 67,

$$\begin{aligned} M_t^2 - [M,M]_t &= \frac{\pi}{2}\{(t-g_t)-g_t\} \\ &= \frac{\pi}{2}t - \pi g_t. \end{aligned}$$

Since the above is a martingale, it remains only to observe that $\frac{\pi}{2}g_t - \frac{\pi}{4}t$ is a martingale. □

Note that if $X_t = \operatorname{sign}(B_t)\sqrt{2}\sqrt{t-g_t}$, then X is a quadratic pure jump martingale with $\langle X,X\rangle_t = t$. Recall that if N is a Poisson process with intensity one, then $N_t - t =$ martingale. Since $[N,N]_t = N_t$, we have $[N,N]_t - t =$ martingale; thus $\langle N,N\rangle_t = t$. Another example of a martingale Y with $\langle Y,Y\rangle_t = t$ is Brownian motion: Since $[B,B]_t = t$ is continuous, $[B,B]_t = \langle B,B\rangle_t = t$.

Bibliographic Notes

The extension by continuity of stochastic integration from processes in $\mathbf{L}$ to predictable processes is presented here for essentially the first time. However the procedure was indicated earlier in Protter [**6, 7**]. Other approaches which are closely related are given by Jacod [**1**] and Chou-Meyer-Stricker [**1**] (see an exposition in Dellacherie-Meyer [**2**, p. 381]).

The "$\mathcal{H}$" in the space $\mathcal{H}^2$ of semimartingales comes from Hardy spaces, and this bears explaining. When the semimartingale $X \in \mathcal{H}^2$ is actually a martingale, the space of $\mathcal{H}^p$ martingales has a theory analogous to that of Hardy spaces. While this is not the case for semimartingales, the $\mathcal{H}^p$ norms for semimartingales are, in a certain sense, a natural generalization of the $\mathcal{H}^p$ norms for martingales, and so the name has been preserved. The $\mathcal{H}^p$ norm was first introduced by Emery [**1**], though it was implicit in Protter [**1**].

Most of the treatment of stochastic integration in Sect. 2 of this chapter is new, though essentially all of the results have been proved elsewhere with different methods. I have benefited greatly throughout by discussions with Svante Janson. Theorem 1 (that $\mathcal{H}^2$ is a Banach space) is due to Emery [**2**]. The equivalence of the two pseudonorms for $\mathcal{H}^2$ (the Corollary of Theorem 24) is originally due to Yor [**7**], while the fact that $\mathcal{H}^2(Q) \subset \mathcal{H}^2(P)$ if $\frac{dQ}{dP}$ is bounded is originally due to Lenglart [**4**]. Theorem 25 is originally due to Jacod [**1**], p. 228. The concept of a predictable σ-algebra and its importance to stochastic integration is due to Meyer [**4**]. The local behavior of the stochastic integral was first investigated by McShane for his integral [**1**], and then independently by Meyer [**8**], who established Theorem 26 and its Corollary for the semimartingale integral. The first dominated convergence theorem for the semimartingale integral appearing in print seems to be in Jacod [**1**].

The theory of martingale representation dates back to Itô's work on multiple stochastic integrals [**6**], and the theory for $\mathbf{M}^2$ presented here is largely due to Kunita-Watanabe [**1**]. A more powerful theory (for $\mathbf{M}^1$) is presented in Dellacherie-Meyer [**2**].

The theory of stochastic integration depending on a parameter is due to the fundamental work of Stricker-Yor [**1**], who used a key idea of Doléans-Dade [**1**]. The Fubini theorems for stochastic integration have their origins in the book of Doob [**1**] and Kallianpur-Striebel [**1**] for the Itô integral, Kailath-Segall-Zakai [**1**] for martingale integrals, and Jacod [**1**] for semimartingales. The counterexample to a general Fubini theorem presented here is due to S. Janson.

The theory of semimartingale local time is of course abstracted from Brownian local time, which is due to Lévy [**1**]. It was related to stochastic integration by Tanaka (see McKean [**1**]) for Brownian motion and the Itô integral. The theory presented here for semimartingale local time is due largely to Meyer [**8**]. See also Millar [**1**]. The measure theory needed for the rigorous development of semimartingale local time was developed by Stricker-Yor [**1**], and the Meyer-Itô formula (Theorem 51) was formally presented in Yor [**6**], as well as in Jacod [**1**]. Theorem 52 is due to Yor [**4**], and has been extended by Ouknine [**1**]. A more general version (but more complicated) is in Çinlar-Jacod-Protter-Sharpe [**1**]. The proof of Kolmogorov's Lemma given here is from Meyer [**14**]. The results proved under Hypothesis A are all due to Yor [**3**] except the Bouleau-Yor formula [**1**]. See Bouleau [**1**] for more on this formula.

Azema's martingale is of course due to Azema [**1**], though our presentation of it is new and it is due largely to Svante Janson. For many more interesting results concerning Azema's martingale, see Azema-Yor [**1**], Emery [**6**], and Meyer [**16**].

CHAPTER V

Stochastic Differential Equations

1. Introduction

A *diffusion* can be thought of as a strong Markov process (in $\mathbb{R}^n$) with continuous paths. Before the development of Itô's theory of stochastic integration for Brownian motion, the primary method of studying diffusions was to study their transition semigroups; this was equivalent to studying the infinitesimal generators of their semigroups, which are partial differential operators. Thus Feller's investigations of diffusions (for example) were actually investigations of partial differential equations, inspired by diffusions.

The primary tool to study diffusions was Kolmogorov's differential equations and Feller's extensions of them. Such approaches did not permit an analysis of the paths of diffusions and their properties. Inspired by Lévy's investigations of sample paths, Itô studied diffusions that could be represented as solutions of stochastic differential equations[1] of the form:

$$(*) \qquad X_t = X_0 + \int_0^t \sigma(X_s)dB_s + \int_0^t b(X_s)ds$$

where B is a Brownian motion in $\mathbb{R}^n$ and σ is an $n \times n$ matrix, b an n-vector, of appropriately smooth functions to ensure the existence and uniqueness of solutions. This gives immediate intuitive meaning: If $(\mathcal{F}_t)_{t\geq 0}$ is the underlying filtration for the Brownian motion B, then for small $\epsilon > 0$:

$$E\{X^i_{t+\epsilon} - X^i_t | \mathcal{F}_t\} = b^i(X_t)\epsilon + o(\epsilon)$$
$$E\{(X^i_{t+\epsilon} - X^i_t - \epsilon b^i(X_t))(X^j_{t+\epsilon} - X^j_t - \epsilon b^j(X_t)) | \mathcal{F}_t\} = (\sigma\sigma')^{ij}(X_t)\epsilon + o(\epsilon),$$

where σ' denotes the transpose of the matrix σ. Much of current research is still concerned with equations of the form $(*)$.

[1] More properly (but less often) called "stochastic integral equations".

Itô's differential "dB" was found to have other interpretations as well. In particular "dB" can be thought of as "white noise" in statistical communication theory, and thus if ξ_t is white noise at time t, $B_t = \int_0^t \xi_s ds$, an equation which can be given a rigorous meaning using the theory of generalized functions (cf., e.g., Arnold [1]). Here the Markov nature of the solutions is not as important, and coefficients that are functionals of the paths of the solutions can be considered.

Finally it is now possible to consider semimartingale driving terms (or "semimartingale noise"), and to study stochastic differential equations in full generality. Since "dB" and "dt" are semimartingale differentials, they are always included in our results as special cases. While our treatment is very general, it is not always the most general available: We have at times preferred to keep proofs simple and non-technical rather than to achieve maximum generality.

The study of stochastic differential equations driven by general semimartingales (rather than just by dB, dt, dN, and combinations thereof, where N is a Poisson process) allows one to see which properties of the solutions are due to certain special properties of Brownian motion, and which are true in general. For example in Sect. 6 we see that the Markov nature of the solutions is due to the independence of the increments of the differentials; in Sects. 8 and 10 we see precisely how the homeomorphic and diffeomorphic nature of the flow of the solution is a consequence of path continuity; in Sect. 5 we study Fisk-Stratonovich equations which reveal that the "correction" term is due to the continuous part of the quadratic variation of the differentials.

2. The $\underline{\underline{H}}^p$ Norms for Semimartingales

We defined an $\mathcal{H}^2$ norm for semimartingales in Chap. IV as follows: if X is a special semimartingale with $X_0 = 0$ and canonical decomposition $X = \overline{N} + \overline{A}$, then

$$\|X\|_{\mathcal{H}^2} = \|[\overline{N}, \overline{N}]_\infty^{1/2}\|_{L^2} + \|\int_0^\infty |d\overline{A}_s|\|_{L^2}.$$

We now use an *equivalent* norm; to avoid confusion, we write $\underline{\underline{H}}^2$ instead of $\mathcal{H}^2$; moreover we will define $\underline{\underline{H}}^p$, $1 \le p \le \infty$.

We begin, however, with a different norm on the space $\mathbb{D}$ (i.e., the space of adapted càdlàg processes). For a process H in $\mathbb{D}$ we define

$$H^* = \sup_t |H_t|,$$

$$\|H\|_{\underline{\underline{S}}^p} = \|H^*\|_{L^p}.$$

Occasionally if H is in $\mathbb{L}$ (adapted and càglàd) we write $\|H\|_{\underline{\underline{S}}^p}$ as well, where the meaning is clear.

If A is a semimartingale with paths of finite variation, a natural definition of a norm would be $\|A\|_p = \| \int_0^\infty |dA_s| \|_{L^p}$, where $|dA_s(\omega)|$ denotes the total variation measure on $\mathbb{R}_+$ induced by $s \to A_s(\omega)$. Since semimartingales do not in general have such nice paths, however, such a norm is not appropriate. *Throughout this chapter, we will let Z denote a semimartingale with $Z_0 = 0$,* a.s. Let Z be an arbitrary semimartingale (with $Z_0 = 0$). By the Bichteler-Dellacherie theorem (Theorem 22 of Chap. III) we know there exists at least one decomposition $Z = N + A$, with N a local martingale and A an adapted, càdlàg process, with paths of finite variation (also $N_0 = A_0 = 0$ a.s.). For $1 \leq p \leq \infty$ we set:

$$j_p(N, A) = \|[N, N]_\infty^{1/2} + \int_0^\infty |dA_s| \|_{L^p}.$$

Definition. Let Z be a semimartingale. For $1 \leq p \leq \infty$ define

$$\|Z\|_{\underline{\underline{H}}^p} = \inf_{Z=N+A} j_p(N, A)$$

where the infimum is taken over all possible decompositions $Z = N + A$ where N is a local martingale, $A \in \mathbb{D}$ with paths of finite variation on compacts, and $A_0 = N_0 = 0$.

The Corollary of Theorem 1 below shows that this norm generalizes the $\underline{\underline{H}}^p$ norm for local martingales, which has given rise to a martingale theory analogous to the theory of Hardy spaces in complex analysis. We do not pursue this topic (cf. eg. Dellacherie-Meyer [**2**]).

Theorem 1. *Let Z be a semimartingale ($Z_0 = 0$). Then* $\|[Z, Z]_\infty^{1/2}\|_{L^p} \leq \|Z\|_{\underline{\underline{H}}^p}$ $(1 \leq p \leq \infty)$.

Proof. Let $Z = M + A$, $M_0 = A_0 = 0$, be a decomposition of Z. Then

$$\begin{aligned}
[Z, Z]_\infty^{1/2} &\leq [M, M]_\infty^{1/2} + [A, A]_\infty^{1/2} \\
&= [M, M]_\infty^{1/2} + (\sum_s (\Delta A_s)^2)^{1/2} \\
&\leq [M, M]_\infty^{1/2} + \sum_s |\Delta A_s| \\
&\leq [M, M]_\infty^{1/2} + \int_0^\infty |dA_s|,
\end{aligned}$$

where the equality above holds because A is a quadratic pure jump semimartingale. Taking L^p norms yields $\|[Z, Z]_\infty^{1/2}\|_{L^p} \leq j_p(M, A)$ and the result follows. □

Corollary. *If Z is a local martingale ($Z_0 = 0$), then* $\|Z\|_{\underline{\underline{H}}^p} = \|[Z,Z]_\infty^{1/2}\|_{L^p}$.

Proof. Since Z is a local martingale, we have that $Z = Z + 0$ is a decomposition of Z. Therefore

$$\|Z\|_{\underline{\underline{H}}^p} \leq j_p(Z,0) = \|[Z,Z]_\infty^{1/2}\|_{L^p}.$$

By Theorem 1 we have $\|[Z,Z]_\infty^{1/2}\|_{L^p} \leq \|Z\|_{\underline{\underline{H}}^p}$, hence we have equality. □

Theorem 2 is analogous to Theorem 5 of Chap. IV. For most of the proofs which follow we need only the case $p = 2$. Since this case does not need Burkholder's inequalities, we distinguish it from the other cases in the proof.

Theorem 2. *For $1 \leq p < \infty$ there exists a constant c_p such that for any semimartingale Z, $Z_0 = 0$, $\|Z\|_{\underline{\underline{S}}^p} \leq c_p \|Z\|_{\underline{\underline{H}}^p}$.*

Proof. A semimartingale Z is in $\mathbf{D}$, so $\|Z\|_{\underline{\underline{S}}^p}$ makes sense. Let $Z = M + A$ be a decomposition with $M_0 = A_0 = 0$. Then

$$\|Z\|_{\underline{\underline{S}}^p}^p = E\{(Z_\infty^*)^p\} \leq E\{(M_\infty^* + \int_0^\infty |dA_s|)^p\} \leq c_p E\{(M_\infty^*)^p + (\int_0^\infty |dA_s|)^p\},$$

using $(a+b)^p \leq 2^{p-1}(a^p + b^p)$.

In the case $p = 2$ we have by Doob's maximal quadratic inequality that

$$E\{(M_\infty^*)^2\} \leq 4E\{M_\infty^2\} = 4E\{[M,M]_\infty\}.$$

For general p, $1 \leq p < \infty$, we need Burkholder's inequalities, which state:

$$E\{(M_\infty^*)^p\} \leq c_p E\{[M,M]_\infty^{p/2}\}$$

for a universal constant c_p which depends only on p and not on the local martingale M. For *continuous* local martingales and $p \geq 2$ we proved this using Itô's formula in Chap. IV (Theorem 54). For general local martingales and for all finite $p \geq 1$ see, for example, Dellacherie-Meyer [**2**, p. 287].

Continuing, letting the constant c_p vary from place to place we have:

$$\begin{aligned}\|Z\|_{\underline{\underline{S}}^p}^p &\leq c_p E\{(M_\infty^*)^p + (\int_0^\infty |dA_s|)^p\} \\ &\leq c_p E\{[M,M]_\infty^{p/2} + (\int_0^\infty |dA_s|)^p\} \\ &\leq c_p [j_p(M,A)]^p,\end{aligned}$$

and taking p-th roots yields the result. □

Corollary. *On the space of semimartingales the $\underline{\underline{H}}^p$ norm is stronger than the $\underline{\underline{S}}^p$ norm, $1 \le p < \infty$.*

Theorem 3 (Emery's Inequality). *Let Z be a semimartingale, $H \in \mathbb{L}$, and $\frac{1}{p} + \frac{1}{q} = \frac{1}{r}$ $(1 \le p \le \infty; 1 \le q \le \infty)$. Then*

$$\| \int_0^\infty H_s dZ_s \|_{\underline{\underline{H}}^r} \le \|H\|_{\underline{\underline{S}}^p} \|Z\|_{\underline{\underline{H}}^q}.$$

Proof. Let $H \cdot Z$ denote $(\int_0^t H_s dZ_s)_{t \ge 0}$. Recall that we always assume $Z_0 = 0$ a.s., and let $Z = M + A$ be a decomposition of Z with $M_0 = A_0 = 0$ a.s. Then $H \cdot M + H \cdot A$ is a decomposition of $H \cdot Z$, hence

$$\|H \cdot Z\|_{\underline{\underline{H}}^r} \le j_r(H \cdot M, H \cdot A).$$

Next recall that $[H \cdot M, H \cdot M] = \int H_s^2 d[M,M]_s$, by Theorem 29 of Chap. II. Therefore

$$\begin{aligned} j_r(H \cdot M, H \cdot A) &= \|(\int_0^\infty H_s^2 d[M,M]_s)^{1/2} + \int_0^\infty |H_s||dA_s|\|_{L^r} \\ &\le \|H_\infty^*([M,M]_\infty^{1/2} + \int_0^\infty |dA_s|)\|_{L^r} \\ &\le \|H_\infty^*\|_{L^p} \|([M,M]_\infty^{1/2} + \int_0^\infty |dA_s|)\|_{L^q} \\ &= \|H\|_{\underline{\underline{S}}^p} j_q(M,A), \end{aligned}$$

where the last inequality above follows from Hölder's inequality. The foregoing implies that

$$\|H \cdot Z\|_{\underline{\underline{H}}^r} \le \|H\|_{\underline{\underline{S}}^p} j_q(M,A)$$

for any such decomposition $Z = M + A$. Taking infimums over all such decompositions yields the result. □

For a process $X \in \mathbb{D}$ and a stopping time T, recall that:

$$\begin{aligned} X^T &= X_t 1_{[0,T)} + X_T 1_{[T,\infty)} \\ X^{T-} &= X_t 1_{[0,T)} + X_{T-} 1_{[T,\infty)}. \end{aligned}$$

A property holding locally was defined in Chap. I, and a property holding prelocally was defined in Chap. IV. Recall that a property π is said to hold **locally** for a process X if $X^{T^n} 1_{\{T^n > 0\}}$ has property π for each n, where T^n is a sequence of stopping times tending to ∞ a.s. If the process X is zero at zero (i.e., $X_0 = 0$ a.s.) then the property π is said to hold **prelocally** if X^{T^n-} has property π for each n.

Definition. A process X **is locally in** $\underline{\underline{S}}^p$ (resp. $\underline{\underline{H}}^p$) if there exist stopping times $(T^n)_{n\geq 1}$ increasing to ∞ a.s. such that $X^{T^n}1_{\{T^n>0\}}$ is in $\underline{\underline{S}}^p$ (resp. $\underline{\underline{H}}^p$) for each $n (1 \leq p \leq \infty)$. If $X_0 = 0$ then X **is said to be prelocally in** $\underline{\underline{S}}^p$ (resp. $\underline{\underline{H}}^p$) if X^{T^n-} is in $\underline{\underline{S}}^p$ (resp. $\underline{\underline{H}}^p$) for each n.

While there are many semimartingales which are not locally in $\underline{\underline{H}}^p$, *all* semimartingales are prelocally in $\underline{\underline{H}}^p$. The proof of Theorem 4 below closely parallels the proof of Theorem 13 of Chap. IV.

Theorem 4. *Let Z be a semimartingale ($Z_0 = 0$). Then Z is prelocally in $\underline{\underline{H}}^p$, $1 \leq p \leq \infty$.*

Proof. By the Fundamental Theorem of Local Martingales (Theorem 13 of Chap. III) and the Bichteler-Dellacherie theorem (Theorem 22 of Chap. III) we know that for given $\epsilon > 0$, Z has a decomposition $Z = M + A$, $M_0 = A_0 = 0$ a.s., such that the jumps of the local martingale M are bounded by ϵ. Define inductively:

$$T_0 = 0;$$

$$T_{k+1} = \inf\{t \geq T_k : [M,M]_t^{1/2} + \int_0^t |dA_s| \geq k+1\}.$$

The sequence $(T_k)_{k\geq 1}$ are stopping times increasing to ∞ a.s. Moreover

$$Z^{T_k-} = (M^{T_k}) + (A^{T_k-} - \Delta M_{T_k} 1_{[T_k,\infty)}) = N + C$$

is a decomposition of Z^{T_k-}. Also, since $[M,M]_{T_k} = [M,M]_{T_k-} + (\Delta M_{T_k})^2$, we conclude

$$\begin{aligned} j_\infty(N,C) &= \| [N,N]_\infty^{1/2} + \int_0^\infty |dC_s| \|_{L^\infty} \\ &= \| ([M,M]_{T_k-} + (\Delta M_{T_k})^2)^{1/2} + \int_0^{T_k} |dC_s| \|_{L^\infty} \\ &\leq \| (k^2 + \epsilon^2)^{1/2} + (k+\epsilon) \|_{L^\infty} < \infty. \end{aligned}$$

Therefore $Z^{T_k-} \in \underline{\underline{H}}^\infty$ and hence it is in $\underline{\underline{H}}^p$ as well, $1 \leq p \leq \infty$. □

Definition. Let Z be a semimartingale in $\underline{\underline{H}}^\infty$ and let $\alpha > 0$. A finite sequence of stopping times $0 = T_0 \leq T_1 \leq \cdots \leq T_k$ **is said to α-slice** Z if $Z = Z^{T_k-}$ and $\|(Z - Z^{T_i})^{T_{i+1}-}\|_{\underline{\underline{H}}^\infty} \leq \alpha$, $0 \leq i \leq k-1$. If such a sequence of stopping times exists, we say Z **is α-sliceable**, and we write $Z \in \mathcal{S}(\alpha)$.

Theorem 5. *Let Z be a semimartingale ($Z_0 = 0$ a.s.).*

(i) *For* $\alpha > 0$, *if* $Z \in \mathcal{S}(\alpha)$ *then for every stopping time* T, $Z^T \in \mathcal{S}(\alpha)$ *and* $Z^{T-} \in \mathcal{S}(2\alpha)$.
(ii) *For every* $\alpha > 0$, *there exists an arbitrarily large stopping time* T *such that* $Z^{T-} \in \mathcal{S}(\alpha)$.

Proof. Always $\|Z^T\|_{\underline{\underline{H}}^\infty} \leq \|Z\|_{\underline{\underline{H}}^\infty}$, and since $Z^{T-} = M^T + (A^{T-} - \Delta M_T 1_{[T,\infty)})$ one concludes $\|Z^{T-}\|_{\underline{\underline{H}}^\infty} \leq 2\|Z\|_{\underline{\underline{H}}^\infty}$, and part (i) follows.

Next consider (ii). If semimartingales Z and Y are α-sliceable, let T_i^z and T_j^y be two sequences of stopping times respectively α-slicing Z and Y. By reordering the points T_i^z and T_j^y and using part (i), we easily conclude that $Z+Y$ is 8α-sliceable. Next let $Z = M + A$, $M_0 = A_0 = 0$ a.s., with the local martingale M having jumps bounded by the constant $\beta = \frac{\alpha}{24}$. By the preceding observation it suffices to consider M and A separately.

For A, let $T_0 = 0$, $T_{k+1} = \inf\{t \geq T_k : \int_{T_k}^t |dA_s| \geq \alpha/8 \text{ or } \int_0^t |dA_s| \geq k\}$. Then $A^{T_k-} \in \mathcal{S}(\alpha/8)$ for each k, and the stopping times (T_k) increase to ∞ a.s.

For M, let $R_0 = 0$, $R_{k+1} = \inf\{t \geq R_k : [M,M]_t - [M,M]_{R_k} \geq \beta^2 \text{ or } [M,M]_t \geq k\}$. Then $M^{R_k-} \in \underline{\underline{H}}^\infty$, each k, and moreover:

$$(M - M^{R_k})^{R_{k+1}-} = M^{R_{k+1}} - M^{R_k} - (\Delta M_{R_{k+1}} 1_{\{R_{k+1} > R_k\}}) 1_{[R_{k+1},\infty)},$$

hence

$$\begin{aligned}
&\|(M - M^{R_k})^{R_{k+1}-}\|_{\underline{\underline{H}}^\infty} \\
&\quad \leq \|([M,M]_{R_{k+1}} - [M,M]_{R_k})^{1/2} + |\Delta M_{R_{k+1}}|\|_{L^\infty} \\
&\quad = \|((\Delta M_{R_{k+1}})^2 + [M,M]_{R_{k+1}-} - [M,M]_{R_k})^{1/2} + |\Delta M_{R_{k+1}}|\|_{L^\infty} \\
&\quad \leq \|(\beta^2 + \beta^2)^{1/2} + \beta\|_{L^\infty} = (1+\sqrt{2})\beta.
\end{aligned}$$

Thus for each k, $M^{R_k-} \in \mathcal{S}((1+\sqrt{2})\beta)$, and since $\beta = \frac{\alpha}{24}$, the result follows. □

3. Existence and Uniqueness of Solutions

In presenting theorems on the existence and uniqueness of solutions of stochastic differential equations, there are many choices to be made. First, we do not present the most general conditions known to be allowed; in exchange we are able to give simpler proofs. Moreover the conditions we do give are extremely general and are adequate for the vast majority of applications. For more general results the interested reader can consult Jacod [1, pp. 451ff]. Second, we consider only Lipschitz-type hypotheses and thus obtain strong solutions. There is a vast literature on weak solutions (cf. eg. Stroock-Varadhan [1]), however weak solutions are more natural (and simpler) when the differentials are the Wiener process and Lebesgue measure, rather than general semimartingales.

A happy consequence of our approach to stochastic differential equations is that it is just as easy to prove theorems for coefficients that depend not only on the state X_t of the solution at time t (the traditional framework), but on the past history of the process X before t as well.

We begin by stating a theorem whose main virtue is its simplicity. It is a trivial corollary of Theorem 7 which follows it. Recall that a process H is in $\mathbb{L}$ if it has càglàd paths and is adapted.

Theorem 6. *Let Z be a semimartingale with $Z_0 = 0$ and let $f : \mathbb{R}_+ \times \Omega \times \mathbb{R} \to \mathbb{R}$ be such that:*

(i) for fixed x, $(t, \omega) \to f(t, \omega, x)$ is in $\mathbb{L}$;
(ii) for each (t, ω), $|f(t, \omega, x) - f(t, \omega, y)| \le K(\omega)|x - y|$ for some finite random variable K.

Let X_0 be finite and $\mathcal{F}_0$-measurable. Then the equation

$$X_t = X_0 + \int_0^t f(s, \cdot, X_{s-}) dZ_s$$

admits a solution. The solution is unique and it is a semimartingale.

Of course one could state such a theorem for a finite number of differentials $dZ^j (1 \le j \le d)$ and for a finite system of equations.

In the theory of (non-random) ordinary differential equations, coefficients are typically Lipschitz continuous, which ensures the existence and the uniqueness of a solution. In stochastic differential equations we are led to consider more general coefficients that arise, for example, in control theory. There are enough different definitions to cause some confusion, so we present all the definitions here in ascending order of generality. Note that we add, for technical reasons, the non-customary hypothesis (ii) below to the definition of Lipschitz which follows.

Definition. A function $f : \mathbb{R}_+ \times \mathbb{R}^n \to \mathbb{R}$ is **Lipschitz** if there exists a (finite) constant k such that

(i) $|f(t, x) - f(t, y)| \le k|x - y|$, each $t \in \mathbb{R}_+$;
(ii) $t \to f(t, x)$ is right continuous with left limits, each $x \in \mathbb{R}^n$.

f is said to be **autonomous** if $f(t, x) = f(x)$, all $t \ge 0$.

Definition. A function $f : \mathbb{R}_+ \times \Omega \times \mathbb{R}^n \to \mathbb{R}$ is **random Lipschitz** if f satisfies conditions (i) and (ii) of Theorem 6.

Let D^n denote the space of processes $\mathbf{X} = (X^1, \ldots, X^n)$ where each $X^i \in \mathsf{D}$ $(1 \leq i \leq n)$.

Definition. An operator F from D^n into $\mathsf{D}^1 = \mathsf{D}$ is said to be **process Lipschitz** if for any $\mathbf{X}$, $\mathbf{Y}$ in D^n the following two conditions are satisfied:

(i) for any stopping time T, $\mathbf{X}^{T-} = \mathbf{Y}^{T-}$ implies $F(\mathbf{X})^{T-} = F(\mathbf{Y})^{T-}$;
(ii) there exists an adapted process $K \in \mathsf{L}$ such that

$$\|F(\mathbf{X})_t - F(\mathbf{Y})_t\| \leq K_t \|\mathbf{X}_t - \mathbf{Y}_t\|.$$

Definition. An operator F mapping D^n to $\mathsf{D}^1 = \mathsf{D}$ is **functional Lipschitz** if for any $\mathbf{X}$, $\mathbf{Y}$ in D^n the following two conditions are satisfied:

(i) for any stopping time T, $\mathbf{X}^{T-} = \mathbf{Y}^{T-}$ implies $F(\mathbf{X})^{T-} = F(\mathbf{Y})^{T-}$;
(ii) there exists an increasing (finite) process $K = (K_t)_{t \geq 0}$ such that $|F(\mathbf{X})_t - F(\mathbf{Y})_t| \leq K_t \|\mathbf{X} - \mathbf{Y}\|_t^*$ a.s., each $t \geq 0$.

Note that if $g(t,x)$ is a Lipschitz function, then $f(t,x) = g(t-,x)$ is random Lipschitz. A Lipschitz, or a random Lipschitz, function induces a process Lipschitz operator, and if an operator is process Lipschitz, then it is also functional Lipschitz.

An autonomous function with a bounded derivative is Lipschitz by the Mean Value Theorem. If a function f has a continuous but not bounded derivative, f will be **locally Lipschitz**; such functions are defined and considered in Sect. 7 of this chapter.

Let $A = (A_t)_{t \geq 0}$ be continuous and adapted. Then a linear coefficient such as $f(t,\omega,x) = A_t(\omega)x$ is an example of a process Lipschitz coefficient. A functional Lipschitz operator F will typically be of the form $F(X) = f(t,\omega; X_s, s \leq t)$, where f is defined on $[0,t] \times \Omega \times D[0,t]$ for each $t \geq 0$; here $D[0,t]$ denotes the space of càdlàg functions defined on $[0,t]$. Another example is a generalization of the coefficients introduced by Itô and Nisio [1]:

$$F(X)_t = \int_0^t g(u,\omega,X_u)\mu(\omega,du)$$

for a random signed measure μ and a bounded Lipschitz function g with constant $C(\omega)$. In this case the Lipschitz process for F is given by $K_t(\omega) = C(\omega)\|\mu(\omega)_t\|$, where $\|\mu(\omega)_t\|$ denotes the total mass of the measure $\mu(\omega,du)$ on $[0,t]$.

Lemmas 1 and 2 which follow are used to prove Theorem 7. We state and prove them in the 1-dimensional case, their generalizations to n dimensions being simple.

Lemma 1. *Let* $1 \leq p < \infty$, *let* $J \in \underline{\underline{S}}^p$, *let* F *be functional Lipschitz and suppose* $F(0) = 0$, *and that* $\sup_t |K_t(\omega)| \leq k$ *a.s. Let* Z *be a semimartingale* *in* $\underline{\underline{H}}^\infty$ *such that* $\|Z\|_{\underline{\underline{H}}^\infty} \le \frac{1}{2c_p k}$. *Then the equation*

$$X_t = J_t + \int_0^t F(X)_{s-} dZ_s$$

has a solution in $\underline{\underline{S}}^p$, *it is unique, and moreover*

$$\|X\|_{\underline{\underline{S}}^p} \le 2\|J\|_{\underline{\underline{S}}^p}.$$

Proof. Define $\Lambda : \underline{\underline{S}}^p \to \underline{\underline{S}}^p$ by $\Lambda(X)_t = J_t + \int_0^t F(X)_{s-} dZ_s$. Then by Theorems 2 and 3 the operator is $\frac{1}{2}$-Lipschitz, and the fixed point theorem gives existence and uniqueness. Indeed

$$\begin{aligned} \|X\|_{\underline{\underline{S}}^p} &\le \|J\|_{\underline{\underline{S}}^p} + \|\int F(X)_{s-} dZ_s\|_{\underline{\underline{S}}^p} \\ &\le \|J\|_{\underline{\underline{S}}^p} + c_p\|F(X)\|_{\underline{\underline{S}}^p}\|Z\|_{\underline{\underline{H}}^\infty} \\ &\le \|J\|_{\underline{\underline{S}}^p} + \frac{1}{2k}\|F(X)\|_{\underline{\underline{S}}^p}. \end{aligned}$$

Since $\|F(X)\|_{\underline{\underline{S}}^p} = \|F(X) - F(0)\|_{\underline{\underline{S}}^p}$, we have $\|X\|_{\underline{\underline{S}}^p} \le \|J\|_{\underline{\underline{S}}^p} + \frac{1}{2}\|X\|_{\underline{\underline{S}}^p}$, which yields the estimate. □

Lemma 2. *Let* $1 \le p < \infty$, *let* $J \in \underline{\underline{S}}^p$, *let* F *be functional Lipschitz with* $F(0) = 0$ *and* $\sup_t |K_t(\omega)| \le k < \infty$ *a.s. Let* Z *be a semimartingale such that* $Z \in S(\frac{1}{2c_p k})$. *Then the equation*

$$X_t = J_t + \int_0^t F(X)_{s-} dZ_s$$

has a solution in $\underline{\underline{S}}^p$, *it is unique, and moreover* $\|X\|_{\underline{\underline{S}}^p} \le C(k, Z)\|J\|_{\underline{\underline{S}}^p}$, *where* $C(k, Z)$ *is a constant depending only on* k *and* Z.

Proof. Let $z = \|Z\|_{\underline{\underline{H}}^\infty}$ and $j = \|J\|_{\underline{\underline{S}}^p}$. Let $0 = T_0, T_1, \ldots, T_\ell$ be the slicing times for Z, and consider the equations

$$X = J^{T_i-} + \int F(X)_{s-} dZ_s^{T_i-}, \qquad i = 0, 1, 2, \ldots. \tag{i}$$

Equation (0) has the trivial solution $X \equiv 0$ since $J^{0-} = Z^{0-} = 0$ for all t, and its $\underline{\underline{S}}^p$ norm is 0. Assume that equation (i) has a unique solution X^i, and let $x^i = \|X^i\|_{\underline{\underline{S}}^p}$. Stopping next at T_i instead of T_i-, let Y^i denote the unique solution of $Y^i = J^{T_i} + \int F(Y^i)_{s-} dZ_s^{T_i}$, and set $y^i = \|Y^i\|_{\underline{\underline{S}}^p}$. Since $Y^i = X^i + \{\Delta J_{T_i} + F(X^i)_{T_i-}\Delta Z_{T_i}\}1_{[T_i,\infty)}$, we conclude that

$$\begin{aligned} \|Y^i\|_{\underline{\underline{S}}^p} &\le \|X^i\|_{\underline{\underline{S}}^p} + 2\|J\|_{\underline{\underline{S}}^p} + \|F(X^i)\|_{\underline{\underline{S}}^p}\|Z\|_{\underline{\underline{H}}^\infty} \\ &\le x^i + 2j + kx^i z \\ &= x^i(1 + kz) + 2j; \end{aligned}$$

hence

$$(*) \qquad y^i \leq 2j + x^i(1+kz).$$

We set for $U \in \mathbb{D}$, $D_iU = (U - U^{T_i})^{T_{i+1}-}$. Since each solution X of $(i+1)$ satisfies $X^{T_i} = Y^i$ on $[0, T_{i+1})$, we can change the unknown by $U = X - (Y^i)^{T_{i+1}-}$, to get the equations: $U = D_iJ + \int F(Y^i + U)_{s-} dD_iZ_s$; however since $F(Y^i + 0)$ need not be 0 we define $G_i(\cdot) = F(Y^i + \cdot) - F(Y^i)$, and thus the above equation can be equivalently expressed as:

$$U = (D_iJ + \int F(Y^i)_{s-} dD_iZ_s) + \int G_i(U)_{s-} dD_iZ_s.$$

We can now apply Lemma 1 to this equation to find that it has a unique solution in $\underline{\underline{S}}^p$, and its norm u^i is majorized by

$$\begin{aligned} u^i &\leq 2\|D_iJ + \int F(Y^i)_{s-} dD_iZ_s\|_{\underline{\underline{S}}^p} \\ &\leq 2(2j + c_pky^i \frac{1}{2c_pk}) \leq 4j + y^i. \end{aligned}$$

We conclude equation $(i+1)$ has a unique solution in $\underline{\underline{S}}^p$ with norm x^{i+1} dominated by (using (*))

$$x^{i+1} \leq u^i + y^i \leq 4j + 2y^i \leq 8j + 2(1+kz)x^i.$$

Next we iterate from $i = 0$ to $\ell - 1$ to conclude that

$$x^\ell \leq 8\{\frac{(2+2kz)^\ell - 1}{1+2kz}\}j.$$

Finally, since $Z = Z^{T_\ell -}$, we have seen that the equation $X = J + \int F(X)_{s-} dZ_s$ has a unique solution in $\underline{\underline{S}}^p$, and moreover $X = X^\ell + J - J^{T_\ell -}$. Therefore $\|X\|_{\underline{\underline{S}}^p} \leq x^\ell + 2j$, hence $C(k, Z) \leq 2 + 8\{\frac{(2+2kz)^\ell - 1}{1+2kz}\}$. □

Theorem 7. *Given a vector of semimartingales $\mathbf{Z} = (Z^1, \ldots, Z^d)$, $\mathbf{Z}_0 = 0$ processes $J^i \in \mathbb{D}(1 \leq i \leq n)$, and operators F^i_j which are functional Lipschitz $(1 \leq i \leq n; 1 \leq j \leq d)$, then the system of equations*

$$X^i_t = J^i_t + \sum_{j=1}^d \int_0^t F^i_j(\mathbf{X})_{s-} dZ^j_s$$

$(1 \leq i \leq n)$ has a solution in $\mathbb{D}^n$, and it is unique. Moreover if $(J^i)_{i \leq n}$ is a vector of semimartingales, then so is $(X^i)_{i \leq n}$.

Proof. The proof for systems is the same as the proof for one equation provided we take F to be matrix valued and X, J and Z to be vector-valued; hence we give here the proof for $n = d = 1$. Thus we will consider the equation

$$(*) \qquad X_t = J_t + \int_0^t F(X)_{s-} dZ_s.$$

Assume that $\max_{i,j} \sup_t K_t^{i,j}(\omega) \le k < \infty$ a.s. Also, by considering the equation:

$$X_t = \{J_t + \int_0^t F(0)_{s-} dZ_s\} + \int_0^t G(X)_{s-} dZ_s,$$

where $G(X) = F(X) - F(0)$, it suffices to consider the case where $F(0) = 0$. We also need Lemmas 1 and 2 only for $p = 2$. In this case $c_2 = \sqrt{8}$.

Let T be an arbitrarily large stopping time such that $J^{T-} \in \underline{\underline{S}}^2$ and such that $Z^{T-} \in \mathcal{S}(\frac{1}{4\sqrt{8}k})$. Then by Lemma 2 there exists a unique solution in $\underline{\underline{S}}^2$ of:

$$X(T)_t = J_t^{T-} + \int_0^t F(X(T))_{s-} dZ_s^{T-}.$$

By the uniqueness in $\underline{\underline{S}}^2$ one has, for $R > T$, that $X(R)^{T-} = X(T)^{T-}$, and therefore we can define a process X on $\Omega \times [0, \infty)$ by $X = X(T)$ on $[0, T)$. Thus we have existence.

Suppose next Y is another solution. Let S be arbitrarily large such that $(X - Y)^{S-}$ is bounded, and let $R = \min(S, T)$, which can also be taken arbitrarily large. Then X^{R-} and Y^{R-} are both solutions of

$$U = J^{R-} + \int_0^t F(U)_{s-} dZ_s^{R-},$$

and since $Z^{R-} \in \mathcal{S}(\frac{1}{2\sqrt{8}k})$, we know that $X^{R-} = Y^{R-}$ by the uniqueness established in Lemma 2. Thus $X = Y$, and we have uniqueness.

We have assumed that $\max_{i,j} \sup_t K_t^{i,j}(\omega) \le k < \infty$ a.s. By proving existence and uniqueness on $[0, t_0]$, for t_0 fixed, we can reduce the Lipschitz processes $K_t^{i,j}$ to the random constants $K_{t_0}^{i,j}(\omega)$, which we replace with $K(\omega) = \max_{i,j} K_{t_0}^{i,j}(\omega)$. Thus without loss of generality we can assume we have a Lipschitz constant $K(\omega) < \infty$ a.s. Then we can choose a constant c such that $P\{K \le c\} > 0$. Let $\Omega_n = \{K \le c + n\}$, each $n = 1, 2, 3, \ldots$. Define a new probability P_n by $P_n(A) = P(A \cap \Omega_n)/P(\Omega_n)$, on the space Ω_n equipped with the filtration $\mathcal{F}_t^n = \mathcal{F}_t|_{\Omega_n}$, the trace of $\mathcal{F}_t$ on Ω_n. Then it is a simple consequence of the definition of a semimartingale in Chap. II that the restriction of Z in equation (*) to $\Omega_n \times [0, \infty)$ is an $(\mathcal{F}_t^n)_{0 \le t < \infty}$-semimartingale. Let Y^n be the unique solution on Ω_n that we have seen exists. For $m > n$ we have $\Omega_m \supset \Omega_n$ and $P_n \ll P_m$. By Theorem 14 of Chap. II we know P_m-stochastic integrals are P_n indistinguishable from P_n-stochastic integrals, whence Y^m restricted to $\Omega_n \times [0, \infty)$ is a P_n solution on $(\Omega_n, \mathcal{F}_t^n)$. Thus Y^m is P_n-indistinguishable from Y^n, and we can define a solution Y on $\mathbb{R}_+ \times \Omega$ by setting $Y = Y^n$ on $\Omega_n \setminus \Omega_{n-1}$, where $\Omega_n = \{\omega : K(\omega) \leq c + n\}$. □

Theorem 7 can be generalized by weakening the Lipschitz hypothesis on the coefficients. If the coefficients are Lipschitz on compact sets, for example, in general one has unique solutions existing only up to a stopping time T; at this time one has $\overline{\lim}_{t \to T} |\mathbf{X}_t| = \infty$. Such times are called **explosion times**, and they can be finite or infinite. Coefficients that are Lipschitz on compact sets are called *locally Lipschitz*. Simple cases are treated in Sect. 7 of this chapter (cf. Theorems 38–40), where they arise naturally in the study of flows.

We end this section with the remark that we have already met a fundamental stochastic differential equation in Chap. II: that of the **stochastic exponential equation** ($Z_0 = 0$):

$$X_t = X_0 + \int_0^t X_{s-} dZ_s,$$

where we obtained a formula for its solution (thus *a fortiori* establishing the existence of a solution):

$$X_t = X_0 \exp(Z_t - \frac{1}{2}[Z,Z]_t) \prod_{0<s\leq t} (1 + \Delta Z_s) \exp(-\Delta Z_s + \frac{1}{2}(\Delta Z_s)^2).$$

The uniqueness of this solution is a consequence of Theorem 7, or of Theorem 6.

A traditional way to show the existence and uniqueness of solutions of ordinary differential equations is the Picard iteration method; one might well wonder if Picard-type iterations converge in the case of stochastic differential equations. As it turns out, the following theorem is quite useful.

Theorem 8. *Let the hypotheses of Theorem 7 be satisfied, and in addition let $(X^0)^i = H^i$ be processes in $\mathbb{D}$ $(1 \leq i \leq n)$, and define inductively*

$$(X_t^{m+1})^i = J_t^i + \sum_{j=1}^d \int_0^t F_j^i(\mathbf{X}^m)_{s-} dZ_s^j$$

and let $\mathbf{X}$ be the solution of

$$X_t^i = J_t^i + \sum_{j=1}^d \int_0^t F_j^i(\mathbf{X})_{s-} dZ_s^j \; (1 \leq i \leq n).$$

Then $\mathbf{X}^m$ converges to $\mathbf{X}$ in ucp.

Proof. We give the proof for $d = n = 1$. It is easy to see that if $\lim_{m\to\infty} X^m = X$ prelocally in $\underline{\underline{S}}^2$, then $\lim_{m\to\infty} X^m = X$ in ucp; in any event this is proved in Theorem 12 in Sect. 4. Thus we will show that $\lim_{m\to\infty} X^m = X$ prelocally in $\underline{\underline{S}}^2$. We first assume $\sup_t K_t \le a < \infty$ a.s. Without loss of generality we can assume $Z \in \mathcal{S}(\alpha)$, with $\alpha = \frac{1}{2\sqrt{8a}}$, and that $J \in \underline{\underline{S}}^2$. Let $0 = T_0 \le T_1 \le \cdots \le T_k$ be the stopping times that α-slice Z. Then $(X^m)^{T_1-}$ and X^{T_1-} are in $\underline{\underline{S}}^2$ by Lemma 1. Then

$$\|(X^{m+1} - X)^{T_1-}\|_{\underline{\underline{S}}^2} \le \sqrt{8}\|(F(X^m) - F(X))^{T_1-}\|_{\underline{\underline{S}}^2}\|Z^{T_1-}\|_{\underline{\underline{H}}^\infty}$$

by Theorems 2 and 3. Therefore

$$\begin{aligned}\|(X^{m+1} - X)^{T_1-}\|_{\underline{\underline{S}}^2} &\le \frac{1}{2}\|(X^m - X)^{T_1-}\|_{\underline{\underline{S}}^2} \\ &\le \frac{1}{2^m}\|(X^1 - X)^{T_1-}\|_{\underline{\underline{S}}^2}\end{aligned}$$

and therefore $\lim_{m\to\infty}\|(X^{m+1} - X)^{T_1-}\|_{\underline{\underline{S}}^2} = 0$. We next analyze the jump at T_1. Since

$$X^{m+1}_{T_1} = X^{m+1}_{T_1-} + F(X^m)_{T_1-}\Delta Z_{T_1},$$

we have

$$\|(X^{m+1} - X)^{T_1}\|_{\underline{\underline{S}}^2} \le \|(X^{m+1} - X)^{T_1-}\|_{\underline{\underline{S}}^2} + az\|(X^m - X)^{T_1-}\|_{\underline{\underline{S}}^2},$$

where $z = \|Z\|_{\underline{\underline{H}}^\infty} < \infty$. Therefore

$$\|(X^{m+1} - X)^{T_1}\|_{\underline{\underline{S}}^2} \le \frac{1}{2^{m-1}}(1 + az)\|(X^1 - X)^{T_1-}\|_{\underline{\underline{S}}^2}.$$

Next suppose we know that

$$\|(X^{m+1} - X)^{T_\ell}\|_{\underline{\underline{S}}^2} \le \frac{(m+1)^{\ell-1}(1+az)^\ell\gamma}{2^{m-1}}$$

where $\gamma = \|(X^1 - X)^{T_k-}\|_{\underline{\underline{S}}^2}$. Then

$$(X^{m+1}_t - X_t)^{T_{\ell+1}-} = (X^{m+1}_t - X_t)^{T_\ell} + \int_{T_\ell}^t (F(X^m)_{s-} - F(X)_{s-})d\hat{Z}^{\ell+1}_s$$

where $\hat{Z}^{\ell+1} = (Z - Z^{T_\ell})^{T_{\ell+1}-}$. Therefore by iterating on ℓ, $0 \le \ell \le k$,

$$\begin{aligned}\|(X^{m+1} - X)^{T_{\ell+1}-}\|_{\underline{\underline{S}}^2} &\le \|(X^{m+1} - X)^{T_\ell}\|_{\underline{\underline{S}}^2} + a\sqrt{8}\alpha\|(X^m - X)^{T_{\ell+1}-}\|_{\underline{\underline{S}}^2} \\ &\le \frac{(m+1)^{\ell-1}(1+az)^\ell}{2^{m-1}} + a\sqrt{8}\alpha\|(X^m - X)^{T_{\ell+1}-}\|_{\underline{\underline{S}}^2} \\ &\le \frac{(m+1)^{\ell-1}(1+az)^\ell\gamma}{2^{m-1}}.\end{aligned}$$

Note that the above expression tends to 0 as m tends to ∞. Therefore X^m tends to X prelocally in $\underline{\underline{S}}^2$ by a (finite) induction, and hence $\lim_{m\to\infty} X^m = X$ in ucp.

It remains to remove the assumption that $\sup_t K_t \le a < \infty$ a.s. Fix a $t < \infty$; we will show ucp on $[0,t]$. Since t is arbitrary, this will imply the result. As we did at the end of the proof of Theorem 7, let $c > 0$ be such that $P(K_t \le c) > 0$. Define $\Omega_n = \{\omega : K_t(\omega) \le c+n\}$, and $P_n(\Lambda) \equiv P(\Lambda|\Omega_n)$. Then $P_n \ll P$, and under P_n, $\lim_{m\to\infty} X^m = X$ in ucp on $[0,t]$. For $\epsilon > 0$, choose N such that $n \ge N$ implies $P(\Omega_n^c) < \epsilon$. Then

$$P((X^m - X)_t^* > \delta) \le P_n((X^m - X)_t^* > \delta) + \epsilon,$$

hence $\lim_{m\to\infty} P((X^m - X)_t^* > \delta) \le \epsilon$. Since $\epsilon > 0$ was arbitrary, the limit must be zero. □

4. Stability of Stochastic Differential Equations

Since one is never exactly sure of the accuracy of a proposed model, it is important to know how robust the model is; that is, if one perturbs the model a bit, how large are the resulting changes? Stochastic differential equations are stable with respect to perturbations of the coefficients, or of the initial conditions. Perturbations of the differentials, however, is a more delicate matter: One must perturb the differentials in the right way to have stability. Not surprisingly, an $\underline{\underline{H}}^p$ perturbation is the right kind of perturbation. In this section we will be concerned with equations of the form:

$$(*n) \qquad X_t^n = J_t^n + \int_0^t F^n(X^n)_{s-} dZ_s^n,$$

$$(*) \qquad X_t = J_t + \int_0^t F(X)_{s-} dZ_s,$$

where J^n, J are in $\mathbf{D}$, Z^n, Z are semimartingales, and F^n, F are functional Lipschitz, with Lipschitz processes K_n, K respectively. *We will assume that the Lipschitz processes K_n, K are each uniformly bounded by the same constant, and that the semimartingale differentials Z^n, Z are always zero at* 0 (that is, $Z_0^n = 0$ a.s., $n \ge 1$, and $Z_0 = 0$ a.s.).

For simplicity we state and prove the theorems in this section for one equation (rather than for finite systems), with one semimartingale driving term (rather than a finite number), and for $p = 2$. The generalizations are obvious, and the proofs are exactly the same except for notation. We say a functional Lipschitz operator F is **bounded** if for all $H \in \mathbf{D}$, there exists a non-random constant $c < \infty$ such that $F(H)^* < c$.

Theorem 9. *Let J, $J^n \in \mathbb{D}$, Z, Z^n be semimartingales; and F, F^n be functional Lipschitz with constants K, K_n respectively. Assume:*

(i) J, J^n are in $\underline{\underline{S}}^2$ (respectively $\underline{\underline{H}}^2$) and $\lim_{n\to\infty} J^n = J$ in $\underline{\underline{S}}^2$ (resp. $\underline{\underline{H}}^2$);

(ii) F^n are all bounded by the same constant c, and $\lim_{n\to\infty} F^n(X) = F(X)$ in $\underline{\underline{S}}^2$, where X is the solution of (*);

(iii) $\max(\sup_n K_n, K) \leq a < \infty$ a.s. (a not random); $Z \in \mathcal{S}(\frac{1}{2\sqrt{8a}})$; $(Z^n)_{n\geq 1}$ are in $\underline{\underline{H}}^2$, and $\lim_{n\to\infty} Z^n = Z$ in $\underline{\underline{H}}^2$.[2]

*Then $\lim_{n\to\infty} X^n = X$ in $\underline{\underline{S}}^2$ (respectively in $\underline{\underline{H}}^2$), where X^n is the solution of (*n) and X is the solution of (*).*

Proof. We use the notation $H \cdot Z_t$ to denote $\int_0^t H_s dZ_s$, and $H \cdot Z$ to denote the process $(H \cdot Z_t)_{t\geq 0}$. We begin by supposing that J, $(J^n)_{n\geq 1}$ are in $\underline{\underline{S}}^2$ and J^n converges to J in $\underline{\underline{S}}^2$. Then

$$X - X^n = J - J^n + (F(X) - F^n(X))_- \cdot Z$$
$$+ (F^n(X) - F^n(X^n))_- \cdot Z + F^n(X^n)_- \cdot (Z - Z^n).$$

Let $Y^n = (F(X) - F^n(X))_- \cdot Z + F^n(X^n) \cdot (Z - Z^n)$. Then

$$(**) \qquad X - X^n = J - J^n + Y^n + (F^n(X) - F^n(X^n))_- \cdot Z.$$

For $U \in \mathbb{D}$ define G^n by:

$$G^n(U) = F^n(X) - F^n(X - U).$$

Then $G^n(U)$ is functional Lipschitz with constant a and $G^n(0) = 0$. Take $U = X - X^n$, and (**) becomes

$$U = (J - J^n + Y^n) + G^n(U)_- \cdot Z.$$

By Lemma 2 preceding Theorem 7 we have

$$\|X - X^n\|_{\underline{\underline{S}}^2} \leq C(a, Z)\|J - J^n + Y^n\|_{\underline{\underline{S}}^2} \leq C(a, Z)\{\|J - J^n\|_{\underline{\underline{S}}^2} + \|Y^n\|_{\underline{\underline{S}}^2}\}.$$

Since $C(a, Z)$ is independent of n and $\lim_{n\to\infty}\|J - J^n\|_{\underline{\underline{S}}^2} = 0$ by hypothesis, it suffices to show $\lim_{n\to\infty}\|Y^n\|_{\underline{\underline{S}}^2} = 0$. But

$$\|Y^n\|_{\underline{\underline{S}}^2} \leq \|(F(X) - F^n(X))_- \cdot Z\|_{\underline{\underline{S}}^2} + \|F^n(X^n)_- \cdot (Z - Z^n)\|_{\underline{\underline{S}}^2}$$
$$(***) \qquad \leq \sqrt{8}\|F(X) - F^n(X)\|_{\underline{\underline{S}}^2}\|Z\|_{\underline{\underline{H}}^\infty}$$
$$+ \sqrt{8}\|F^n(X^n)_-\|_{\underline{\underline{S}}^\infty}\|Z - Z^n\|_{\underline{\underline{H}}^2}$$

[2] $\mathcal{S}(\alpha)$ is defined in Sect. 2 of this chapter (page 192).

by Theorem 2 and Emery's inequality (Theorem 3). Since $\|Z\|_{\underline{\underline{H}}^\infty} < \infty$ by hypothesis and since

$$\lim_{n\to\infty} \|F(X) - F^n(X)\|_{\underline{\underline{S}}^2} = \lim_{n\to\infty} \|Z - Z^n\|_{\underline{\underline{H}}^2} = 0,$$

again by hypothesis, we are done.

Note that if we knew J^n, $J \in \underline{\underline{H}}^2$ and that J^n converged to J in $\underline{\underline{H}}^2$, then

$$\|X - X^n\|_{\underline{\underline{H}}^2} \leq \|J - J^n\|_{\underline{\underline{H}}^2} + \|Y^n\|_{\underline{\underline{H}}^2} + C\|X - X^n\|_{\underline{\underline{S}}^2}\|Z\|_{\underline{\underline{H}}^\infty};$$

we have seen already that $\lim_{n\to\infty}\|X - X^n\|_{\underline{\underline{S}}^2} = 0$, hence it suffices to show $\lim_{n\to\infty}\|Y^n\|_{\underline{\underline{H}}^2} = 0$. Proceeding as in (***) we obtain the result. □

Comment. Hypothesis (ii) in Theorem 9 seems very strong; it is the perturbation of the semimartingale differentials that make it necessary. Indeed, the hypothesis cannot be relaxed in general, as the following example shows. We take $\Omega = [0,1]$, P to be Lebesgue measure on $[0,1]$, and $(\mathcal{F}_t)_{t\geq 0}$ equal to $\mathcal{F}$, the Lebesgue sets on $[0,1]$. Let $\phi(t) = \min(t,1)$, $t \geq 0$. Let $f_n(\omega) \geq 0$ and set $Z_t^n(\omega) = \phi(t)f_n(\omega)$, $\omega \in [0,1]$, and $Z_t(\omega) = \phi(t)f(\omega)$. Let $F^n(X) = F(X) \equiv X$, and finally let $J_t^n = J_t = 1$, all $t \geq 0$. Thus the equations (*n) and (*) become respectively

$$X_t^n = 1 + \int_0^t X_{s-}^n dZ_s^n$$

$$X_t = 1 + \int_0^t X_{s-} dZ_s,$$

which are elementary continuous exponential equations and have solutions:

$$X_t^n = \exp(Z_t^n) = \exp(f_n(\omega)\phi(t)),$$

$$X_t = \exp(Z_t) = \exp(f(\omega)\phi(t)).$$

We can choose f_n such that $\lim_{n\to\infty} E\{f_n^2\} = 0$ but $\lim_{n\to\infty} E\{f_n^p\} \neq 0$ for $p > 2$. Then the Z^n converge to 0 in $\underline{\underline{H}}^2$ but X^n does not converge to $X = 1$ (since $f = 0$) in $\underline{\underline{S}}^p$, for any $p \geq 1$. Indeed, $\lim_{n\to\infty} E\{f_n^p\} \neq 0$ for $p > 2$ implies $\lim_{n\to\infty} E\{e^{tf_n}\} \neq 1$ for any $t > 0$.

The next result *does not require that the coefficients be bounded*, because there is only one, fixed, semimartingale differential. Theorems 10, 11 and 13 all have $\underline{\underline{H}}^2$ as well as $\underline{\underline{S}}^2$ versions as in Theorem 9, but we state and prove only the $\underline{\underline{S}}^2$ versions.

Theorem 10. *Let $J, J^n \in \mathbb{D}$; Z be a semimartingale, F, F^n be functional Lipschitz with constants K, K_n, and let X^n, X be the unique solutions of equations (*n) and (*). Assume:*

(i) J^n, J are in $\underline{\underline{S}}^2$ and $\lim_{n\to\infty} J^n = J$ in $\underline{\underline{S}}^2$;

(ii) $\lim_{n\to\infty} F^n(X) = F(X)$ in $\underline{\underline{S}}^2$, where X is the solution of (*);

(iii) $\max(\sup_n K_n, K) \le a < \infty$ a.s. for a non-random constant a, and $Z \in \mathcal{S}(\frac{1}{2\sqrt{8a}})$.

*Then $\lim_{n\to\infty} X^n = X$ in $\underline{\underline{S}}^2$ where X^n is the solution of (*n) and X is the solution of (*).*

Proof. Let X^n and X be the solutions of equations (*n) and (*), respectively. Then

$$X - X^n = J - J^n + (F(X) - F^n(X))_- \cdot Z + (F^n(X) - F^n(X^n))_- \cdot Z.$$

We let $Y^n = (F(X) - F^n(X))_- \cdot Z$, and we define a new functional Lipschitz operator G^n by

$$G^n(U) = F^n(X) - F^n(X - U).$$

Then $G^n(0) = 0$. If we set $U = X - X^n$, we obtain the equation

$$U = J - J^n + Y^n + G^n(U)_- \cdot Z.$$

Since $Z \in \underline{\underline{H}}^\infty$, by Emery's inequality (Theorem 3) we have Y^n tends to 0 in $\underline{\underline{H}}^2$, hence also in $\underline{\underline{S}}^2$ (Theorem 2). In particular $\|Y^n\|_{\underline{\underline{S}}^2} < \infty$, and therefore by Lemma 2 in Sect. 3 we have

$$\|U\|_{\underline{\underline{S}}^2} \le C(a, Z)\|J - J^n + Y^n\|_{\underline{\underline{S}}^2},$$

where $C(a, Z)$ is independent of n, and where the right side tends to zero as n tends to ∞. Since $U = X - X^n$, we are done. □

We now wish to localize the results of Theorems 9 and 10 so that they hold for general semimartingales and exogenous processes J^n, J. We first need a definition, which is consistent with our previous definitions of properties holding locally and prelocally (defined in Chap. IV, Sect. 2).

Definition. Processes M^n are said to converge locally (respectively prelocally) in $\underline{\underline{S}}^p$ (respectively $\underline{\underline{H}}^p$) to M if M^n, M are in $\underline{\underline{S}}^p$ (resp. $\underline{\underline{H}}^p$) and if there exists a sequence of stopping times T_k increasing to ∞ a.s. such that $\lim_{n\to\infty}\|(M^n - M)^{T_k} 1_{\{T_k>0\}}\|_{\underline{\underline{S}}^p} = 0$ (resp. $\lim_{n\to\infty}\|(M^n - M)^{T_k-}\|_{\underline{\underline{S}}^p} = 0$) for each $k \ge 1$ (resp. $\underline{\underline{S}}^p$ replaced by $\underline{\underline{H}}^p$).

Theorem 11. *Let J, $J^n \in \mathbb{D}$; Z be a semimartingale ($Z_0 = 0$); and F, F^n be functional Lipschitz with Lipschitz processes K, K_n respectively. Let X^n, X be solutions of*

$$(*\text{n}) \qquad X_t^n = J_t^n + \int_0^t F^n(X^n)_{s-} dZ_s$$

$$(*) \qquad X_t = J_t + \int_0^t F(X)_{s-} dZ_s.$$

Assume:

(i) J^n *converge to* J *prelocally in* $\underline{\underline{S}}^2$;

(ii) $F^n(X)$ *converges to* $F(X)$ *prelocally in* $\underline{\underline{S}}^2$ *where* X *is the solution of* (*);

(iii) $\max(\sup_n K_n, K) \leq a < \infty$ *a.s.* (*a not random*);

Then $\lim_{n\to\infty} X^n = X$ *prelocally in* $\underline{\underline{S}}^2$ *where* X^n *is the solution of* (*n) *and* X *is the solution of* (*).

Proof. By stopping at $T-$ for an arbitrarily large stopping time T we can assume without loss of generality that $Z \in \mathcal{S}(\frac{1}{2\sqrt{8a}})$ by Theorem 5, and that J^n converges to J in $\underline{\underline{S}}^2$ and $F(X^n)$ converges to $F(X)$ in $\underline{\underline{S}}^2$, by hypothesis. Next we need only to apply Theorem 10. □

We can recast Theorem 11 in terms of convergence in ucp (uniform convergence on compacts, in probability), which we introduced in Sect. 4 of Chap. II in order to develop the stochastic integral.

Corollary. *Let* J^n, $J \in \mathbb{D}$; Z *be a semimartingale* ($Z_0 = 0$); *and* F, F^n *be functional Lipschitz with Lipschitz processes* K, K_n *respectively. Let* X, X^n *be as in Theorem 11. Assume:*

(i) J^n *converges to* J *in ucp;*

(ii) $F^n(X)$ *converges to* $F(X)$ *in ucp;*

(iii) $\max(\sup_n K_n, K) \leq a < \infty$ *a.s.* (*a not random*).

Then $\lim_{n\to\infty} X^n = X$ *in ucp.*

Proof. Recall that convergence in ucp is metrizable; let d denote a distance compatible with it. If X^n does not converge to 0 in ucp, we can find a subsequence n' such that $\inf_{n'} d(X^{n'}, 0) > 0$. Therefore no sub-subsequence $(X^{n''})$ can converge to 0 in ucp, and hence $X^{n''}$ cannot converge to 0 prelocally in $\underline{\underline{S}}^2$ as well. Therefore to establish the result we need to show only that for any subsequence n', there exists a further subsequence n'' such that $X^{n''}$ converges prelocally to 0 in $\underline{\underline{S}}^2$. This is the content of Theorem 12 which follows, so the proof is complete. □

Theorem 12. *Let* H^n, $H \in \mathbb{D}$. *For* H^n *to converge to* H *in ucp it is necessary and sufficient that there exist a subsequence* n' *such that* $\lim_{n'\to\infty} H^{n'} = H$, *prelocally in* $\underline{\underline{S}}^2$.

Proof. We first show the necessity. Without loss of generality, we assume that $H = 0$. We construct by iteration a decreasing sequence of subsets $(\mathbb{N}_k)$ of $\mathbb{N} = \{1, 2, 3, \dots\}$, such that

$$\lim_{\substack{n\to\infty \\ n\in\mathbb{N}_k}} \sup_{0\leq s\leq k} |H_s^n| = 0 \quad \text{a.s.}$$

By Cantor's diagonalization procedure we can find an infinite subset $\mathbb{N}'$ of $\mathbb{N}$ such that

$$\lim_{\substack{n\to\infty \\ n\in\mathbb{N}'}} \sup_{0\le s\le k} |H_s^n| = 0 \quad \text{a.s.,}$$

each integer $k > 0$. By replacing $\mathbb{N}$ with $\mathbb{N}'$ we can assume without loss of generality that H^n tends to 0 uniformly on compacts, *almost surely*. We next define:

$$T_n = \inf\{t \ge 0 : |H_t^n| \ge 1\}$$
$$S_n = \inf_{m\ge n} T_m.$$

Then T_n and S_n are stopping times and the S_n increase to ∞ a.s. Indeed, for each k there exists $N(\omega)$ such that for $n \ge N(\omega) \sup_{0\le s\le k}|H_s^n(\omega)| < 1$; hence $T_n(\omega) \ge k$ and $S_{N(\omega)}(\omega) \ge k$. Next, define

$$L^m = (H^m)^{(S_n\wedge n)-};$$

then

$$(L^m)^* \le \sup_{0\le s\le n} |H_s^m|,$$

which tends to 0 a.s. Moreover, when $m \ge n$,

$$(L^m)^* \le \sup_{0\le s<S_n} |H_s^m| \le \sup_{0\le s<T_m} |H_s^m| \le 1,$$

hence by the dominated convergence theorem we have that $(L^m)^*$ tends to 0 in L^2, which implies the result.

For the sufficiency, suppose that H^n does not converge to H in ucp. Then there exist $t_0 > 0$, $\delta > 0$, $\epsilon > 0$, and a subsequence n' such that $P\{(H^{n'} - H)_{t_0}^* > \delta\} > \epsilon$ for each n'. Then $\|(H^{n'} - H)_{t_0}^*\|_{L^2} \ge \delta\sqrt{\epsilon}$ for all n'. Let T^k tend to ∞ a.s. such that $(H^{n'} - H)^{T^k-}$ tends to 0 in $\underline{\underline{S}}^2$, each k. Then there exists $K > 0$ such that $P(T^k < t_0) < \frac{\delta\sqrt{\epsilon}}{2}$ for all $k > K$. Hence

$$\varlimsup_{n'\to\infty} \|(H^n - H)_{t_0}^*\|_{L^2} \le \varlimsup_{n'\to\infty} \|(H^{n'} - H)^{T^k-}\|_{\underline{\underline{S}}^2} + \frac{\delta\sqrt{\epsilon}}{2}.$$

We conclude $\delta\sqrt{\epsilon} \le \frac{\delta\sqrt{\epsilon}}{2}$, a contradiction; whence H^n converges to H in ucp. □

Recall that we have stated and proven our theorems for the simplest case of one equation (rather than finite systems) and one semimartingale driving term (rather than a finite number). The extensions to systems and several driving terms is simple and essentially only an exercise in notation. We leave this to the reader.

An interesting consequence of the preceding results is that prelocal $\underline{\underline{S}}^p$ and prelocal $\underline{\underline{H}}^p$ convergence *are not topological* in the usual sense. If they were, then one would have that a sequence converged to zero if and only if every subsequence had a sub-subsequence that converged to zero. To see that this is not the case for $\underline{\underline{S}}^2$ for instance, consider the example given in the Comment following Theorem 9. In this situation, the solutions X^n converge to X in ucp. By Theorem 12, this implies the existence of a subsequence n' such that $\lim_{n'\to\infty} X^{n'} = X$, prelocally in $\underline{\underline{S}}^2$. However we saw in the comment that X^n does not converge to X in $\underline{\underline{S}}^2$. It is still *a priori* possible that X^n converges to X *prelocally* in $\underline{\underline{S}}^2$, however. In the framework of the example a stopping time is simply a nonnegative random variable. Thus our counter-example is complete with the following real analysis result (see Protter [**4**, p. 344] for a proof): There exist nonnegative functions f_n on $[0,1]$ such that $\lim_{n\to\infty} \int_0^1 f_n(x)^2 dx = 0$ and $\limsup_{n\to\infty} \int_\Lambda (f_n(x))^p dx = +\infty$ for all $p > 2$ and all Lebesgue sets Λ with strictly positive Lebesgue measure. In conclusion, this counterexample gives a sequence of semimartingales X^n such that every subsequence has a sub-subsequence converging prelocally in $\underline{\underline{S}}^p$, but the sequence itself does not converge prelocally in $\underline{\underline{S}}^p$ $(1 \le p < \infty)$.

Finally we observe that such non-topological convergence is not as unusual as one might think at first: Indeed, let X_n be random variables which converge to zero in probability but not a.s. Then every subsequence has a sub-subsequence which converges to zero a.s., and thus almost sure convergence is also not topological in the usual sense.

As an example of Theorem 10, let us consider the equations

$$X_t^n = J_t^n + \int_0^t F^n(X^n)_s dW_s + \int_0^t G^n(X^n)_s ds$$

$$X_t = J_t + \int_0^t F(X)_s dW_s + \int_0^t G(X)_s ds$$

where W is a standard Wiener process. If $(J^n - J)_t^*$ converges to 0 in L^2, each $t > 0$; and if F^n, G^n, F, G are all functional Lipschitz with constant $K < \infty$ and are such that $(F^n(X) - F(X))_t^*$ and $(G^n(X) - G(X))_t^*$ converge to 0 in L^2, each $t > 0$; then $(X^n - X)_t^*$ converges to 0 in L^2 as well, each $t > 0$. Note that we require only that $F^n(X)$ and $G^n(X)$ converge respectively to $F(X)$ and $G(X)$ for the one X that is the solution, and not for all processes in $\mathbb{D}$.

One can weaken the hypothesis of Theorem 9 and still let the differentials vary, provided the coefficients stay bounded, as the next theorem shows.

Theorem 13. *Let J^n, $J \in \mathbb{D}$; Z, Z^n be semimartingales ($Z_0^n = Z_0 = 0$ a.s.); and F, F^n be functional Lipschitz with Lipschitz processes K, K_n respectively. Let X^n, X be solutions of $(*^n)$ and $(*)$. Assume:*

(i) J^n converges to J prelocally in $\underline{\underline{S}}^2$;

(ii) $F^n(X)$ converges to $F(X)$ prelocally in $\underline{\underline{S}}^2$, and the coefficients F^n, F are all bounded by $c < \infty$;

(iii) Z^n *converges to* Z *prelocally in* $\underline{\underline{H}}^2$;
(iv) $\max(\sup_n K_n, K) \leq a < \infty$ *a.s. (a not random).*

Then $\lim_{n\to\infty} X^n = X$ *prelocally in* $\underline{\underline{S}}^2$.

Proof. By stopping at $T-$ for an arbitrarily large stopping time T we can assume without loss that $Z \in \mathcal{S}(\frac{1}{2\sqrt{8a}})$ by Theorem 5, and that: J^n converges to J in $\underline{\underline{S}}^2$, $F^n(X)$ converges in $\underline{\underline{S}}^2$ to $F(X)$, and Z^n converges to Z in $\underline{\underline{H}}^2$, all by hypothesis. We then invoke Theorem 9, and the proof is complete. □

The assumptions of prelocal convergence are a bit awkward. This type of convergence, however, leads to a topology on the space of semimartingales which is the natural topology for convergence of semimartingale differentials, just as ucp is the natural topology for processes related to stochastic integration. This is exhibited in Theorem 15.

Before defining a topology on the space of semimartingales, let us recall that we can define a "distance" on $\mathbb{D}$ by setting, for Y, $Z \in \mathbb{D}$:

$$r(Y) = \sum_{n>0} 2^{-n} E\{1 \wedge \sup_{0\leq t\leq n} |Y_t|\},$$

and $d(Y, Z) = r(Y - Z)$. This distance is compatible with uniform convergence on compacts in probability, and it was previously defined in Sect. 4 of Chap. II.

Using stochastic integration we can define, for a semimartingale X,

$$\hat{r}(X) = \sup_{|H|\leq 1} r(H \cdot X)$$

where the supremum is taken over all predictable processes bounded by one; then the **semimartingale topology** is defined by the distance $\hat{d}(X, Y) = \hat{r}(X - Y)$.

The semimartingale topology can be shown to make the space of semimartingales a topological vector space which is complete. Furthermore, the following theorem relates the semimartingale topology to convergence in $\underline{\underline{H}}^p$. For its proof and a general treatment of the semimartingle topology, see Émery [2] or Protter [6].

Theorem 14. *Let* $1 \leq p < \infty$, *let* X^n *be a sequence of semimartingales, and let* X *be a semimartingale.*

(i) If X^n *converges to* X *in the semimartingale topology, then there exists a subsequence which converges prelocally in* $\underline{\underline{H}}^p$;
(ii) If X^n *converges to* X *prelocally in* $\underline{\underline{H}}^p$, *then it converges to* X *in the semimartingale topology.*

In Chap. IV we established the equivalence of the norms $\|X\|_{\underline{\underline{H}}^2}$ and $\sup_{|H|\leq 1}\|H \cdot X\|_{\underline{\underline{S}}^2}$ in the Corollary to Theorem 24 in Sect. 2 of Chap. IV.

Given this result, Theorem 14 can be seen as a uniform version of Theorem 12.

We are now able once again to recast a result in terms of ucp convergence. Theorem 13 has the following corollary.

Corollary. *Let J^n, $J \in \mathbb{D}$; Z^n, Z be semimartingales ($Z_0^n = Z_0 = 0$); and F^n, F be functional Lipschitz with Lipschitz processes K, K_n respectively. Let X^n, X be solutions of (*n) and (*), respectively. Assume:*

(i) J^n converges to J in ucp;
(ii) $F^n(X)$ converges to $F(X)$ in ucp where X is the solution of (*), and moreover all the coefficients F^n are bounded by a random $c < \infty$;
(iii) Z^n converges to Z in the semimartingale topology;
(iv) $\max(\sup_n K_n, K) \le a < \infty$ a.s.;

Then $\lim_{n\to\infty} X^n = X$ in ucp.

Proof. Since Z^n converges to Z in the semimartingale topology, by Theorem 14 there exists a subsequence n' such that $Z^{n'}$ converges to Z prelocally in $\underline{\underline{H}}^2$. Then by passing to further subsequences if necessary, by Theorem 12 we may assume without loss that J^n converges to J and $F^n(X)$ converges to $F(X)$ both prelocally in $\underline{\underline{S}}^2$, where X is the solution of (*). Therefore by Theorem 13 X^n converges to X prelocally in $\underline{\underline{S}}^2$ for this subsequence. We have shown that for the sequence (X^n) there is always a subsequence that converges prelocally in $\underline{\underline{S}}^2$. We conclude by Theorem 12 that X^n converges to X in ucp. □

The next theorem extends Theorem 9 and the preceding corollary by relaxing the hypotheses on convergence and especially the hypothesis that all the coefficients be bounded.

Theorem 15. *Let J^n, $J \in \mathbb{D}$; Z^n, Z be semimartingales ($Z_0^n = Z_0 = 0$); and F^n, F be functional Lipschitz with Lipschitz process K, the same for all n. Let X^n, X be solutions respectively of*

$$(*n) \qquad X_t^n = J_t^n + \int_0^t F^n(X^n)_{s-} dZ_s^n$$

$$(*) \qquad X_t = J_t + \int_0^t F(X)_{s-} dZ_s.$$

Assume:

(i) J^n converges to J in ucp;
(ii) $F^n(X)$ converges to $F(X)$ in ucp, where X is the solution of (*);
(iii) Z^n converges to Z in the semimartingale topology.

Then X^n converges to X in ucp.

Proof. First we assume that $\sup_t K_t(\omega) \leq a < \infty$; we remove this hypothesis at the end of the proof. By Theorem 12, it suffices to show that there exists a subsequence n' such that $X^{n'}$ converges to X prelocally in $\underline{\underline{S}}^2$. Then by Theorem 12 we can assume with loss of generality, by passing to a subsequence if necessary, that J^n converges to J and $F^n(X)$ converges to $F(X)$ both prelocally in $\underline{\underline{S}}^2$. Moreover by Theorem 14 we can assume without loss, again by passing to a subsequence if necessary, that Z^n converges to Z prelocally in $\underline{\underline{H}}^2$, and that $Z \in \mathcal{S}(\frac{1}{4\sqrt{8}a})$. Thus all the hypotheses of Theorem 13 are satisfied except one: We do not assume that the coefficients F^n are bounded. However by pre-stopping we can assume without loss that $|F(X)|$ is uniformly bounded by a constant $c < \infty$.

Let us introduce *truncation operators* T^x defined by (for $x \geq 0$):

$$T^x(Y) = \min(x, \sup(-x, Y)).$$

Then T^x is functional Lipschitz with Lipschitz constant 1, for each $x \geq 0$. Consider the equations

$$Y_t^n = J_t^n + \int_0^t (T^{a+c+1}F^n)(Y^n)_{s-} dZ_s^n.$$

Then by Theorem 13 Y^n converges to X prelocally in $\underline{\underline{S}}^2$. By passing to yet another subsequence, if necessary, we may assume that $F^n(X)$ tends to $F(X)$ and Y^n tends to X uniformly on compacts almost surely. Next we define:

$$S^k = \inf\{t \geq 0 : \text{ there exists } m \geq k : |Y_t^m - X_t| + |F^m(X)_t - F(X)_t| \geq 1\}.$$

The stopping times S^k increase a.s. to ∞. By stopping at S^k-, we have for $n \geq k$ that $(Y^n - X)^*$ and $(F^n(X) - F(X))^*$ are a.s. bounded by 1. (Note that stopping at S^k- changes Z to being in $\mathcal{S}(\frac{1}{2\sqrt{8}a})$ instead of $\mathcal{S}(\frac{1}{4\sqrt{8}a})$ by Theorem 5.) Observe that:

$$\begin{aligned} |F^n(Y^n)| &\leq |F^n(Y^n) - F^n(X)| + |F^n(X) - F(X)| + |F(X)| \\ &\leq a(Y^n - X)^* + (F^n(X) - F(X))^* + F(X)^* \\ &\leq a + 1 + c, \end{aligned}$$

whence $(T^{a+c+1}F^n)(Y^n) = F^n(Y^n)$. We conclude that, for an arbitrarily large stopping time R, with J^n and Z^n stopped at R, Y^n is a solution of

$$Y_t^n = J_t^n + \int_0^t F^n(Y^n)_{s-} dZ_s^n,$$

which is equation ($^*n$). By the uniqueness of solutions we deduce $Y^n = X^n$ on $[0, R)$. Since Y^n converges to X prelocally in $\underline{\underline{S}}^2$, we thus conclude X^n converges to X prelocally in $\underline{\underline{S}}^2$.

It remains only to remove the hypothesis that $\sup_t K_t(\omega) \leq a < \infty$. Since we are dealing with local convergence, it suffices to consider $\sup_{s\leq t} K_s \leq K_t$, for a fixed t. Since $K_t < \infty$ a.s., let $\alpha > 0$ be such that $P(K_t \leq \alpha) > 0$, and define

$$\Omega_m = \{\omega : K_t(\omega) \leq \alpha + m\}.$$

Then Ω_m increase to Ω a.s. and as in the proof of Theorem 7 we define P_m by $P_m(A) = P(A|\Omega_m)$, and define $\mathcal{F}_t^m = \mathcal{F}_t|_{\Omega_m}$, the trace of $\mathcal{F}_t$ on Ω_m. Then $P_m \ll P$, hence by Lemma 2 preceding Theorem 25 in Chap. IV, if Z^n converges to Z prelocally in $\underline{\underline{H}}^2(P)$, then Z^n converges to Z prelocally in $\underline{\underline{H}}^2(P_m)$ as well. Therefore by the first part of this proof, X^n converges to X in ucp under P_m, each $m \geq 1$.

Choose $\epsilon > 0$ and m so large that $P(\Omega_m^c) < \epsilon$. Then

$$\begin{aligned}\lim_{n\to\infty} P((X^n - X)_t^* > \delta) &\leq \lim_{n\to\infty} P_m((X^n - X)_t^* > \delta) + P(\Omega_m^c) \\ &\leq \lim_{n\to\infty} P_m((X^n - X)_t^* > \delta) + \epsilon \\ &\leq \epsilon,\end{aligned}$$

and since $\epsilon > 0$ was arbitrary, we conclude that X^n converges to X in ucp on $[0,t]$. Finally since t was arbitrary, we conclude X^n converges to X in ucp. □

Another important topic is how to approximate solutions by difference solutions. Our preceding convergence results yield two consequences. (Theorem 16 and its Corollary).

The next lemma is a type of dominated convergence theorem for stochastic integrals and it is used in the proof of Theorem 16.

Lemma (Dominated Convergence). *Let p, q, r be given such that $\frac{1}{p}+\frac{1}{q} = \frac{1}{r}$, where $1 < r < \infty$. Let Z be a semimartingale in $\underline{\underline{H}}^q$, and let $H^n \in \underline{\underline{S}}^p$ such that $|H^n| \leq Y \in \underline{\underline{S}}^p$, all $n \geq 1$. Suppose $\lim_{n\to\infty} H_{t-}^n(\omega) = 0$, all (t,ω). Then $\lim_{n\to\infty} \| \int H_{s-}^n dZ_s \|_{\underline{\underline{H}}^r} = 0$.*

Proof. Since $Z \in \underline{\underline{H}}^q$, there exists a decomposition of Z, $Z = N + A$, such that

$$j_q(N,A) = \| [N,N]_\infty^{1/2} + \int_0^\infty |dA_s| \|_{L^q} < \infty.$$

Let C^n be the random variable given by

$$C^n = (\int_0^\infty (H_{s-}^n)^2 d[N,N]_s)^{1/2} + \int_0^\infty |H_{s-}^n||dA_s|.$$

The hypothesis that $|H^n| \leq Y$ implies

$$C^n \leq Y^*([N,N]_\infty^{1/2} + \int_0^\infty |dA_s|) \quad \text{a.s.}$$

However

$$\|Y^*([N,N]_\infty^{1/2}+\int_0^\infty |dA_s|)\|_{L^r}\le \|Y^*\|_{L^p}\|[N,N]_\infty^{1/2}+\int_0^\infty |dA_s|\|_{L^q}$$
$$=\|Y\|_{\underline{\underline{S}}^p}j_q(N,A)<\infty.$$

Thus C^n is dominated by a random variable in L^r and hence by the dominated convergence theorem C^n tends to 0 in L^r. □

We let σ_n *denote a sequence of random partitions tending to the identity.*[3] Recall that for a process Y and a random partition $\sigma=\{0=T_0\le T_1\le \cdots\le T_{k_n}\}$, we define:

$$Y^\sigma\equiv Y_0 1_{\{0\}}+\sum_k Y_{T_k}1_{(T_k,T_{k+1}]}.$$

Note that if Y is adapted, càdlàg (i.e, $Y\in \mathbb{D}$), then $(Y_s^\sigma)_{s\ge 0}$ is left continuous with right limits (and adapted, of course). It is convenient to have a version of $Y^\sigma\in\mathbb{D}$, occasionally, so we define

$$Y^{\sigma+}=\sum_k Y_{T_k}1_{[T_k,T_{k+1})}.$$

Theorem 16. *Let $J\in\underline{\underline{S}}^2$, let F be process Lipschitz with Lipschitz process $K\le a<\infty$ a.s. and $F(0)\in\underline{\underline{S}}^2$. Let Z be a semimartingale in $\mathcal{S}(\frac{1}{2\sqrt{8a}})$, and let $X(\sigma)$ be the solution of*

$$(*\sigma)\qquad X_t=J_t+\int_0^t F(X^{\sigma+})_s^\sigma dZ_s,$$

for a random partition σ. If σ_n is sequence of random partitions tending to the identity, then $X(\sigma_n)$ tends to X in $\underline{\underline{S}}^2$, where X is the solution of (*).

Proof. For the random partition σ_n (n fixed), define an operator G^n on $\mathbb{D}$ by:

$$G^n(H)=F(H^{\sigma_n+})^{\sigma_n+}.$$

Note that $G^n(H)\in\mathbb{D}$ for each $H\in\mathbb{D}$ and that $G^n(H)_-=F(H^{\sigma_n+})^{\sigma_n}$. Then G^n is functional Lipschitz with constant K and sends $\underline{\underline{S}}^2$ into itself, as the reader can easily verify. Since $F(0)\in\underline{\underline{S}}^2$, so also are $G^n(0)\in\underline{\underline{S}}^2$, $n\ge 1$, and an argument analogous to the proof of Theorem 10 (though a bit simpler) shows that it suffices to show that $\int_0^t G^n(X)_{s-}dZ_s$ converges to $\int_0^t F(X)_{s-}dZ_s$ in $\underline{\underline{S}}^2$, where X is the solution of

$$X_t=J_t+\int_0^t F(X)_{s-}dZ_s.$$

[3] Random partitions tending to the identity are defined in Chap. II, preceding Theorem 21 (page 57).

Towards this end, fix (t,ω) with $t > 0$, and choose $\epsilon > 0$. Then there exist $\delta > 0$ such that $|X_u(\omega) - X_{t-}(\omega)| < \epsilon$ for all $u \in [t-2\delta, t)$. If mesh$(\sigma) < \delta$, then also $|X_u^{\sigma+}(\omega) - X_u(\omega)| < 2\epsilon$ for all $u \in [t-\delta, t)$. This then implies that $|F(X^{\sigma+})(\omega) - F(X)(\omega)| < 2a\epsilon$. Therefore $(F(X^{\sigma_n+}) - F(X))_t^{\sigma_n}(\omega)$ tends to 0 as mesh(σ_n) tends to 0. Since, on $(0,\infty)$,

$$\lim_{\text{mesh}(\sigma_n)\to 0} F(X)^{\sigma_n} = F(X)_-,$$

we conclude that

$$\lim_{\text{mesh}(\sigma_n)\to 0} F(X^{\sigma_n+})^{\sigma_n} = F(X)_-,$$

where convergence is pointwise in (t,ω). Thus

$$\lim_{\text{mesh}(\sigma_n)\to 0} G^n(X)_- = F(X)_-.$$

However we also know that

$$\begin{aligned}|G^n(X)_-| &= |F(X^{\sigma_n+})^{\sigma_n}| \le F(X^{\sigma_n+})^* \\ &\le F(0)^* + K(X^{\sigma_n+})^* \\ &\le F(0)^* + aX^*\end{aligned}$$

which is in $\underline{\underline{S}}^2$ and is independent of σ_n. Therefore using the preceding dominated convergence lemma we obtain the convergence in $\underline{\underline{S}}^2$ of $\int_0^t G^n(X)_{s-}dZ_s$ to $\int_0^t F(X)_{s-}dZ_s$, and the proof is complete. □

Remark. In Theorem 16 and its corollary which follows, we have assumed that F is *process Lipschitz*, and not functional Lipschitz. Indeed, Theorem 16 is not true in general for functional Lipschitz coefficients: Let $J_t = 1_{\{t\ge 1\}}$, $Z_t = t \wedge 2$, and $F(Y) = Y_1 1_{\{t\ge 1\}}$. Then X, the solution of (*) is given by $X_t = (t\wedge 2)1_{\{t\ge 1\}}$, but if σ is any random partition such that $T_k \neq 1$ a.s., then $(X(\sigma)^{\sigma+})_t = 0$ for $t \le 1$, and therefore $F(X(\sigma)^{\sigma+}) = 0$, and $X(\sigma)_t = J_t = 1_{\{t\ge 1\}}$. (Here $X(\sigma)$ denotes the solution to equation $(*\sigma)$ of Theorem 16.)

Corollary. *Let $J \in \mathbb{D}$; F be process Lipschitz; Z be a semimartingale; and let σ_n be a sequence of random partitions tending to the identity. Then*

$$\lim_{n\to\infty} X(\sigma_n) = X \quad \textit{in ucp}$$

where $X(\sigma_n)$ is the solution of $(*^{\sigma_n})$ and X is the solution of (*), as in Theorem 16.

Proof. First assume $K \le a < \infty$, a.s. Fix $t > 0$ and $\epsilon > 0$. By Theorem 5 we can find a stopping time T such that $Z^{T-} \in \mathcal{S}(\frac{1}{8\sqrt{8a}})$, and $P(T < t) < \epsilon$. Thus without loss of generality we can assume that $Z \in \mathcal{S}(\frac{1}{8\sqrt{8a}})$. By letting $S^k = \inf\{t \ge 0 : |J_t| > k\}$, we have that S^k is a stopping time, and $\lim_{k\to\infty} S^k = \infty$ a.s. By now stopping at S^k- we have that J is bounded, hence also in $\underline{\underline{S}}^2$, and $Z \in \mathcal{S}(\frac{1}{4\sqrt{8a}})$. An analogous argument gives us that $F(0)$ can be assumed bounded (and hence in $\underline{\underline{S}}^2$) as well; hence $Z \in \mathcal{S}(\frac{1}{2\sqrt{8a}})$. We now can apply Theorem 16 to obtain the result.

To remove the assumption that $K \le a < \infty$ a.s., we need only apply an argument like the one used at the end of the proofs of Theorems 7, 8, and 15. □

Theorem 16 and its Corollary give us a way to approximate the solution of a general stochastic differential equation with finite differences. Indeed, let X be the solution of

$$(*) \qquad X_t = J_t + \int_0^t F(X)_{s-} dZ_s$$

where Z is a semimartingale and F is process Lipschitz. For each random partition $\sigma_n = \{0 = T_0^n \le T_1^n \le \cdots \le T_{k_n}^n\}$, we see that the random variables $X(\sigma_n)_{T_k^n}$ verify the relations (writing σ for σ_n, X for $X(\sigma_n)$, T_k for T_k^n):

$$X_{T_0} = J_0$$
$$X_{T_{k+1}} = X_{T_k} + J_{T_{k+1}} - J_{T_k} + F(X^{\sigma+})_{T_k}(Z_{T_{k+1}} - Z_{T_k}).$$

Then the solution of the finite difference equation above converges to the solution of (*), under the appropriate hypotheses.

As an example we give a means to approximate the stochastic exponential.

Theorem 17. *Let Z be a semimartingale and let $X = \mathcal{E}(Z)$, the stochastic exponential of Z. That is, X is the solution of*

$$X_t = 1 + \int_0^t X_{s-} dZ_s.$$

Let σ_n be a sequence of random partitions tending to the identity. Let

$$X^n = \prod_{i=1}^{k_n - 1} (1 + (Z^{T_{i+1}^n} - Z^{T_i^n})).$$

Then $\lim_{n\to\infty} X^n = X$ *in ucp.*

Proof. Let Y^n be the solution of:

$$Y_t = 1 + \int_0^t Y_s^{\sigma_n} dZ_s,$$

equation $(*^{\sigma_n})$ of Theorem 16. By the Corollary of Theorem 16 we know that Y^n converges to $X = \mathcal{E}(Z)$ in ucp. Thus it suffices to show $Y^n = X^n$.

Let $\sigma_n = \{0 = T_0^n \leq T_1^n \leq \cdots \leq T_{k_n}^n\}$. On $(T_i^n, T_{i+1}^n]$ we have

$$\begin{aligned} Y_t^n &= Y_{T_i^n}^n + Y_{T_i^n}^n (Z^{T_{i+1}^n} - Z^{T_i^n}) \\ &= Y_{T_i^n}^n (1 + (Z^{T_{i+1}^n} - Z^{T_i^n})). \end{aligned}$$

Inducting on i down to 0 we have

$$Y_t^n = \prod_{j \leq i} (1 + (Z^{T_{j+1}^n} - Z^{T_j^n})),$$

for $T_i^n < t \leq T_{i+1}^n$. Since $Z^{T_{j+1}^n} - Z^{T_j^n} = 0$ for all $j > i$ when $T_i^n < t \leq T_{i+1}^n$, we have that $Y^n = X^n$, and the theorem is proved. □

5. Fisk-Stratonovich Integrals and Differential Equations

In this section we extend the notion of the Fisk-Stratonovich integral given in Chap. II, Sect. 7, and we develop a theory of stochastic differential equations with Fisk-Stratonovich differentials. We begin with some results on the quadratic variation of stochastic processes.

Definition. Let H, J be adapted, càdlàg processes. The **quadratic covariation process of** H, J denoted $[H, J] = ([H, J]_t)_{t \geq 0}$, if it exists, is defined to be the adapted, càdlàg process of finite variation on compacts, such that for any sequence σ_n of random partitions tending to the identity:

$$\begin{aligned} \lim_{n \to \infty} S_{\sigma_n}(H, J) &= \lim_{n \to \infty} H_0 J_0 + \sum_i (H^{T_{i+1}^n} - H^{T_i^n})(J^{T_{i+1}^n} - J^{T_i^n}) \\ &= [H, J] \end{aligned}$$

with convergence in ucp, where σ_n is the sequence $0 = T_0^n \leq T_1^n \leq \cdots \leq T_{k_n}^n$. A process H in $\mathbb{D}$ **is said to have finite quadratic variation** if $[H, H]_t$ exists and is finite a.s., each $t \geq 0$.

If H, J and $H + J$ in $\mathbb{D}$ have finite quadratic variation, then the *polarization identity* holds:

$$[H, J] = \frac{1}{2}([H + J, H + J] - [H, H] - [J, J]).$$

For X a semimartingale, in Chap. II we defined the quadratic variation of X using the stochastic integral. However Theorem 22 of Chap. II shows every semimartingale is of finite quadratic variation and that the two definitions are consistent.

Notation. For H of finite quadratic variation we let $[H,H]^c$ denote the continuous part of the (non-decreasing) paths of $[H,H]$. Thus:

$$[H,H]_t = [H,H]^c_t + \sum_{0\leq s\leq t} \Delta[H,H]_s,$$

where $\Delta[H,H]_t = [H,H]_t - [H,H]_{t-}$, the jump at t.

The next definition extends the definition of the Fisk-Stratonovich integral given in Chap. II, Sect. 7.

Definition. Let $H \in \mathbf{D}$, X be a semimartingale, and assume $[H,X]$ exists. **The Fisk-Stratonovich integral of H with respect to X**, denoted $\int_0^t H_{s-} \circ dX_s$, is defined to be:

$$\int_0^t H_{s-} \circ dX_s \equiv \int_0^t H_{s-} dX_s + \frac{1}{2}[H,X]^c_s.$$

To consider properly general Fisk-Stratonovich differential equations, we need a generalization of Itô's formulas (Theorems 32 and 33 of Chap. II). Since Itô's formula is proved in detail there, we only sketch the proof of this generalization.

Theorem 18 (Generalized Itô's Formula). *Let $\mathbf{X} = (X^1, \ldots, X^n)$ be an n-tuple of semimartingales, and let f map $\mathbf{R}_+ \times \Omega \times \mathbf{R}^n \to \mathbf{R}$ be such that:*

(i) there exists an adapted FV process A and a function g such that

$$f(t,\omega,\mathbf{x}) = \int_0^t g(s,\omega,\mathbf{x})dA_s,$$

$(s,\omega) \to g(s,\omega,\mathbf{x})$ is an adapted, jointly measurable process for each $\mathbf{x}$, and $\int_0^t \sup_{\mathbf{x}\in K} |g(s,\omega,\mathbf{x})||dA_s| < \infty$ a.s. for compact sets K;

(ii) the function g of part (i) is C^2 in $\mathbf{x}$ uniformly in s on compacts: that is,

$$\sup_{s\leq t}\{|g(s,\omega,\mathbf{y}) - \sum_{i=1}^n g_{x_i}(s,\omega,\mathbf{x}))(y_i - x_i)$$
$$- \sum_{1\leq i,j\leq n} g_{x_i x_j}(s,\omega,\mathbf{x})(y_i - x_i)(y_j - x_j)|\}$$
$$\leq r_t(\omega, \|\mathbf{x}-\mathbf{y}\|)\|\mathbf{x}-\mathbf{y}\|^2$$

a.s., where $r_t : \Omega \times \mathbf{R}_+ \to \mathbf{R}_+$ is an increasing function with $\lim_{u\downarrow 0} r_t(u) = 0$ a.s., provided $\mathbf{x}$ ranges through a compact set (r_t depends on the compact set chosen);

(iii) the partial derivatives f_{x_i}, $f_{x_i x_j}$, $1 \leq i, j \leq n$ all exist and are continuous and moreover

$$f_{x_i}(t,\omega,\mathbf{x}) = \int_0^t g_{x_i}(s,\omega,\mathbf{x})dA_s;$$

$$f_{x_i x_j}(t,\omega,\mathbf{x}) = \int_0^t g_{x_i x_j}(s,\omega,\mathbf{x})dA_s.$$

Then

$$\begin{aligned} f(t,\omega,\mathbf{X}_t) = f(0,\omega,\mathbf{X}_0) &+ \int_0^t g(s,\omega,\mathbf{X}_s)dA_s \\ &+ \sum_{i=1}^n \int_0^t f_{x_i}(s-,\omega,\mathbf{X}_{s-})dX_s^i \\ &+ \frac{1}{2}\sum_{i,j=1}^n \int_{0+}^t f_{x_i x_j}(s-,\omega,\mathbf{X}_{s-})d[X^i,X^j]_s^c \\ &+ \sum_{0<s\leq t} \{f(s,\omega,\mathbf{X}_s) - f(s-,\omega,\mathbf{X}_{s-}) \\ &\qquad - g(s,\omega,\mathbf{X}_s)\Delta A_s - \sum_{i=1}^n f_{x_i}(s-,\omega,\mathbf{X}_{s-})\Delta X_s^i\}. \end{aligned}$$

Proof. We sketch the proof for $n = 1$. We have, letting $0 = t_0 \leq t_1 \leq \cdots \leq t_m = t$ be a partition of $[0,t]$, and assuming temporarily $|X| \leq k$ for all $s \leq t$, k a constant:

$$\begin{aligned} f(t,\omega,X_t) - f(0,\omega,X_0) &= \sum_{k=0}^{m-1}\{f(t_{k+1},\omega,X_{t_{k+1}}) - f(t_k,\omega,X_{t_{k+1}})\} \\ &\quad + \sum_{k=0}^{m-1}\{f(t_k,\omega,X_{t_{k+1}}) - f(t_k,\omega,X_{t_k})\} \\ (*) \qquad &= \sum_{k=0}^{m-1}\int_{t_k}^{t_{k+1}} g(u,\omega,X_{t_{k+1}})dA_u \\ &\quad + \sum_{k=0}^{m-1}\{f(t_k,\omega,X_{t_{k+1}}) - f(t_k,\omega,X_{t_k})\} \end{aligned}$$

Consider first the term $\sum_{k=0}^{m-1}\int_{t_k}^{t_{k+1}} g(u,\omega,X_{t_{k+1}})dA_u$. The integrand is not adapted, however one can interpret this integral as a path by path Stieltjes integral since A is an FV process. Expanding the integrand for fixed (u,ω) by the Mean Value Theorem yields:

$$g(u,\omega,X_{t_{k+1}}) = g(u,\omega,X_u) + g_x(u,\omega,\hat{X}_u)(X_{t_{k+1}} - X_u)$$

where $\hat{X}_u$ is in between X_u and $X_{t_{k+1}}$. Therefore

$$\sum_k \int_{t_k}^{t_{k+1}} g(u,\omega,X_{t_{k+1}})dA_u = \sum_k \int_{t_k}^{t_{k+1}} g(u,\omega,X_u)dA_u + \sum_k \int_{t_k}^{t_{k+1}} g_x(u,\omega,\hat{X}_u)(X_{t_{k+1}} - X_u)dA_u,$$

and since A is of finite variation and X is right continuous, the second sum tends to zero as the mesh of the partitions tends to zero. Therefore letting π_n denote a sequence of partitions of $[0,t]$ with $\lim_{n\to\infty}$ mesh$(\pi_n) = 0$,

$$\lim_{n\to\infty} \sum_{t_k\in\pi_n} \int_{t_k}^{t_{k+1}} g(u,\omega,X_{t_{k+1}})dA_u = \int_0^t g(u,\omega,X_u)dA_u(\omega) \ a.s.$$

Next consider the second term on the right side of (*):

$$\sum_{k=0}^{m-1} \{f(t_k,\omega,X_{t_{k+1}}) - f(t_k,\omega,X_{t_k})\}.$$

Here we proceed analogously to the proof of Theorem 32 of Chap. II. Given $\epsilon > 0$, $t > 0$, let $A(\epsilon,t)$ be a set of jumps of X that has a.s. a finite number of times s, and let $B = B(\epsilon,t)$ be such that $\sum_{s\in B}(\Delta X_s)^2 \le \epsilon^2$, where $A \cup B$ exhaust the jumps of X on $(0,t]$. Then

$$\sum_{k=0}^{m-1} \{f(t_k,\omega,X_{t_{k+1}}) - f(t_k,\omega,X_{t_k})\} = \sum_{k,A} \{f(t_k,\omega,X_{t_{k+1}}) - f(t_k,\omega,X_{t_k})\} + \sum_{k,B} \{f(t_k,\omega,X_{t_{k+1}}) - f(t_k,\omega,X_{t_k})\}$$

where $\sum_{k,A}$ denotes $\sum_{t_k\in\pi_n} 1_{\{A\cap(t_k,t_{k+1}]\neq\emptyset\}}$. Then

$$\lim_{n\to\infty} \sum_{k,A} \{f(t_k,\omega,X_{t_{k+1}}) - f(t_k,\omega,X_{t_k})\} = \sum_{s\in A} \{f(s-,\omega,X_s) - f(s-,\omega,X_{s-})\}$$

By Taylor's formula, and letting $\Delta_k X$ denote $X_{t_{k+1}} - X_{t_k}$,

(**)
$$\sum_{k,B} \{f(t_k,\omega,X_{t_{k+1}}) - f(t_k,\omega,X_{t_k})\} = \sum_k f_x(t_k,\omega,X_{t_k})\Delta_k X + \frac{1}{2}\sum_k f_{xx}(t_k,\omega,X_{t_k})(\Delta_k X)^2 - \sum_{k,A} \{f_x(t_k,\omega,X_{t_k})\Delta_k X + \frac{1}{2} f_{xx}(t_k,\omega,X_{t_k})(\Delta_k X)^2\} + \sum_{k,B} R(t_k,\omega,X_{t_k},X_{t_{k+1}}).$$

By Theorems 21 and 30 of Chap. II, the first sums on the right above converge in ucp to $\int_0^t f_x(s-,\omega,X_{s-})dX_s$ and $\frac{1}{2}\int_0^t f_{xx}(s-,\omega,X_{s-})d[X,X]_s$ respectively. The third sum converges a.s. to

$$-\sum_{s\in A}\{f_x(s-,\omega,X_{s-})\Delta X_s+\frac{1}{2}f_{xx}(s-,\omega,X_{s-})(\Delta X_s)^2\}.$$

By the hypothesis (ii) on the function g, we have

$$\limsup_n \sum_{t_k\in\pi_n,B} R(t_k,\omega,X_{t_k},X_{t_{k+1}})\leq r_t(\omega,\epsilon+)[X,X]_t$$

where

$$r_t(\omega,\epsilon+)=\limsup_{\delta\downarrow\epsilon} r_t(\omega,\delta).$$

Next we let ϵ tend to 0; then $r_t(\omega,\epsilon+)[X,X]_t$ tends to 0 a.s., and finally combining the two series indexed by A we see that

$$\sum_{s\in A(\epsilon,t)}\{f(s-,\omega,X_s)-f(s-,\omega,X_{s-})-f_x(s-,\omega,X_{s-})\Delta X_s$$
$$-\frac{1}{2}f_{xx}(s-,\omega,X_{s-})(\Delta X_s)^2\}$$

tends to the series

$$\sum_{0<s\leq t}\{f(s-,\omega,X_s)-f(s-,\omega,X_{s-})-f_x(s-,\omega,X_{s-})\Delta X_s$$
$$-\frac{1}{2}f_{xx}(s-,\omega,X_{s-})(\Delta X_s)^2\}$$

which is easily seen to be an absolutely convergent series (cf. the proof of Theorem 32 of Chap. II). Incorporating $-\sum_{0<s\leq t}\frac{1}{2}f_{xx}(s-,\omega,X_{s-})(\Delta X_s)^2$ into $\frac{1}{2}\int_0^t f_{xx}(s-,\omega,X_{s-})d[X,X]_s$ yields the term $\frac{1}{2}\int_0^t f_{xx}(s-,\omega,X_{s-})d[X,X]_s^c$.

We further note that $f(s,\omega,X_s)-f(s-,\omega,X_s)=g(s,\omega,X_s)\Delta A_s$, and since $\sum_{0<s\leq t}g(s,\omega,X_s)\Delta A_s$ is absolutely convergent, the theorem is proved. The assumption that $|X_s|\leq k$ for $s\leq t$ is removed as it was in Theorem 32 of Chap. II. □

Note that a consequence of Theorem 18 is that for f satisfying the hypotheses, we have $f(t,\cdot,\mathbf{X}_t)$ is a semimartingale when $\mathbf{X}=(X^1,\ldots,X^n)$ is an n-tuple of semimartingales. Also, an important special case is when the process $A_s\equiv s$; in this case the hypotheses partly reduce to assuming f is absolutely continuous in t.

The next theorem allows an improvement of the Fisk-Stratonovich change of variables formula given in Chap. II (Theorem 34).

Theorem 19. *Let $\mathbf{X} = (X^1, \dots, X^d)$ be a vector of semimartingales and let $f : \mathbb{R}_+ \times \Omega \times \mathbb{R}^d \to \mathbb{R}$ be such that*

(i) there exists an adapted FV process A and a function g such that

$$f(t, \omega, \mathbf{x}) = \int_0^t g(s, \omega, \mathbf{x}) dA_s$$

where $(s, \omega) \to g(s, \omega, \mathbf{x})$ is an adapted, jointly measurable process for each $\mathbf{x}$;

(ii) for each compact set K, $\int_0^t \sup_{\mathbf{x} \in K} |g(s, \omega, \mathbf{x})||dA_s| < \infty$ a.s.;

(iii) f_{x_i} exists and is continuous in $\mathbf{x}$ and $f_{x_i}(t, \omega, \mathbf{x}) = \int_0^t g_{x_i}(s, \omega, \mathbf{x}) dA_s$.

Let

$$Y_t = f(t, \omega, X^1, \dots, X^d).$$

Then Y is an adapted process of finite quadratic variation, and moreover

$$[Y,Y]_t = \sum_{1 \le i,j \le d} \int_0^t \frac{\partial f}{\partial x_i}(s-, \omega, \mathbf{X}_{s-}) \frac{\partial f}{\partial x_j}(s-, \omega, \mathbf{X}_{s-}) d[X^i, X^j]^c_s + \sum_{0 \le s \le t} \Delta Y_s^2.$$

Proof. First assume that for fixed (t, ω) the function f and all its first partials are bounded functions of $\mathbf{x}$. Then by optional stopping at times $T-$ we can assume without loss of generality that f and all its first partials (in $\mathbf{x}$) are bounded functions. Let f_k be a sequence of functions on $\mathbb{R}_+ \times \Omega \times \mathbb{R}^d$ which are C^2 on $\mathbb{R}^d$ such that f_k and its first partials converge uniformly on $\mathbb{R}^d$ respectively to f and its corresponding first partials. Moreover we can take all the f_k bounded and Lipschitz continuous with Lipschitz constant c, independent of k. For simplicity take $d = 1$. Let σ_n be a sequence of random partitions tending to the identity. We write, for a process Z,

$$S_{\sigma_n}(Z) = \sum_i (Z^{T^n_{i+1}} - Z^{T^n_i})^2$$

where

$$\sigma_n = \{0 = T^n_0 \le T^n_1 \le \dots \le T^n_{k_n}\}.$$

Since f_k and their first partials converge to f and its first partials uniformly, we have $f_k - f_\ell$ is Lipschitz with constant $\epsilon_{k\ell}$, which tends to 0 as k, ℓ tend to ∞. Therefore, letting τ denote σ_n for a given n and writing $f_k(X)$ for $f_k(t, \omega, X_t)$, we have

$$\begin{aligned} &|S_\tau(f_k(X)) - S_\tau(f_\ell(X))| \\ (*) \qquad &\le S_\tau((f_k - f_\ell)(X)) + 2\{S_\tau(f_k(X)) S_\tau((f_k - f_\ell)(X))\}^{1/2} \\ &\le (\epsilon_{k\ell}^2 + 2c\epsilon_{k\ell}) S_\tau(X), \end{aligned}$$

by the Lipschitz properties. Since $f_k(X)$ and $f_\ell(X)$ are semimartingales, we know (by Theorem 22 of Chap. II) that $S_{\sigma_n}(f_k(X))$ and $S_{\sigma_n}(f_\ell(X))$ converge in ucp respectively to $[f_k(X), f_k(X)]$ and $[f_\ell(X), f_\ell(X)]$. By restricting our attention to an interval $(s,t]$, we then have from (*) that:

$$|\{[f_k(X), f_k(X)]_t - [f_k(X), f_k(X)]_s\} - \{[f_\ell(X), f_\ell(X)]_t - [f_\ell(X), f_\ell(X)]_s\}| \leq (\epsilon_{k\ell}^2 + 2c\epsilon_{k\ell})\{[X,X]_t - [X,X]_s\}.$$

Since this is true for all $0 < s < t < \infty$, we deduce:

$$\int_0^t |d[f_k(X), f_k(X)]_s - d[f_\ell(X), f_\ell(X)]_s| \leq (\epsilon_{k\ell}^2 + 2c\epsilon_{k\ell})[X,X]_t.$$

Therefore $d[f_k(X), f_k(X)]$ is a Cauchy sequence of random measures, converging in total variation norm on $(0,\infty)$, a.s. In addition, by Itô's formula (Theorem 18) and Theorem 29 of Chap. II we have that

$$[f_k(X), f_k(X)]_t^c = \int_0^t \frac{\partial f_k}{\partial x}(s-,\omega,X_{s-})^2 d[X,X]_s^c;$$

from Theorem 23 of Chap. II we also know that

$$(\Delta f_k(X))^2 = \Delta[f_k(X), f_k(X)].$$

Therefore

$$[f_k(X), f_k(X)]_t = \int_0^t \frac{\partial f_k}{\partial x}(s-,\omega,X_{s-})^2 d[X,X]_s^c + \sum_{0\leq s\leq t} (\Delta f_k(X_s))^2,$$

and since we have a.s. convergence in total variation norm, we can pass to the limit to obtain:

$$V_t = \int_0^t f'(X_{s-})^2 d[X,X]_s^c + \sum_{0\leq s\leq t} (\Delta f(X_s))^2.$$

Our identification of the limit removes the dependence on the representation of Y by f, because

$$|S_{\sigma_n}(f(X))_t - V_t| \leq |S_{\sigma_n}(f(X))_t - S_{\sigma_n}(f_k(X))_t| + |V_t - [f_k(X), f_k(X)]_t| + |S_{\sigma_n}(f_k(X))_t - [f_k(X), f_k(X)]_t|,$$

and by taking k large enough the first two terms on the right side can be taken small in probability independently of n; one then takes n large enough in the third term to make it small in probability, and we have $\lim_{n\to\infty} S_{\sigma_n}(f(X))_t = V_t$, in probability. Therefore $V_t = [Y,Y]_t$ and this completes the proof for f and its first partials bounded functions of $\mathbf{x}$, and $d = 1$. The proof for general d is exactly analogous. For general f satisfying the hypotheses, let $g_m : \mathbf{R}^d \to \mathbf{R}^d$ be $\mathcal{C}^\infty$ such that $g_m(\mathbf{x}) = \mathbf{x}$ if $|\mathbf{x}| \leq m$, and g_m has compact support. Defining $Y_t^m = f(t,\omega,g_m(X_t))$, let $T^m(\omega) = \inf\{t \geq 0 : Y_t(\omega) \neq Y_t^m(\omega)\}$. Then T^m increase to ∞ a.s. Also, since the quadratic variation is a path property, $[Y^m, Y^m]_t(\omega) = [Y,Y]_t(\omega)$, for all $t < T^m(\omega)$. Therefore Y is of finite quadratic variation, and since by the preceding (for $d = 1$)

$$[Y^m, Y^m]_t = \int_0^t \frac{\partial f}{\partial x}(s-, \omega, g_m(X_{s-}))g'_m(X_{s-})d[X,X]^c_s + \sum_{0<s\leq t} \Delta f(s, \omega, g_m(X_s))^2$$

which, on $[0, T_m)$, is equal to:

$$= \int_0^t \frac{\partial f}{\partial x}(s-, \omega, X_{s-})d[X,X]^c_s + \sum_{0<s\leq t} \Delta f(s, X_s)^2,$$

whence the result. □

Note that the subspace of processes Y that have a representation $Y = f(X^1, \ldots, X^d)$, where $\mathbf{X} = (X^1, \ldots, X^d)$ is a finite dimensional semimartingale (d is not fixed) and $f \in C^1$, is a vector subspace of the space of processes of finite quadratic variation. As an immediate consequence of Theorem 19 we have the following extension of Theorem 34 of Chap. II.

Theorem 20. *Let X be a semimartingale and let f be C^2. Then*

$$f(X_t) - f(X_0) = \int_{0+}^t f'(X_{s-}) \circ dX_s + \sum_{0<s\leq t} \{f(X_s) - f(X_{s-}) - f'(X_{s-})\Delta X_s\}.$$

Proof. Note that f' is C^1, so that $f'(X)$ is in the domain of definition of the $F-S$ integral by Theorem 19. Also by definition we have

$$\int_{0+}^t f'(X_{s-}) \circ dX_s = \int_{0+}^t f'(X_{s-})dX_s + \frac{1}{2}[f'(X), X]^c_t.$$

By Theorem 19 and polarization we know that

$$[f'(X), X]^c_t = \int_0^t f''(X_{s-})d[X,X]^c_s,$$

and thus the result follows by Itô's formula (Theorem 32, Chap. II). □

Theorem 20 has an obvious generalization to the multidimensional case. We omit the proof which is an analogous consequence of Theorem 33 of Chap. II.

Theorem 21. *Let $\mathbf{X} = (X^1, \ldots, X^n)$ be an n-tuple of semimartingales and let $f : \mathbb{R}^n \to \mathbb{R}$ have second order continuous partial derivatives. Then $f(\mathbf{X})$ is a semimartingale and the following formula holds:*

$$f(\mathbf{X}_t) - f(\mathbf{X}_0) = \sum_{i=1}^n \int_{0+}^t \frac{\partial f}{\partial x_i}(\mathbf{X}_{s-}) \circ dX^i_s + \sum_{0<s\leq t} \{f(\mathbf{X}_s) - f(\mathbf{X}_{s-}) - \sum_{i=1}^n \frac{\partial f}{\partial x^i}(\mathbf{X}_{s-})\Delta X^i_s\}.$$

As a corollary of Theorem 21 we have the **Stratonovich integration by parts formula:**

Corollary 1. *Let X and Y be semimartingales. Then*

$$X_tY_t - X_0Y_0 = \int_{0+}^t X_{s-} \circ dY_s + \int_{0+}^t Y_{s-} \circ dX_s + \sum_{0<s\le t} \Delta X_s \Delta Y_s.$$

Proof. Let $f(x,y) = xy$ and apply Theorem 21. □

Corollary 2. *Let X and Y be semimartingales, with at least one of X or Y continuous. Then*

$$X_tY_t - X_0Y_0 = \int_{0+}^t X_{s-} \circ dY_s + \int_{0+}^t Y_{s-} \circ dX_s.$$

Recall that since $X_{0-} = 0$ by convention for a càdlàg process X, we did not really need to write $\int_{0+}^t$; the formula also holds for $\int_0^t$.

For stochastic differential equations with Fisk-Stratonovich differentials we are limited as to the coefficients we can consider, because the integrands must be of finite quadratic variation. Nevertheless we can still obtain reasonably general results. We first describe the coefficients.

Definition. A function $f : \mathbf{R}_+ \times \Omega \times \mathbf{R}^d \to \mathbf{R}$ is said to be **Fisk-Stratonovich acceptable** if

(i) there exists an adapted FV process A and a function g such that

$$f(t,\omega,\mathbf{x}) = \int_0^t g(s,\omega,\mathbf{x})dA_s$$

where $(s,\omega) \to g(s,\omega,\mathbf{x})$ is an adapted, jointly measurable process for each $\mathbf{x}$;

(ii) for each compact set K, $\int_0^t \sup_{x\in K}|g(s,\omega,\mathbf{x})||dA_s| < \infty$ a.s.;

(iii) f_{x_i} is C^1 and

$$f_{x_i}(t,\omega,\mathbf{x}) = \int_0^t g_{x_i}(s,\omega,\mathbf{x})dA_s;$$

(iv) for each fixed (t,ω), the functions $\mathbf{x} \to f(t,\omega,\mathbf{x})$ and $\mathbf{x} \to (\frac{\partial f}{\partial x^i})(f)(t,\omega,\mathbf{x})$ are all Lipschitz continuous with Lipschitz constant $K(\omega)$, $K < \infty$ a.s. $(1 \le i \le d)$.

We will often write "$F - S$ acceptable" in place of "Fisk-Stratonovich acceptable".

Theorem 22. *Given a vector of semimartingales $Z = (Z^1, \dots, Z^k)$, semimartingales J^i $(1 \le i \le d)$, and $F - S$ acceptable functions f^i_j $(1 \le i \le$

$d, 1 \leq j \leq k$), then the system of equations

$$X_t^i = J_t^i + \sum_{j=1}^{k} \int_0^t f_j^i(s-, \omega, \mathbf{X}_{s-}) \circ dZ_s^j \tag{*}$$

has a unique semimartingale solution. Moreover the solution $\mathbf{X}$ *of (*) is also the (unique) solution of*

$$\begin{aligned} X_t^i = J_t^i &+ \sum_{j=1}^{k} \int_0^t f_j^i(s-, \omega, \mathbf{X}_{s-}) dZ_s^j \\ &+ \frac{1}{2} \sum_{j,n=1}^{k} \sum_{m=1}^{d} \int_0^t (\frac{\partial f_j^i}{\partial x^m})(f_n^m)(s-, \omega, \mathbf{X}_{s-}) d[Z^n, Z^j]_s^c \\ &+ \frac{1}{2} \sum_{m=1}^{d} \int_0^t \frac{\partial f_j^i}{\partial x^m}(s-, \omega, X_{s-}) d[J^m, Z^j]_s^c. \end{aligned} \tag{**}$$

Proof. We note that equation (**) has a unique solution as a trivial consequence of Theorem 7. Since $\mathbf{X}$ is a d-dimensional semimartingale, we know that $f_j^i(s-, \cdot, \mathbf{X}_{s-})$ is in the domain of definition of the $F-S$ integral by Theorem 19. Further, as a consequence of Theorem 19, we have that[4]

$$\begin{aligned} [f_j^i(\cdot, \omega, \mathbf{X}_\cdot), Z_\cdot^j]_t^c &= \sum_{m=1}^{d} \int_0^t \frac{\partial f_j^i}{\partial x^m}(s, \omega, \mathbf{X}_s)(d[X^m, Z^j]_s^c + d[J^m, Z^j]_s^c) \\ &= \sum_{n=1}^{k} \sum_{m=1}^{d} \int_0^t \frac{\partial f_j^i}{\partial x^m}(s, \omega, \mathbf{X}_s) f_n^m(s, \omega, \mathbf{X}_s) d[Z^n, Z^j]_s^c, \end{aligned}$$

and the equivalence of (*) and (**) follows. Therefore the existence of a unique semimartingale solution of (**) is equivalent to the existence of a unique semimartingale solution of (*), and we are done. Note that if J^m is of finite variation, the terms involving $[J^m, Z^j]^c$ disappear. □

In Chap. II we studied the stochastic exponential of a semimartingale. The $F-S$ integral allows us to give a version that has a more natural appearance.

Theorem 23. *Let Z be a semimartingale, $Z_0 = 0$. The unique solution of the equation*

$$X_t = X_0 + \int_0^t X_{s-} \circ dZ_s$$

[4] Since we are taking the continuous parts of the quadratic variations, we need not write $f(s-, \omega, X_{s-})$, etcetera.

is given by

$$X_t = X_0 \exp(Z_t) \prod_{0<s\leq t} (1+\Delta Z_s)e^{-\Delta Z_s},$$

and it is called the Fisk-Stratonovich exponential, $X_0\mathcal{E}_{F-S}(Z)$.

Proof. By Theorem 22 the equation above is equivalent to

$$\begin{aligned} X_t &= X_0 + \int_0^t X_{s-}dZ_s + \frac{1}{2}\int_0^t X_{s-}d[Z,Z]_s^c \\ &= X_0 + \int_0^t X_{s-}d(Z_s + \frac{1}{2}[Z,Z]_s^c). \end{aligned}$$

Therefore

$$\begin{aligned} \mathcal{E}_{F-S}(Z) &= \mathcal{E}(Z + \frac{1}{2}[Z,Z]^c) \\ &= \mathcal{E}(Z)\mathcal{E}(\frac{1}{2}[Z,Z]^c) \\ &= \mathcal{E}(Z)e^{\frac{1}{2}[Z,Z]^c}, \end{aligned}$$

where the second equality is by Theorem 37 of Chap. II. Using the formula for the stochastic exponential established in Theorem 36 of Chap. II, the result follows. □

The most interesting case is when the semimartingale is continuous:

Corollary 1. *Let* Z *be a continuous semimartingale,* $Z_0 = 0$. *Then the unique solution,* $\mathcal{E}_{F-S}(Z)$, *of the exponential equation*

$$X_t = 1 + \int_0^t X_s \circ dZ_s$$

is given by $\mathcal{E}_{F-S}(Z)_t = \exp(Z_t)$.

Corollary 2. *Let* B *be a Brownian motion with* $B_0 = 0$. *Then the unique solution of*

$$X_t = X_0 + \int_0^t X_s \circ dB_s$$

is given by $X_t = X_0e^{B_t}$.

The simplicity gained by using the $F-S$ integral can be surprisingly helpful. As an example let $\mathbf{Z} = (Z^1,\dots,Z^n)$ be an n-dimensional *continuous* semimartingale and let $\mathbf{x} = (x^1,\dots,x^n)'$ be a column vector in $\mathbb{R}^n$. (Here "$'$" denotes transpose.) Let I be the identity $n\times n$ matrix and define

$$a(\mathbf{x}) = I - \frac{\mathbf{x}\mathbf{x}'}{|\mathbf{x}|^2}.$$

(The matrix $a(\mathbf{x})$ represents projection onto the hyperplane normal to $\mathbf{x}$). Note that $\mathbf{x}'a(\mathbf{x}) = 0$. We want to study the system of F-S differential equations:

$$(***) \qquad \mathbf{X}_t = \mathbf{x}_0 + \int_0^t a(\mathbf{X}_s) \circ d\mathbf{Z}_s, \qquad 1 \le i \le n.$$

where $\mathbf{X} = (X^1, \ldots, X^n)'$ and $\mathbf{Z} = (Z^1, \ldots, Z^n)'$.

Theorem 24. *The solution* $\mathbf{X}$ *of equation (***) above exists, is unique, and it always stays on the sphere of center* 0 *and radius* $\|\mathbf{x}_0\|$.

Proof. Since a is singular at the origin, we need to modify it slightly. Let $g(\mathbf{x})$ be a $\mathcal{C}^\infty$ function equal to $a(\mathbf{x})$ outside of a ball centered at the origin, N_0, such that $\mathbf{x}_0 \notin N_0$. Let $\mathbf{Y}$ be the solution of

$$\mathbf{Y}_t = \mathbf{x}_0 + \int_0^t g(\mathbf{Y}_s) \circ dZ_s.$$

Then $\mathbf{Y}$ exists and is unique by Theorem 22. Let $T = \inf\{t > 0 : Y_t \in N_0\}$. Since $\mathbf{Y}_0 = \mathbf{x}_0$ and $\mathbf{Y}$ is continuous, $P(T > 0) = 1$. However for $s < T$, $g(\mathbf{Y}_s) = a(\mathbf{Y}_s)$. Therefore it suffices to show that the function $t \to \|\mathbf{Y}_t\|$ is constant, since $\|\mathbf{Y}_0\| = \|\mathbf{x}_0\|$. This would imply that $\mathbf{Y}$ always stays on the sphere of center 0 and radius $\|\mathbf{x}_0\|$, and hence $P(T = \infty) = 1$ and $g(\mathbf{Y}) = a(\mathbf{Y})$ always. To this end, let $f(\mathbf{x}) = \sum_{i=1}^n (x^i)^2$, where $\mathbf{x} = (x^1, \ldots, x^n)'$. Then it suffices to show that $df(\mathbf{Y}_t) = 0$, $(t \ge 0)$. Note that f is C^2. We have that for $t < T$,

$$\begin{aligned} df(\mathbf{Y}_t) &= 2\mathbf{Y}_t' \circ d\mathbf{Y} \\ &= 2\mathbf{Y}_t' \cdot a(\mathbf{Y}_t) \circ d\mathbf{Z}_t \\ &= 0. \end{aligned}$$

Therefore $\|Y_t\| = \|x_0\|$ for $t < T$ and thus by continuity $T = \infty$ a.s. □

Corollary. *Let* $\mathbf{B} = (B^1, \ldots, B^n)'$ *be* n*-dimensional Brownian motion, let* $a(\mathbf{x}) = I - \frac{\mathbf{x}\mathbf{x}'}{|\mathbf{x}|^2}$, *and let* $\mathbf{X}$ *be the solution of*

$$\mathbf{X}_t = \mathbf{x}_0 + \int_0^t a(\mathbf{X}_s) \circ d\mathbf{B}_s.$$

Then $\mathbf{X}$ *is a Brownian motion on the sphere of radius* $\|\mathbf{x}_0\|$.

Proof. In Theorem 36 of Sect. 6 we show that the solution $\mathbf{X}$ is a diffusion. By Theorem 24 we know that $\mathbf{X}$ always stays on the sphere of center 0 and radius $\|\mathbf{x}_0\|$. It thus remains to show only that $\mathbf{X}$ is a *rotation invariant* diffusion, since this characterizes Brownian motion on a sphere. Let U be an orthogonal matrix. Then $U\mathbf{B}$ is again a Brownian motion, and thus it suffices to show:

$$d(U\mathbf{X}) = a(U\mathbf{X}) \circ d(U\mathbf{B});$$

the above equation shows that $U\mathbf{X}$ is statistically the same diffusion as is $\mathbf{X}$, and hence $\mathbf{X}$ is rotation invariant. The coefficient a satisfies:

$$a(U\mathbf{x}) = Ua(\mathbf{x})U',$$

and therefore

$$\begin{aligned} d(U\mathbf{X}) = U \circ d\mathbf{X} &= Ua(\mathbf{X}) \circ d\mathbf{B} \\ &= Ua(\mathbf{X})U'U \circ d\mathbf{B} \\ &= a(U\mathbf{X}) \circ d(U\mathbf{B}), \end{aligned}$$

and we are done. □

The F-S integral can also be used to derive explicit formulas for solutions of stochastic differential equations in terms of solutions of (nonrandom) ordinary differential equations. As an example, consider the equation

$$(*4) \qquad X_t = x_0 + \int_0^t f(X_s) \circ dZ_s + \int_0^t g(X_s)ds,$$

where Z is a continuous semimartingale, $Z_0 = 0$. (Note that $\int_0^t g(X_s) \circ ds = \int_0^t g(X_s)ds$, so we have not included "Itô's circle" in this term.) Assume that f is $\mathcal{C}^2$ and that f, g, and ff' are all Lipschitz continuous. Let $u = u(x, z)$ be the unique solution of

$$\frac{\partial u}{\partial z}(x, z) = f(u(x, z))$$
$$u(x, 0) = x.$$

Then $\frac{\partial}{\partial z}\frac{\partial u}{\partial x} = f'(u(x, z))\frac{\partial u}{\partial x}$, and $\frac{\partial u}{\partial x}(x, 0) = 1$, from which we conclude that

$$\frac{\partial u}{\partial x}(x, z) = \exp\{\int_0^z f'(u(x, v))dv\}.$$

Let $Y = (Y_t)_{t \geq 0}$ be the solution of

$$Y_t = x_0 + \int_0^t \exp\{-\int_0^{Z_s} f'(u(Y_s, v))dv\}g(u(Y_s, Z_s))ds$$

which we assume exists. For example if $\dfrac{g(u(x, Z_s))}{\frac{\partial u}{\partial x}(x, Z_s)}$ is Lipschitz, this would suffice.

Theorem 25. *With the notation and hypotheses given above, the solution X of (*4) is given by*

$$X_t = u(Y_t, Z_t).$$

Proof. Using the F-S calculus we have

$$\begin{aligned} u(Y_t, Z_t) &= u(x_0, 0) + \int_0^t \frac{\partial u}{\partial x}(Y_s, Z_s) \circ dY_s + \int_0^t \frac{\partial u}{\partial z}(Y_s, Z_s) \circ dZ_s \\ &= x_0 + \int_0^t \exp\{\int_0^{Z_s} f'(u(Y_s, v))dv\} \circ dY_s + \int_0^t f(u(Y_s, Z_s)) \circ dZ_s. \end{aligned}$$

Since

$$\frac{dY_s}{ds} = \exp\{-\int_0^{Z_s} f'(u(Y_s,v))dv\}g(u(Y_s,Z_s)),$$

we deduce

$$u(Y_t,Z_t) = x_0 + \int_0^t g(u(Y_s,Z_s))ds + \int_0^t f(u(Y_s,Z_s)) \circ dZ_s.$$

By the uniqueness of the solution, we conclude that $X_t = u(Y_t, Z_t)$. □

The Fisk-Stratonovich integrals also have an interpretation as limits of sums, as Theorems 26 through 29 illustrate. These theorems are then useful in turn for approximating solutions of stochastic differential equations.

Theorem 26. *Let H be càdlàg, adapted, and let X be a semimartingale. Assume $[H,X]$ exists. Let $\sigma_n = \{0 = T_0^n \le T_1^n \le \cdots \le T_{k_n}^n\}$ be a sequence of random partitions tending to the identity. If H and X have no jumps in common (i.e., $\sum_{0<s\le t} \Delta H_s \Delta X_s = 0$, all $t \ge 0$), then*

$$\lim_{n\to\infty} \sum_i \frac{1}{2}(H_{T_i^n} + H_{T_{i+1}^n})(X^{T_{i+1}^n} - X^{T_i^n})$$

equals the F-S integral $\int_0^\cdot H_{s-} \circ dX_s$ in ucp.

Proof. It follows easily from the definition of $[H,X]$ at the beginning of this section that $\Delta[H,X]_t = \Delta H_t \Delta X_t$ and $\lim_{n\to\infty} \sum (H_{T_{i+1}^n} - H_{T_i^n})(X^{T_{i+1}^n} - X^{T_i^n}) = [H,X] - H_0X_0$. Thus if H and X have no jumps in common we conclude $[H,X] = [H,X]^c + H_0X_0$. Observing that

$$\begin{aligned}\sum_i \frac{1}{2}&(H_{T_i^n} + H_{T_{i+1}^n})(X^{T_{i+1}^n} - X^{T_i^n}) \\ &= \sum_i H_{T_i^n}(X^{T_{i+1}^n} - X^{T_i^n}) + \frac{1}{2}\sum_i (H_{T_{i+1}^n} - H_{T_i^n})(X^{T_{i+1}^n} - X^{T_i^n}),\end{aligned}$$

the result follows from Theorems 21 and 22 of Chap. II. □

Corollary 1. *If either H or X in Theorem 21 is continuous, then*

$$\lim_{n\to\infty} \sum_i \frac{1}{2}(H_{T_i^n} + H_{T_{i+1}^n})(X^{T_{i+1}^n} - X^{T_i^n}) = \int_0^\cdot H_{s-} \circ dX_s,$$

where $H_{0-} = 0$, in ucp.

Corollary 2. *Let X and Y be continuous semimartingales, and let $\sigma_n = \{0 = T_0^n \le T_1^n \le \cdots \le T_{k_n}^n\}$ be a sequence of random partitions tending to the identity. Then*

$$\lim_{n\to\infty} \sum_i \frac{1}{2}(Y_{T_i^n} + Y_{T_{i+1}^n})(X^{T_{i+1}^n} - X^{T_i^n}) = \int_0^\cdot Y_{s-} \circ dX_s,$$

with convergence in ucp.

Proof. By Theorem 22 of Chap. II, any semimartingale Y has finite quadratic variation. Thus Corollary 2 is a special case of Theorem 26 (and of Corollary 1). □

Theorem 27. *Let H be càdlàg, adapted, of finite quadratic variation, and suppose $\sum_{0<s\leq t}|\Delta H_s| < \infty$ a.s., each $t > 0$. Let X be a semimartingale, and let $\sigma_n = \{T_i^n\}_{0\leq i\leq k_n}$ be a sequence of random partitions tending to the identity. Assume $[H,X]$ exists. Then*

$$\lim_{n\to\infty}\{\sum_i \frac{1}{2}(H_{T_i^n}+H_{T_{i+1}^n}-\sum_{T_i^n<s\leq T_{i+1}^n}\Delta H_s)(X^{T_{i+1}^n}-X^{T_i^n})\} = \int_0^{\cdot} H_{s-}\circ dX_s,$$

with convergence in ucp.

Proof. First observe that $\hat{H}_t = H_t - \sum_{0<s\leq t}\Delta H_s$ defines a continuous process of finite quadratic variation and that $[\hat{H},X] = [H,X]^c + H_0X_0$. Next we note that

$$\begin{aligned}\frac{1}{2}(H_{T_i^n}+H_{T_{i+1}^n}-\sum_{T_i^n<s\leq T_{i+1}^n}\Delta H_s) &= H_{T_i^n}+\frac{1}{2}(H_{T_{i+1}^n}-H_{T_i^n}-\sum_{T_i^n<s\leq T_{i+1}^n}\Delta H_s)\\ &= H_{T_i^n}+\frac{1}{2}(\hat{H}_{T_{i+1}^n}-\hat{H}_{T_i^n}).\end{aligned}$$

Therefore by Theorems 21 and 22 of Chap. II, or directly from the definition as in the proof of Theorem 26, we have

$$\begin{aligned}&\lim_{n\to\infty}\{\sum_i\frac{1}{2}(H_{T_i^n}+H_{T_{i+1}^n}-\sum_{T_i^n<s\leq T_{i+1}^n}\Delta H_s)(X^{T_{i+1}^n}-X^{T_i^n})\}\\ &= \lim_{n\to\infty}\sum_i H_{T_i^n}(X^{T_{i+1}^n}-X^{T_i^n})+\lim_{n\to\infty}\frac{1}{2}\sum_i(\hat{H}_{T_{i+1}^n}-\hat{H}_{T_i^n})(X^{T_{i+1}^n}-X^{T_i^n})\\ &= \int_{0+}^{\cdot} H_{s-}dX_s+\frac{1}{2}([\hat{H},X]_\cdot+H_0X_0)\\ &= \int_0^{\cdot} H_{s-}\circ dX_s+\frac{1}{2}[H,X]^c_\cdot = \int_0^{\cdot} H_{s-}\circ dX_s, \quad \text{where } H_{0-}=H_0,\end{aligned}$$

and the result follows. □

For the general case we have a more complicated result. Let H be a càdlàg, adapted process of finite quadratic variation. For each $\epsilon > 0$ we define

$$J_t^\epsilon = \sum_{0<s\leq t}\Delta H_s 1_{\{|\Delta H_s|>\epsilon\}}.$$

Then J^ϵ is also a càdlàg, adapted process, and it has paths of finite variation on compacts.

Theorem 28. *Let H be càdlàg, adapted, of finite quadratic variation, and let X be a semimartingale. Assume $[H,X]$ exists. Let $\sigma_n = \{T_i^n\}$ be a sequence of random partitions tending to the identity. Let J^ϵ be as defined above. Then:*

$$\lim_{\epsilon\to 0}\lim_{n\to\infty}\{\sum_i \frac{1}{2}(H_{T_i^n} + H_{T_{i+1}^n} + (J^\epsilon_{T_{i+1}^n} - J^\epsilon_{T_i^n}))(X^{T_{i+1}^n} - X^{T_i^n})\}$$
$$= \int_0^\cdot H_{s-}\circ dX_s,$$

with convergence in ucp.

Proof. Let $H^\epsilon = H - J^\epsilon$. Then the jumps of H^ϵ are bounded by ϵ. Moreover:

$$\frac{1}{2}(H_{T_i^n} + H_{T_{i+1}^n} + (J^\epsilon_{T_{i+1}^n} - J^\epsilon_{T_i^n})) = H_{T_i^n} + \frac{1}{2}(H^\epsilon_{T_{i+1}^n} - H^\epsilon_{T_i^n}).$$

Therefore the first limit at $t > 0$ is:

$$\int_{0+}^t H_{s-}dX_s + \frac{1}{2}[H,X]_t^c + \frac{1}{2}\sum_{0<s\leq t}\Delta H_s^\epsilon \Delta X_s = \int_0^t H_{s-}\circ dX_s + \frac{1}{2}\sum_{0<s\leq t}\Delta H_s^\epsilon \Delta X_s,$$

and letting ϵ tend to 0 gives the result, since

$$\sum_{0<s\leq t}|\Delta H_s^\epsilon \Delta X_s| \leq [H^\epsilon, H^\epsilon]_t^{1/2}[X,X]_t^{1/2} < \infty,$$

and since $[H^\epsilon, H^\epsilon]_t = \sum_{0<s\leq t}(\Delta H_s^\epsilon)^2$, we have $\lim_{\epsilon\to 0}\sum_{0<s\leq t}|\Delta H_s^\epsilon \Delta X_s| = 0$. □

The next result is useful, for example, for approximating integrals of the form $\int_0^t f(B_s)\circ dB_s$, where B is a Brownian motion. It can be generalized to Y a continuous, adapted process of finite quadratic variation, but we do not do so here.

Theorem 29. *Let X be a semimartingale and Y a continuous semimartingale, and let f be C^1. Let μ be a probability measure on $[0,1]$ let $\alpha = \int \lambda\mu(d\lambda)$, and let $\sigma_n = \{T_i^n\}_{i\geq 0}$ be a sequence of random partitions tending to the identity. Then*

$$\lim_{n\to\infty}\sum_i \int_0^1 f(Y_{T_i^n} + \lambda(Y_{T_{i+1}^n} - Y_{T_i^n}))\mu(d\lambda)(X^{T_{i+1}^n} - X^{T_i^n})$$
$$= \int_{0+}^\cdot f(Y_s)dX_s + \alpha\int_{0+}^\cdot f'(Y_s)d[Y,X]_s,$$

with convergence in ucp. In particular if $\alpha = \frac{1}{2}$ then the limit is the F-S integral $\int_{0+}^{\cdot} f(Y_s) \circ dX_s$.

Proof. We begin by observing that

$$\sum_i \int_0^1 f(Y_{T_i^n} + \lambda(Y_{T_{i+1}^n} - Y_{T_i^n}))\mu(d\lambda)(X^{T_{i+1}^n} - X^{T_i^n})$$

$$= \sum_i f(Y_{T_i^n})(X^{T_{i+1}^n} - X^{T_i^n})$$

$$+ \sum_i \int_0^1 \mu(d\lambda)\{f(Y_{T_i^n} + \lambda(Y_{T_{i+1}^n} - Y_{T_i^n})) - f(Y_{T_i^n})\}(X^{T_{i+1}^n} - X^{T_i^n}).$$

The first sum on the right side of the above equation tends to $\int_{0+}^{\cdot} f(Y_{s-})dX_s$ in ucp. Using the fundamental theorem of calculus, the second sum on the right above equals

$$\sum_i \int_0^1 \mu(d\lambda) \int_0^1 ds \lambda f'(Y_{T_i^n} + \lambda s(Y_{T_{i+1}^n} - Y_{T_i^n}))(Y_{T_{i+1}^n} - Y_{T_i^n})(X^{T_{i+1}^n} - X^{T_i^n}),$$

which in turn equals

$$\alpha \sum_i f'(Y_{T_i^n})(Y_{T_{i+1}^n} - Y_{T_i^n})(X^{T_{i+1}^n} - X^{T_i^n})$$

$$+ F_{\sigma_n}(\omega) \sum_i |(Y_{T_{i+1}^n} - Y_{T_i^n})(X^{T_{i+1}^n} - X^{T_i^n})|,$$

where

$$F_{\sigma_n}(\omega) \leq \sup_i \{|f'(Y_{T_i^n} + \lambda s(Y_{T_{i+1}^n} - Y_{T_i^n})) - f'(Y_{T_i^n})|\}.$$

Since f' and Y are continuous, $F_{\sigma_n}(\omega)$ tends to 0 on compact time intervals. Also, on $[0,t]$,

$$\limsup_{n\to\infty} \sum_i |(Y_{T_{i+1}^n} - Y_{T_i^n})(X^{T_{i+1}^n} - X^{T_i^n})| \leq [Y,Y]_t^{1/2}[X,X]_t^{1/2},$$

and the result follows by Theorem 30 of Chap. II. □

Corollary. *Let X be a semimartingale, Y be a continuous semimartingale, and $f \in C^1$. Let $\sigma_n = \{T_i^n\}_{0 \leq i \leq k_n}$ be a sequence of random partitions tending to the identity. Then*

$$\lim_{n\to\infty} \sum_i f(\frac{1}{2}(Y_{T_i^n} + Y_{T_{i+1}^n}))(X^{T_{i+1}^n} - X^{T_i^n}) = \int_0^{\cdot} f(Y_{s-}) \circ dX_s$$

with convergence in ucp.

Proof. Let $\mu(d\lambda) = \epsilon_{\frac{1}{2}}(d\lambda)$, point mass at $\frac{1}{2}$. Then $\alpha = \int_0^1 \lambda\mu(d\lambda) = \frac{1}{2}$, and we need only apply Theorem 29. □

For Brownian motion the Fisk-Stratonovich integral is sometimes defined as a limit of the form:

$$\lim \sum_i f(B_{\frac{t_i+t_{i+1}}{2}})(B_{t_{i+1}} - B_{t_i}) = \int_0^t f(B_s) \circ dB_s;$$

that is, the sampling times are averaged. Such an approximation does not hold in general even for continuous semimartingales (see Yor [2], p. 524ff), but it does hold with a supplementary hypothesis on the quadratic covariation, as Theorem 30 reveals.

Theorem 30. *Let X be a semimartingale and Y be a continuous semimartingale. Let μ be a probability measure on $[0,1]$ and let $\alpha = \int \lambda\mu(d\lambda)$. Further suppose that $[X,Y]_t = \int_0^t J_s ds$; that is, the paths of $[X,Y]$ are absolutely continuous. Let $\sigma_n = \{t_i^n\}$ be a sequence of non-random partitions tending to the identity. Let f be C^1. Then*

$$\lim_{n\to\infty} \sum_i \int_0^1 f(Y_{t_i+\lambda(t_{i+1}-t_i)})\mu(d\lambda)(X^{t_{i+1}} - X^{t_i})$$

$$= \int_{0+}^{\cdot} f(Y_s)dX_s + \alpha \int_{0+}^{\cdot} f'(Y_s)d[Y,X]_s,$$

with convergence in ucp. In particular if $\alpha = \frac{1}{2}$ then the limit is the $F-S$ integral $\int_{0+}^{\cdot} f(Y_s) \circ dX_s$.

Proof. We begin with a real analysis result. We let t_i^λ denote $t_i + \lambda(t_{i+1} - t_i)$, where the t_i are understood to be in σ_n. Suppose a is continuous on $[0,t]$. Then

$$\left|\sum_i \int_{t_i}^{t_i^\lambda} a(s)ds - \sum_i a(t_i)\lambda(t_{i+1} - t_i)\right| \le \sum_i \int_{t_i}^{t_i^\lambda} |a(s) - a(t_i)|ds$$

which tends to 0. Therefore

$$(*) \qquad \lim_{n\to\infty} \sum_i \int_{t_i}^{t_i^\lambda} a(s)ds = \lambda \int_0^t a(s)ds.$$

Moreover since continuous functions are dense in $L^1([0,t],ds)$, the limiting result(*) holds for all a in $L^1([0,t],ds)$.

Next suppose H is a continuous, adapted process. Set $H^{\lambda,n} = \sum_i H_{t_i} 1_{(t_i,t_i^\lambda]}$. Then taking limits in ucp we have

$$\lim_n \sum_i H_{t_i}\{[X,Y]^{t_i^\lambda} - [X,Y]^{t_i}\} = \lim_n \int_0^t H_s^{\lambda,n} d[X,Y]_s,$$

and using integration by parts:

$$= \lim_n \{H^{\lambda,n} \cdot (XY) - (H^{\lambda,n} X_-) \cdot Y - (H^{\lambda,n} Y) \cdot X\},$$

and by Theorem 21 of Chap. II

$$= \lim_n \{\sum_i H_{t_i}(X^{t_i^\lambda} Y^{t_i^\lambda} - X^{t_i} Y^{t_i}) - \sum_i H_{t_i} X_{t_i}(Y^{t_i^\lambda} - Y^{t_i})$$

$$- \sum_i H_{t_i} Y_{t_i}(X^{t_i^\lambda} - X^{t_i})\}$$

$$= \lim_n \sum_i H_{t_i}(X^{t_i^\lambda} - X^{t_i})(Y^{t_i^\lambda} - Y^{t_i}).$$

However since $[X,Y]_t = \int_0^t J_s ds$, by the result (*) we conclude
(**)

$$\lim_n \sum_i H_{t_i}(X^{t_i^\lambda} - X^{t_i})(Y^{t_i^\lambda} - Y^{t_i}) = \lim_n \sum_i H_{t_i}\{[X,Y]^{t_i^\lambda} - [X,Y]^{t_i}\}$$

$$= \lim_n \sum_i H_{t_i} \int_{t_i}^{t_i^\lambda} J_s ds$$

$$= \lambda \int_0^t H_s d[X,Y]_s.$$

We now turn our attention to the statement of the theorem. Using the mean value theorem we have:
(***)

$$\sum_i \int_0^1 f(Y_{t_i^\lambda})\mu(d\lambda)(X^{t_{i+1}} - X^{t_i}) = \sum_i f(Y_{t_i})(X^{t_{i+1}} - X^{t_i})$$

$$+ \sum_i \int_0^1 \mu(d\lambda) f'(Y_{t_i})(Y_{t_i^\lambda} - Y_{t_i})(X^{t_{i+1}} - X^{t_i})$$

$$+ \sum_i \int_0^1 \mu(d\lambda) \int_0^1 ds\{f'(Y_{t_i} + s(Y_{t_i^\lambda} - Y_{t_i})) - f'(Y_{t_i})\}$$

$$\times (Y_{t_i^\lambda} - Y_{t_i})(X^{t_{i+1}} - X^{t_i}).$$

The first sum tends in ucp to $\int f(Y_s)dX_s$ by Theorem 21 of Chap. II. The second sum on the right side of (***) can be written as

$$\sum_i \int_0^1 \mu(d\lambda) f'(Y_{t_i})(Y_{t_i^\lambda} - Y_{t_i})(X^{t_i^\lambda} - X^{t_i})$$

$$+ \sum_i \int_0^1 \mu(d\lambda) f'(Y_{t_i})(Y_{t_i^\lambda} - Y_{t_i})(X^{t_{i+1}} - X^{t_i^\lambda}).$$

The first sum above converges to

$$\int_0^1 \mu(d\lambda)\lambda \int_0^t f'(Y_s)d[Y,X]_s$$

by (**), and the second sum can be written as

$$K^{\lambda,n} \cdot X,$$

where

$$K_s^{\lambda,n} = \sum_i f'(Y_{t_i})(Y_{t_i^\lambda} - Y_{t_i})1_{(t_i^\lambda, t_{i+1}]}(s).$$

Then $\lim_n K^{\lambda,n} \cdot X$ converges to 0 locally in ucp by the Dominated Convergence Theorem (Theorem 32 of Chap. IV).

Finally consider the third sum on the right side of (***). Let

$$F_n(\omega) = \sup_{t_i \in \sigma_n} \sup_{s \in [0,1]} |f'(Y_{t_i} + s(Y_{t_i^\lambda} - Y_{t_i})) - f'(Y_{t_i})|.$$

Then

$$\begin{aligned}
&\lim_n \sum_i \int_0^1 \mu(d\lambda) F_n(\omega)|Y_{t_i^\lambda} - Y_{t_i}||X^{t_{i+1}} - X^{t_i}| \\
&\leq \lim_n \int_0^1 \mu(d\lambda) F_n \{\sum_i (Y_{t_i^\lambda} - Y_{t_i})^2\}^{1/2} \{\sum_i (X_{t_{i+1}} - X_{t_i})^2\}^{1/2} \\
&= 0
\end{aligned}$$

since $\lim F_n = 0$ a.s. and the summations stay bounded in probability. This completes the proof. Since Y is continuous, $[Y,X] = [Y,X]^c$, whence if $\alpha = \frac{1}{2}$ we obtain the Fisk-Stratonovich integral. □

Corollary. *Let X be a semimartingale, Y a continuous semimartingale, and f be C^1. Let $[X,Y]$ be absolutely continuous and let $\sigma_n = \{t_i^n\}$ be a sequence of non-random partitions tending to the identity. Then*

$$\lim_{n\to\infty} \sum_i f(Y_{\frac{t_i+t_{i+1}}{2}})(X^{t_{i+1}} - X^{t_i}) = \int_{0+}^{\cdot} f(Y_s) \circ dX_s$$

with convergence in ucp.

Proof. Let $\mu(d\lambda) = \epsilon_{\{1/2\}}(d\lambda)$, point mass at $\frac{1}{2}$. Then $\int \lambda\mu(d\lambda) = \frac{1}{2}$, and apply Theorem 30. □

Note that if Y is a continuous semimartingale and B is standard Brownian motion, then $[Y,B]$ is absolutely continuous as a consequence of the Kunita-Watanabe inequality. Therefore if f is C^1, σ_n are partitions of $[0,t]$:

$$\lim_{n\to\infty} \sum f(Y_{\frac{t_i+t_{i+1}}{2}})(B_{t_{i+1}} - B_{t_i}) = \int_0^t f(Y_s) \circ dB_s,$$

with convergence in probability.

6. The Markov Nature of Solutions

One of the original motivations for the development of the stochastic integral was to study continuous strong Markov processes (that is, **diffusions**), as solutions of stochastic differential equations. Let $B = (B_t)_{t\geq 0}$ be a standard Brownian motion in $\mathbf{R}^n$. K. Itô studied systems of differential equations of the form

$$X_t = X_0 + \int_0^t f(s, X_s)dB_s + \int_0^t g(s, X_s)ds,$$

and under appropriate hypotheses on the coefficients f, g he showed that a unique continuous solution exists and that it is strong Markov.

Today we have semimartingale differentials, and it is therefore natural to replace dB and ds with general semimartingales and to study any resulting Markovian nature of the solution. If we insist that the solution itself be Markov then the semimartingale differentials should have independent increments (see Theorem 32); but if we need only to relate the solution to a Markov process, then more general results are available.

We begin our brief treatment of Markov processes with a naive definition. Assume as given a filtered probability space $(\Omega, \mathcal{F}, (\mathcal{F}_t)_{t\geq 0}, P)$ satisfying the *usual hypotheses*.[5]

Definition. A process Z with values in $\mathbf{R}^d$ and adapted to $(\mathcal{F}_t)_{t\geq 0}$ is a **simple Markov process** with respect to $(\mathcal{F}_t)_{t\geq 0}$ if for each $t \geq 0$ the σ-fields $\mathcal{F}_t$ and $\sigma(Z_u; u \geq t)$ are conditionally independent given Z_t.

Thus one can think of the Markov property as a weakening of the property of independent increments. It is easy to see that the simple Markov property is equivalent to the following: For $u \geq t$ and for every f bounded, Borel measurable,

$$E\{f(Z_u)|\mathcal{F}_t\} = E\{f(Z_u)|\sigma(Z_t)\}. \tag{*}$$

One thinks of this as "the best prediction of the future given the past and the present is the same as the best prediction of the future given the present".

Using the equivalent relation (*), one can define a **transition function** for a Markov process as follows, for $s < t$ and f bounded, Borel measurable:

$$P_{s,t}(Z_s, f) = E\{f(Z_t)|\mathcal{F}_s\}.$$

Note that if $f(x) = 1_A(x)$, the indicator function of a set A, then the preceding equality reduces to

$$P(Z_t \in A|\mathcal{F}_s) = P_{s,t}(Z_s, 1_A).$$

---

[5] See Chap. I, Sect. 1 for a definition of the "usual hypotheses" (page 3).

Identifying 1_A with A, we often write $P_{s,t}(Z_s, A)$ on the right side above. When we speak of a Markov process without specifying the filtration of σ-algebras $(\mathcal{F}_t)_{t\geq 0}$, we mean implicitly that $\mathcal{F}_t^0 = \sigma(Z_s; s \leq t)$, the natural filtration generated by the process.

It often happens that the transition function satisfies the relationship

$$P_{s,t} = P_{t-s}$$

for $t \geq s$. In this case we say the Markov process is **time homogeneous**, and the transition functions are a semigroup of operators, known as the **transition semigroup** $(P_t)_{t\geq 0}$. In the time homogeneous case, the Markov property becomes

$$P(Z_{t+s} \in A|\mathcal{F}_t) = P_s(Z_t, A).$$

A stronger requirement that is often satisfied is that the Markov property hold for stopping times.

Definition. A time homogeneous simple Markov process is **strong Markov** if for any stopping time T with $P(T < \infty) = 1$, $s \geq 0$,

$$P(Z_{T+s} \in A|\mathcal{F}_T) = P_s(Z_T, A)$$

or equivalently

$$E\{f(Z_{T+s})|\mathcal{F}_T\} = P_s(Z_T, f),$$

for any bounded, Borel measurable function f.

The fact that we defined the strong Markov property only for time homogeneous processes is not much of a restriction, since if X is an $\mathbf{R}^d$ valued simple Markov process, then it is easy to see that the process $Z_t = (X_t, t)$ is an $\mathbf{R}^{d+1}$ valued time homogeneous simple Markov process.

Examples of strong Markov processes (with respect to their natural filtrations of σ-algebras) are Brownian motion, the Poisson process, and indeed any Lévy process by Theorem 32 of Chap. I. The results of this section will give many more examples as the solutions of stochastic differential equations.

Since we have defined strong Markov processes for time homogeneous processes only, it is convenient to take the coefficients of our equations to be autonomous. We could let them be non-autonomous, however, and then with an extra argument we can conclude that if X is the solution then the process $Y_t = (X_t, t)$ is strong Markov.

We recall a definition from section three of this chapter.

Definition. A function $f : \mathbf{R}_+ \times \mathbf{R}^n \to \mathbf{R}$ is said to be **Lipschitz** if there exists a finite constant k such that

(i) $|f(t,x) - f(t,y)| \leq k|x - y|$, each $t \in \mathbf{R}_+$;

(ii) $t \to f(t,x)$ is right continuous with left limits, each $x \in \mathbf{R}^n$.

f is said to be **autonomous** if $f(t,x) = f(x)$.

In order to allow arbitrary initial conditions, we need (in general) a larger probability space than the one on which Z is defined. We therefore define:

$$\overline{\Omega} = \mathbf{R}^n \times \Omega$$
$$\overline{\mathcal{F}}_t^0 = \mathcal{B} \otimes \mathcal{F}_t$$
$$\overline{P}^y = \epsilon_y \times P$$

where $\mathcal{B}$ denotes the Borel sets of $\mathbf{R}^n$, and ϵ_y denotes the Dirac point mass measure at y. For a point $\overline{\omega} = (y,\omega) \in \overline{\Omega}$, we further define

$$(*) \qquad X_0(\overline{\omega}) = y, \qquad \text{when } \overline{\omega} = (y,\omega).$$

Finally let $\overline{\mathcal{F}}_t = \cap_{u>t}\overline{\mathcal{F}}_u^0$. A random variable Z defined on Ω is considered to be extended automatically to $\overline{\Omega}$ by the rule $Z(\overline{\omega}) = Z(\omega)$, when $\overline{\omega} = (y,\omega)$.

We begin with a measurability result which is an easy consequence of Sect. 4 of Chap. IV.

Theorem 31. *Let Z^j be semimartingales $(1 \leq j \leq d)$, H^x a vector of adapted processes in $\mathbb{D}$ for each $x \in \mathbf{R}^n$, and suppose $(x,t,\omega) \mapsto H_t^x(\omega)$ is $\mathcal{B}\otimes\mathcal{B}_+\otimes\mathcal{F}$ measurable.*[6] *Let F_j^i be functional Lipschitz and for each $x \in \mathbf{R}^n$, X^x is the unique solution of*

$$(X_t^x)^i = (H_t^x)^i + \sum_{j=1}^d \int_0^t F_j^i(X^x)_{s-}dZ_s^j.$$

There exists a version of X^x such that $(x,t,\omega) \mapsto X_t^x(\omega)$ is $\mathcal{B}\otimes\mathcal{B}_+\otimes\mathcal{F}$ measurable, and for each x, X_t^x is a càdlàg solution of the equation.

Proof. Let $X^0(x,t,\omega) = H_t^x(\omega)$ and define inductively

$$X^{n+1}(x,t,\omega)^i = H_t^x + \sum_{j=1}^d \int_0^t F_j^i(X^n(x,\cdot,\cdot))_{s-}dZ_s^j.$$

The integrands above are in $\mathbb{L}$, hence by Theorem 44 in Chap. IV there exists measurable, càdlàg versions of the stochastic integrals. By Theorem 8 the processes X^n converge ucp to the solution X for each x. Then an application of Theorem 43 of Chap. IV yields the result. □

We state and prove the next theorem for one equation. An analogous result (with a perfectly analogous proof) holds for finite systems of equations.

[6] $\mathcal{B}$ denotes the Borel sets on $\mathbf{R}^n$; $\mathcal{B}_+$ the Borel sets on $\mathbf{R}_+$.

Theorem 32. *Let $\mathbf{Z} = (Z^1, \dots, Z^d)$ be a vector of independent Lévy processes, $\mathbf{Z}_0 = 0$, and let (f^i_j) $1 \le j \le d$, $1 \le i \le n$, be Lipschitz functions. Let X_0 be as in (*) and let X be the solution of*

$$(**) \qquad X^i_t = X_0 + \sum_{j=1}^d \int_0^t f^i_j(s-, X_{s-}) dZ^j_s.$$

Then X is a Markov process, under each $\overline{P}^y$ and X is strong Markov if the f^i_j are autonomous.

Proof. We treat only the case $n = 1$. Let T be an $(\mathcal{F}_t)$-stopping time, $T < \infty$ a.s. Define $\mathcal{G}^T = \sigma\{Z^j_{T+u} - Z^j_T; u \ge 0, 1 \le j \le d\}$. Then $\mathcal{G}^T$ is independent of $\mathcal{F}_T$ under $\overline{P}^y$, since the Z^j are Lévy processes, as a consequence of Theorem 32 of Chap. I. Choose a stopping time $T < \infty$ a.s. and let it be fixed. For $u \ge 0$ define inductively

$$Y^0(x, T, u) = x$$

$$Y^{n+1}(x, T, u) = x + \sum_{j=1}^d \int_T^{T+u} f_j(v-, Y^n(x, T, v-)) dZ^j_v.$$

Also, let $X(x, T, u)$ denote the unique solution of

$$X(x, T, u) = x + \sum_{j=1}^d \int_T^{T+u} f_j(v-, X(x, T, v-) dZ^j_v,$$

taking the jointly measurable version (cf. Theorem 31). By Theorem 8 we know that $X(x, T, u)$ is $\mathcal{G}^T$ measurable. By approximating the stochastic integral as a limit of sums, we see by induction that $Y^n(x, T, u)$ is $\mathcal{G}^T$ measurable as well. Under $\overline{P}^x$ we have $X(X_0, T, u) = X(x, T, u)$ a.s., and $Y^n(X_0, T, u) = Y^n(x, T, u)$ $\overline{P}^x$ - a.s., also. By uniqueness of solutions and using Theorem 31, for all $u \ge 0$ a.s.

$$X(X_0, 0, T + u) = X(X(X_0, 0, T), T, u).$$

There is no problem with sets of probability zero, due to (for example) the continuity of the flows. (See Theorem 37, page 246.) Writing E^x to denote expectation on $\overline{\Omega}$ with respect to $\overline{P}^x$, and using the independence of $\mathcal{F}_T$ and $\mathcal{G}^T$ (as well as of $\overline{\mathcal{F}}_T$ and $\overline{\mathcal{G}}^T$), we have for any bounded, Borel function h:

$$\begin{aligned} E^x\{h(X(X_0, 0, T + u))|\overline{\mathcal{F}}_T\} &= E\{h(X(x, 0, T + u))|\mathcal{F}_T\}1_{\mathbf{R}} \\ &= E\{h(X(X(x, 0, T), T, u))\}1_{\mathbf{R}} \\ &= j(X(x, 0, T))1_{\mathbf{R}}, \end{aligned}$$

where $j(y) = E\{h(X(y,T,u)\}$. The last equality follows from the elementary fact that $E\{F(H,\cdot)|\mathcal{H}\} = f(H)$, where $f(h) = E\{F(h,\cdot)\}$, if F is independent of $\mathcal{H}$ and H is $\mathcal{H}$ measurable. This completes the proof, since the fact that $E^x\{h(X(X_0,0,T+u))|\overline{\mathcal{F}}_T\}$ is a function only of $X(x,0,T)$ implies that

$$E^x\{h(X(X_0,0,T+u))|\overline{\mathcal{F}}_T\} = E^x\{h(X(X_0,0,T+u))|X(X_0,0,T)\}. \quad \square$$

It is interesting to note that Theorem 32 remains true with Fisk-Stratonovich differentials. To see this we need a preliminary result.

Theorem 33. *Let $\mathbf{Z} = (Z^1,\dots,Z^d)$ be a vector of independent Lévy processes, $\mathbf{Z}_0 = 0$. Then $[Z^i,Z^j]^c = 0$ if $i \neq j$, and $[Z^i,Z^i]^c_t = \alpha t$, where $\alpha = E\{[Z^i,Z^i]^c_1\}$.*

Proof. First assume that the jumps of each Z^i are bounded. Then the moments of Z^i of all orders exist (Theorem 34 of Chap. I), and in particular $M^i_t \equiv Z^i_t - E(Z^i_t)$ is an L^2 martingale for each i, with $E(Z^i_t) = tE(Z^i_1)$. By independence M^iM^j is also a martingale and hence $[M^i,M^j] = 0$ by Corollary 2 of Theorem 27 of Chap. II. Therefore $[Z^i,Z^j]^c_t = [M^i,M^j]^c_t = 0$ as well.

Next consider $A^i_t \equiv [Z^i,Z^i]_t = [M^i,M^i]_t$. It is an immediate consequence of approximation by sums (Theorem 22 of Chap. II) that A^i also has independent increments. Since

$$A^i_t = [M^i,M^i]^c_t + \sum_{0<s\leq t} (\Delta M^i_s)^2,$$

and the process $J^i_t = \sum_{0<s\leq t}(\Delta M^i_s)^2$ also clearly has independent increments, we deduce that $[M^i,M^i]^c_t$ has independent increments as well. Therefore $[M^i,M^i]^c_t - E\{[M^i,M^i]^c_t\}$ is a finite variation continuous martingale, hence constant by Theorem 27 of Chap. II. Since $[M^i,M^i]^c_0 = 0$, we deduce that $[Z^i,Z^i]^c_t = [M^i,M^i]^c_t = E\{[M^i,M^i]^c_t\}$. Since $[M^i,M^i]_t$ is also a Lévy process we have $E\{[M^i,M^i]_t\} = tE\{[M^i,M^i]_1\}$ (by the stationarity of the increments), and also

$$E\{\sum_{0<s\leq t} (\Delta M^i_s)^2\} = E\{(\sum_{0<s\leq t} \Delta M^i_s)^2\} = tE\{\sum_{0<s\leq 1} (\Delta M^i_s)^2\}$$

by the same reasoning. Therefore $[Z^i,Z^i]^c_t = tE\{[Z^i,Z^i]^c_1\}$ by subtraction.

If the jumps of $\mathbf{Z}$ are not bounded, let $J^i_t = \sum_{0<s\leq t} \Delta Z^i_s 1_{\{|\Delta Z^i_s|\geq 1\}}$. Then $\hat{Z}^i \equiv Z^i - J^i$ and J^i are independent Lévy processes, and J^i is a quadratic pure jump semimartingale. It therefore follows that $[Z^i,Z^j] = [\hat{Z}^i,\hat{Z}^j] = 0$ and that $[Z^i,Z^i]^c = [\hat{Z}^i,\hat{Z}^i]^c$, and the theorem is proved. $\square$

Theorem 34. *Let $\mathbf{Z} = (Z^1, \dots, Z^d)$ be a vector of independent Lévy processes, $\mathbf{Z}_0 = 0$, and let $(f^i_j)_{1\leq j\leq d,\,1\leq i\leq n}$, be non-random $F - S$ acceptable functions.*[7] *Let X_0 be as in Theorem 32 and X be the solution of*

$$X^i_t = X^i_0 + \sum_{j=1}^{d} \int_0^t f^i_j(s-, X_{s-}) \circ dZ^j_s.$$

Then X is a Markov process and X is strong Markov if the coefficients f^i_j are autonomous.

Proof. We treat only the case $n = 1$. By Theorem 33 $[Z^i, Z^j]^c_t = 0$ if $i \neq j$ and $[Z^i, Z^i]^c_t = \alpha_i t$, $\alpha_i \geq 0$. Therefore by Theorem 22 the equation is equivalent to:

$$X_t = X_0 + \sum_{j=1}^{d} \int_0^t f_j(s-, X_{s-}) dZ^j_s$$
$$+ \frac{1}{2} \sum_{j=1}^{d} \alpha_j \int_0^t \left(\frac{\partial f_j}{\partial x}\right)(f_j)(s-, X_{s-}) ds,$$

and since the process $Y_t = t$ is a Lévy process we need only to invoke Theorem 32 to complete the proof. □

If the differentials are not Lévy processes but only strong Markov processes which are semimartingales, then the solutions of equations such as (**) in Theorem 32 need not be Markov processes. One does, however, have the following result.

Theorem 35. *Let Z be a strong Markov processes with values in $\mathbf{R}$, $Z_0 = 0$, such that Z is a semimartingale. Let f and g be Lipschitz functions. Let X_0 be as in Theorem 32 and X be the solution of*

$$X_t = X_0 + \int_0^t f(s-, X_{s-}) dZ_s + \int_0^t g(s, X_{s-}) ds.$$

Then the vector process (X, Z) is Markov under $\overline{P}^y$, each $y \in \mathbf{R}$, and strong Markov if f and g are autonomous.

Proof. First recall that X is defined on $\overline{\Omega} = \mathbf{R} \times \Omega$ and Z is automatically extended to $\overline{\Omega}$ as explained at the beginning of this section. The probability $\overline{P}^y = \epsilon_y \times P$ is such that $\overline{P}^y(X_0 = y) = 1$, each $y \in \mathbf{R}$. As in the proof of Theorem 32 for a fixed stopping time T ($T < \infty$ a.s.) and $x \in \mathbf{R}$ define

[7] $F - S$ acceptable functions are defined on page 223.

inductively for $u \geq 0$:

$$Y^0(x,T,u) = x$$

$$Y^{n+1}(x,T,u) = x + \int_T^{T+u} f(v-, Y^n(x,T,v-))dZ_v + \int_T^{T+u} g(v-, Y^n(x,T,v-))dv,$$

and let $X(x,T,u)$ denote the unique solution of

$$X(x,T,u) = x + \int_T^{T+u} f(v-, X(x,T,v-))dZ_v + \int_T^{T+u} g(v-, X(x,T,v-))dv.$$

Next define $\mathcal{G}^T = \sigma\{Z_{T+u} - Z_T; u \geq 0\}$ for the same stopping time T. As in the proof of Theorem 32 we see that $Y^n(x,T,u)$ is $\mathcal{G}^T$ measurable for each $n \geq 1$ and that $X(x,T,u)$ is $\mathcal{G}^T$ measurable as well. Next let h be Borel measurable and bounded, and let $G \in \mathcal{G}^T$ and bounded.

Observe that

$$\begin{aligned} E\{h(X_T)G|\mathcal{F}_T\} &= h(X_T)E\{G|\mathcal{F}_T\} \\ &= h(X_T)E\{G|Z_T\} \\ &= j(X_T, Z_T), \end{aligned}$$

where $j : \mathbf{R}^2 \to \mathbf{R}$ is Borel. Since $X_{T+u} = X(X_T, T, u)$ by the uniqueness of the solution, the theorem now follows from the monotone class theorem (Theorem 8 of Chap. I). □

Theorem 35 can also be shown to be true with Stratonovich differentials, but the proof is more complicated since the quadratic variation process is an additive functional of Z, rather than a deterministic process (as is the case when Z is a Lévy process). For this type of result we refer the reader to Çinlar-Jacod-Protter-Sharpe [**1**].

Each of Theorems 32, 34, 35 can be interpreted with X having an arbitrary initial distribution μ. Indeed for μ a probability measure on $(\mathbf{R}, \mathcal{B})$, define $\overline{P}^\mu$ on $\overline{\Omega}$ by $\overline{P}^\mu(\Lambda) = \int_{\mathbf{R}} \overline{P}^x(\Lambda)\mu(dx)$, for $\Lambda \in \mathcal{B} \otimes \mathcal{F}$. Then the conclusions of the three theorems are trivially still valid for $\overline{P}^\mu$, and the distribution of X_0 is μ.

Traditionally the most important Markovian solutions of stochastic differential equations are diffusions. Suppose as given a space $(\Omega, \mathcal{F}, (\mathcal{F}_t)_{t \geq 0}, P)$ satisfying the usual hypotheses.

Definition. An adapted process X with values in $\mathbf{R}^n$ is a **diffusion**[8] if it has continuous sample paths and if it satisfies the strong Markov property.

[8] The definition of a diffusion is not yet standardized. We give a general definition.

A restatement of Theorems 32 and 34 yields the following.

Theorem 36. *Let* $\mathbf{B} = (B^1, \dots, B^d)$ *be a standard Brownian motion on* $\mathbb{R}^d$, $\mathbf{B}_0 = 0$, *and let* $(f^i_j)_{1 \le j \le d, 1 \le i \le n}$, g^i *be autonomous Lipschitz functions. Let* X^i_0 *be as in Theorem 32 and let* X *be the solution of*

$$(***) \qquad X^i_t = X^i_0 + \sum_{j=1}^{d} \int_0^t f^i_j(X_s)dB^j_s + \int_0^t g^i(X_s)ds.$$

Then X *is a diffusion. If the coefficients* $(f^i_j)_{1 \le j \le d}$, $1 \le i \le n$, *are non-random* $F - S$ *acceptable functions and if* Y *is the solution of*

$$Y^i_t = Y^i_0 + \sum_{j=1}^{d} \int_0^t f^i_j(Y_s) \circ dB^j_s + \int_0^t g^i(Y_s)ds,$$

then Y *is a diffusion.*

In equation (***) of Theorem 36 the coefficients f^i_j are called the **diffusion coefficients** and the coefficients g^i are called the **drift coefficients**.

Note that a diffusion need not be semimartingale, even though of course the solutions of equations such as (***) are semimartingales. Indeed any deterministic continuous function with paths of unbounded variation is a diffusion which is not a semimartingale. Another interesting example is provided by Tanaka's formula. The process

$$|B_t| = \int_0^t \operatorname{sign}(B_s)dB_s + L^0_t$$

is a diffusion and it is a semimartingale, where B is standard Brownian motion on $\mathbb{R}$. Since the paths of L^0_t are singular with respect to Lebesgue measure, and since a semimartingale decomposition is unique for continuous processes (the Corollary of Theorem 19 of Chap. III), we see that $|B_t|$ cannot be represented as a solution of (***). Another example is that of $|B_t|^{\frac{1}{3}}$ which is a time homogeneous diffusion but not a semimartingale (Theorem 52 of Chap. IV).

An intuitive notion of a diffusion is to imagine a pollen grain floating downstream in a river. The grain is subject to two forces: the current of the river (drift), and the aggregate bombardment of the grain by the surrounding water molecules (diffusion). The coefficient $f(t,x)$ then represents the sensitivity of the particle at time t and place x to the diffusion forces. For example if part of the river water is warmer at certain times and places (due to sunlight or industrial effluents, for example), then f might be larger. Analogously g would be larger when the river was flowing faster due to a steeper incline.

We give three simple examples of diffusions.

Example 1. The stochastic exponential $e^{B_t - \frac{1}{2}t}$ is a diffusion where $f(t,x) = x$ and $g(t,x) = 0$. □

Example 2. Consider the simple system:

$$V_t = V_0 + \int_0^t \sigma dB_s + \int_0^t \alpha V_s ds$$

$$X_t = X_0 + \int_0^t V_s ds.$$

The process X can be used as a model of Brownian motion alternative to Einstein's; it is called the Ornstein-Uhlenbeck Brownian motion, or simply the **Ornstein-Uhlenbeck process**. Note that here the process X has paths of finite variation and hence the process $(V_t)_{t \geq 0}$ is a true velocity process for X. Using integration by parts we can verify that

$$V_t = e^{\alpha t}(V_0 + \int_0^t e^{-\alpha s}\sigma dB_s)$$

is an explicit solution for V. Indeed:

$$\begin{aligned} e^{-\alpha t}V_t &= V_0 + \int_0^t V_s(-\alpha e^{-\alpha s})ds + \int_0^t e^{-\alpha s} dV_s \\ &= V_0 - \alpha \int_0^t V_s e^{-\alpha s} ds + \int_0^t e^{-\alpha s}\alpha V_s ds + \int_0^t e^{-\alpha s}\sigma dB_s \\ &= V_0 + \int_0^t e^{-\alpha s}\sigma dB_s, \end{aligned}$$

and we are done. Since V_0 and the Brownian motion B are independent (by construction), we see that when V_0 has a Gaussian distribution then V is a Gaussian process. If α is negative and we take $\sigma^2 = -\frac{1}{2\alpha}$, then V is a *stationary* Gaussian process. □

Example 3. Consider next the equation

$$X_t = \int_0^t \frac{-X_s}{1-s} ds + B_t, \qquad (0 \leq t < 1)$$

for B a standard Brownian motion.

For each t_0, $0 < t_0 < 1$, there is a solution which is a diffusion on $[0, t_0]$. By the uniqueness of solutions if $t_1 < t_0$ then the solution for $[0, t_0]$ agrees with the solution for $[0, t_1]$ on the interval $[0, t_1]$. Thus we have a solution on $[0, 1)$. If we can show that $\lim_{t \to 1} X_t = 0$, then the solution extends by continuity to $[0, 1]$ and we will have constructed a diffusion X on $[0, 1]$ with $X_0 = X_1 = 0$, known as the **Brownian bridge**.[9]

[9] The Brownian bridge is also known as *tied down Brownian motion*, and alternatively as *pinned Brownian motion*.

An application of integration by parts shows that the solution of the Brownian bridge equation is given by:

$$X_t = (1-t)\int_0^t \frac{1}{1-s}dB_s \qquad (0 \le t < 1).$$

Write $X_t = f(t)M_t$, where $f(t) = (1-t)$ and $M_t = \int_0^t \frac{1}{1-s}dB_s$. To see that $\lim_{t\to 1} X_t = 0$, we study M_t, making the change of variables $t = \frac{u}{1+u}$. Then $0 \le t < 1$ corresponds to $0 \le u < \infty$, and define

$$N_u = M_{\frac{u}{1+u}}, \quad \mathcal{G}_u = \mathcal{F}_{\frac{u}{1+u}}, \ 0 \le u < \infty.$$

N is clearly a continuous $\mathcal{G}_u$ martingale, and moreover $[N,N]_u = u$; hence N is a standard Brownian motion by Lévy's theorem (Theorem 38 of Chap. II). It is then easy to show that, as a consequence of the Strong Law of Large Numbers, that

$$\lim_{t\to\infty} \frac{N_t}{t} = 0 \quad \text{a.s.}^{10}$$

Let $g(t,\omega) = \frac{N_t(\omega)}{t}$. Then $\lim_{t\to 0} g(\frac{1}{t},\omega) = 0$ a.s., and therefore $\lim_{t\to 0} tN_{1/t} = 0$ a.s. We then have, replacing t with $(1-t)$

$$\lim_{t\to 1}(1-t)N_{1/1-t} = \lim_{t\to 1}(1-t)M_{1/(2-t)} = 0 \quad a.s.,$$

and therefore $\lim_{u\to 1}(1-u)M_u = 0$ a.s., and hence $\lim_{t\to 1} X_t = 0$ a.s.

We now know that X is a continuous diffusion on $[0,1]$, and that $X_0 = X_1 = 0$ a.s. Also X is clearly a semimartingale on $[0,1)$, but it is not obvious that X is a semimartingale on $[0,1]$: One needs to show that the integral $\int_0^1 \frac{|X_s|}{1-s}ds < \infty$ a.s. To see this calculate $E\{X_t^2\}$, $0 \le t < 1$:

$$\begin{aligned} E\{X_t^2\} &= f(t)^2 E\{M_t^2\} \\ &= f(t)^2 E\{[M,M]_t\} \\ &= (1-t)^2 \int_0^t \frac{1}{(1-s)^2}ds \\ &= (1-t)^2\left(\frac{1}{1-t} - 1\right). \end{aligned}$$

By the Cauchy-Schwarz inequality

$$E\{|X_t|\} \le E\{X_t^2\}^{1/2} = \sqrt{t(1-t)}.$$

Therefore

$$\begin{aligned} E\{\int_0^1 \frac{|X_s|}{1-s}ds\} &= \int_0^1 \frac{E\{|X_s|\}}{1-s}ds \\ &\le \int_0^1 \frac{\sqrt{s(1-s)}}{1-s}ds < \infty, \end{aligned}$$

[10] See, for example, Breiman [1, p. 265].

whence $\int_0^1 \frac{|X_s|}{1-s}ds < \infty$ a.s. Therefore the solution X is a semimartingale on $[0,1]$.

Finally we remark that one can construct a similar Brownian bridge from any value a to any value b, on an interval of arbitrary finite length τ. Using the interval $[0,\tau]$ one has the equation:

$$X_t = a + \int_0^t \frac{b-X_s}{\tau - s}ds + B_t, \qquad 0 \leq t < \tau,$$

with solution:

$$X_t = \begin{cases} a(1-\frac{t}{\tau}) + \frac{bt}{\tau} + (\tau - t)\int_0^t \frac{1}{\tau - s}dB_s, & 0 \leq t < \tau, \\ X_\tau = b. \end{cases} \qquad \square$$

7. Flows of Stochastic Differential Equations: Continuity and Differentiability

Consider a stochastic differential equation of the form:

$$X_t = x + \int_0^t F(X)_{s-}dZ_s.$$

Obviously there is a dependence on the initial condition, and we can write the solution in the form $X(t,\omega,x)$, or $X_t^x(\omega)$. The study of the *flow of a stochastic differential equation* is the study of the functions $\phi : x \mapsto X(t,\omega,x)$ which can be considered as mapping $\mathbf{R}^n \to \mathbf{R}^n$ for (t,ω) fixed, or as mapping $\mathbf{R}^n \to \mathcal{D}^n$, where $\mathcal{D}^n$ denotes the space of càdlàg functions from $\mathbf{R}_+$ to $\mathbf{R}^n$, equipped with the topology of uniform convergence on compacts. It is important to distinguish between $\mathcal{D}^n$ and $\mathbf{D}^n$: The former is a *function space*, and it is associated in the literature with weak convergence results (see, e.g., Billingsley [**1**], or Ethier-Kurtz [**1**], or Kurtz-Protter [**1**] for recent results in a semimartingale context); the latter is the space of *stochastic processes* with càdlàg paths, and which are adapted to the underlying filtration.

Note that we have already encountered flows in Sect. 6 (cf. Theorem 31), where we proved measurability of the solution with respect to a parameter (which can be taken to be, of course, the initial condition). We will be interested in several properties of the flows: continuity, differentiability, injectivity, and when the flows are diffeomorphisms of $\mathbf{R}^n$.

We begin with continuity. We consider a general system of equations of the form:

$$(*) \qquad X_t^x = H_t^x + \int_0^t F(X^x)_{s-}dZ_s,$$

where X_t^x and H_t^x are $\mathbf{R}^n$ column vectors, Z is a column vector of m semimartingales with $\mathbf{Z}_0 = 0$, and F is an $n \times m$ matrix with elements (F_α^i). For x fixed, for each y we have that $\overline{X}_t = X_t^y - X_t^x$ is a solution of the equation:

$$(**) \qquad \overline{X}_t = H_t^y - H_t^x + \int_0^t \overline{F}(\overline{X})_{s-} dZ_s,$$

where $\overline{F}(Y) = F(X^x + Y) - F(X^x)$.

Theorem 37. *Let H^x be processes in $\mathbf{D}^n$, and let $x \mapsto H^x : \mathbf{R}^n \mapsto \mathbf{D}^n$ be prelocally Lipschitz continuous in $\underline{\underline{S}}^p$, some $p > n$. Let F be an $n \times m$ matrix of functional Lipschitz operators (F_α^i), $1 \le i \le n$, $1 \le \alpha \le m$.*[11] *Then there exists a function $X(t, \omega, x)$ on $\mathbf{R}_+ \times \Omega \times \mathbf{R}^n$ such that:*

(i) For each x the process $X_t^x(\omega) = X(t, \omega, x)$ is a solution of (*);
(ii) For almost all ω, the flow $x \mapsto X(\cdot, \omega, x)$ from $\mathbf{R}^n$ into $\mathcal{D}^n$ is continuous in the topology of uniform convergence on compacts.

Proof. We recall the method of proof used to show the existence and uniqueness of a solution (Theorem 7). By stopping at a fixed time t_0, we can assume the Lipschitz process is just a random variable K which is finite a.s. Then by conditioning[12] we can assume without loss of generality that this Lipschitz constant is non random, and we call it $c < \infty$. By replacing H_t^x with $X_t^x + \int_0^t F(0)_{s-} dZ_s$, and then by replacing F with G given by: $G(Y)_t = F(Y)_t - F(0)_t$, we can further assume without loss of generality that $F(0) = 0$. Then for $\beta = C_p(c)$, by Theorem 5 we can find an arbitrarily large stopping time T such that $Z^{T-} \in \mathcal{S}(\beta)$, and H^x is Lipschitz continuous in $\underline{\underline{S}}^p$ on $[0, T)$. Then by Lemma 2 (preceding Theorem 7) we have that for the solution $\overline{X}$ of (**):

$$\|\overline{X}^{T-}\|_{\underline{\underline{S}}^p} \le C_p(c, Z) \|(H^x - H^y)^{T-}\|_{\underline{\underline{S}}^p}, \qquad \text{for any } p \ge 2,$$

and some (finite) constant $C_p(c, Z)$. Choose $p > n$, and we have:

$$(***) \qquad E\{\sup_{s<T} |X_s^x - X_s^y|^p\} \le C_p(c, Z) K \|x - y\|^p,$$

due to the Lipschitz hypothesis on $x \to H^x$. By Kolmogorov's Lemma (Theorem 53 of Chap. IV) we have the result on $\mathbf{R}^n \times [0, T)$. However since T was arbitrarily large, the result holds as well on $\mathbf{R}^n \times \Omega \times \mathbf{R}_+$. □

For the remainder of this section we will consider less general equations. Indeed, the following will be our basic hypotheses, which we will supplement as needed.

---

[11] See page 195 for the definition of functional Lipschitz.

[12] See the proofs of Theorem 7, 8, or 15 for this argument.

Hypotheses.

(H1) Z^α *are given semimartingales with* $Z_0^\alpha = 0$, $1 \le \alpha \le m$;
(H2) $f_\alpha^i : \mathbf{R}^n \to \mathbf{R}$ *are given functions,* $1 \le i \le n$, $1 \le \alpha \le m$, *and* $f(x)$ *denotes the* $n \times m$ *matrix* $(f_\alpha^i(x))$.

We will study the system of equations:

$$(\otimes) \qquad X_t^i = x^i + \sum_{\alpha=1}^{m} \int_0^t f_\alpha^i(X_{s-})dZ_s^\alpha, \quad 1 \le i \le n$$

which we also write

$$(\otimes\otimes) \qquad X_t = x + \int_0^t f(X_{s-})dZ_s$$

where it is understood that X_t and x are column vectors in $\mathbf{R}^n$, $f(X_{s-})$ is an $n \times m$ matrix, and Z is a column vector of m semimartingales.

To study the differentiability of the flow we will need a more general theorem on existence and uniqueness of solutions: Indeed, we wish to replace our customary (global) Lipschitz condition with a local Lipschitz condition.

Definition. A function $f : \mathbf{R}^n \to \mathbf{R}$ is said to be **locally Lipschitz** if there exists an increasing sequence of open sets Λ_k such that $\bigcup_k \Lambda_k = \mathbf{R}^n$ and f is Lipschitz with a constant K_k on each Λ_k.

For example, if f has continuous first partial derivatives, then it is locally Lipschitz, while if its continuous first partials are *bounded*, then it is Lipschitz. If f, g, are both Lipschitz, then their product fg is locally Lipschitz. These coefficients arise naturally in the study of Fisk-Stratonovich equations (see Sect. 5).

Theorem 38. *Let* Z *be as in (H1) and let the functions* (f_α^i) *in (H2) be locally Lipschitz. Then there exists a function* $\zeta(x,\omega) : \mathbf{R}^n \times \Omega \to [0,\infty]$ *such that for each* x $\zeta(x,\cdot)$ *is a stopping time, and there exists a unique solution of*

$$(\otimes\otimes) \qquad X_t = x + \int_0^t f(X_{s-})dZ_s$$

up to $\zeta(x,\cdot)$ *with* $\limsup_{t\to\zeta(x,\cdot)} \|X_t\| = \infty$ *a.s. on* $\{\zeta < \infty\}$. *Moreover* $x \to \zeta(x,\omega)$ *is lower semi-continuous, strictly positive, and the flow of* X *is continuous on* $[0,\zeta(x,\cdot))$.

Remark. Before proving the theorem we comment that for each x fixed the stopping time $T(\omega) = \zeta(x,\omega)$ is called an **explosion time**. Thus Theorem 38 assures the existence and uniqueness of a solution *up to an explosion time*; and at that time, the solution does indeed explode in the sense that

$\overline{\lim}_{t\to T}\|X_t\| = +\infty$ on $\{T < \infty\}$. Note however that if the coefficients (f^i_α) in (H2) are (globally) Lipschitz, then a.s. $\zeta = \infty$ for all x (Theorem 7).

Proof. Let Λ_ℓ be open sets increasing to $\mathbf{R}^n$ such that there exist (h_ℓ) a sequence of $\mathcal{C}^\infty$ functions with compact support mapping $\mathbf{R}^n$ to $[0,1]$ such that $h_\ell = 1$ on Λ_ℓ. For each ℓ we let $X_\ell(t,\omega,x)$ denote the solution (continuous in x) of the equation

$$X_t = x + \int_0^t g_\ell(X_{s-})dZ_s$$

where the matrix $g_\ell(x)$ is defined by $h_\ell(x)f(x)$.

Define stopping times, for x fixed:

$$S_\ell(\omega,x) = \inf\{t > 0 : X_\ell(t,\omega,x) \notin \Lambda_\ell \text{ or } X_\ell(t-,\omega,x) \notin \Lambda_\ell\}.$$

By Theorem 37 the flow is uniformly continuous on compact sets, hence the functions S_ℓ are lower semi continuous. Then on $[0, S_\ell)$ we have that $X_\ell = X_{\ell+1}$ by the uniqueness of the solutions, since they both satisfy equation $(\otimes\otimes)$. We wish to show that the relation

$$\text{(L)} \qquad X_\ell(\cdot,\omega,x) = X_{\ell+1}(\cdot,\omega,x) \quad \text{on } [0, S_\ell(\omega,x))$$

holds for all $x \in \mathbf{R}^n$ simultaneously. Choose an x and let:

$$A^x = \{\omega : X_\ell(\cdot,\omega,x) = X_{\ell+1}(\cdot,\omega,x) \text{ on } [0, S_\ell(\omega,x))\}.$$

Then $P(A^x) = 1$. We set $A = \bigcup_{x\in\mathbf{Q}^n} A^x$, where $\mathbf{Q}^n$ denotes the rationals in $\mathbf{R}^n$. Then $P(A) = 1$ as well, and without loss of generality we take $A = \Omega$. Next for arbitrary $y \in \mathbf{R}^n$, let $y_n \to y$, with $y_n \in \mathbf{Q}^n$. Then

$$S_\ell(\omega,y) \le \liminf_{n\to\infty} S_\ell(\omega,y_n),$$

and therefore the relation (L) holds for y as well, by the previously established continuity.

Observe that $S_\ell(\omega,x) \le S_{\ell+1}(\omega,x)$, and let $\zeta(x,\omega) = \sup_\ell S_\ell(\omega,x)$, for each x. Then ζ is lower semi-continuous because the S_ℓ are, and it is strictly positive. Further, we have shown that there exists a unique function $X(\cdot,\omega,x)$ on $[0,\zeta(x,\omega))$ which is a solution of $(\otimes\otimes)$, and it is equal to $X_\ell(\cdot,\omega,x)$ on $[0,S_\ell(\omega,x))$, each $\ell \ge 1$. Indeed, on $[0,S_\ell(\omega,x))$ we have $X(\cdot,\omega,x) = X_\ell(\cdot,\omega,x)$, and since $X_\ell(\cdot,\omega,x) \in \Lambda_\ell$, we have $h_\ell(X_\ell(\cdot,\omega,x)) = 1$, wherefore X_ℓ is a solution of $(\otimes\otimes)$ on $[0,S_\ell)$.

By our construction we have that $X(S_\ell(\omega,x),\omega,x)$ belongs to Λ_ℓ^c: Indeed, letting $S_\ell(\omega,x) = s$, then $X(s-,\omega,x) = X_\ell(s-,\omega,x) = u \in \overline{\Lambda_\ell}$, for some value u. Since $u \in \overline{\Lambda_\ell}$ we have $h_\ell(u) = 1$, and thus X and X_ℓ have the same jump at s and we can conclude $X(s,\omega,x) = X_\ell(s,\omega,x)$. But $X_\ell(s,\omega,x)$ or $X_\ell(s-,\omega,x)$ must be in Λ_ℓ^c by right continuity and the definition of S_ℓ;

therefore $X(s,\omega,x)$ or $X(s-,\omega,x)$ is in Λ_ℓ^c. This shows that $\overline{\lim}_{t\to\zeta}\|X_t\| = \infty$ on $\{\zeta < \infty\}$. □

It is tempting to conclude that if there are no explosions at a finite time for each initial condition x a.s. (null set depending on x), then also $\zeta(x,\cdot) = \infty$ a.s. (with the same null set for all initial values x). Unfortunately this is not true as the next example shows.

Example. Let B be a complex Brownian motion: That is, let B^1 and B^2 be two independent Brownian motions on $\mathbb{R}$ and let $B_t \equiv B_t^1 + iB_t^2$, where $i^2 = -1$. Consider the stochastic differential equation:

$$Z_t = z - \int_0^t Z_s^2 dB_s,$$

where the initial value z is complex. This equation has a closed form solution:

$$Z(t,\omega,z) = \frac{z}{1 + zB_t(\omega)}.$$

Indeed, if we set $f(x) = \frac{z}{1+zx}$ and $Z_t = f(B_t)$, then $f'(B_t) = -Z_t^2$, and since $B_t = B_t^1 + iB_t^2$, we have by Itô's formula:

$$Z_t = z - \int_0^t Z_s^2 dB_s + \frac{1}{2}\int_0^t f''(B_s)(d[B^1,B^1]_s + i^2 d[B^2,B^2]_s);$$

since $d[B^1,B^1]_s = d[B^2,B^2]_s = ds$, we see that $Z_t = z - \int_0^t Z_s^2 dB_s$.

For z fixed we know that $P(\exists t : B_t = -\frac{1}{z}) = 0$, since Brownian motion in $\mathbb{R}^2$ a.s. does not hit a specified point. Therefore Z does not have explosions in a finite time for each fixed initial condition $Z_0 = z$. On the other hand for *any* ω_0 and t_0 fixed we have $B_{t_0}(\omega_0) = z_0$ for some $z_0 \in \mathbb{C}$, and thus for the initial value $z = -\frac{1}{z_0}$ we have an explosion at the chosen finite time t_0. Thus each trajectory has initial conditions such that it will explode at a finite time, and $P(\{\omega : \exists z \text{ with } \zeta(\omega,z) < \infty\}) = 1$. □

We next turn our attention to the differentiability of the flows. To this end we consider the system of $n + n^2$ equations (assuming that the coefficients (f_α^i) are at least C^1):

$$\text{(D)}\qquad \begin{aligned} X_t^i &= x_i + \sum_{\alpha=1}^m \int_0^t f_\alpha^i(X_{s-})dZ_s^\alpha \\ D_{kt}^i &= \delta_k^i + \sum_{\alpha=1}^m \sum_{j=1}^n \int_0^t \frac{\partial f_\alpha^i}{\partial x_j}(X_{s-})D_{ks-}^j dZ_s^\alpha, \end{aligned}$$

$(1 \le i \le n)$ where D denotes an $n \times n$ matrix valued process and $\delta_k^i = 1$ if $i = k$ and 0 otherwise (Kronecker's delta). A convenient convention, sometimes called *Einstein's convention*, is to leave the summations implicit. Thus the system of equations (D) can be alternatively written:

$$
\text{(D)} \qquad
\begin{aligned}
X_t^i &= x_i + \int_0^t f_\alpha^i(X_{s-})dZ_s^\alpha \\
D_{kt}^i &= \delta_k^i + \int_0^t \frac{\partial f_\alpha^i}{\partial x_j}(X_{s-})D_{ks-}^j dZ_s^\alpha.
\end{aligned}
$$

We will use the Einstein convention when there is no ambiguity. Note that in equations (D) if X is already known, then the second system is *linear* in D. Also note that the coefficients for the system (D) are not *globally* Lipschitzian, but if the first partials of the (f_α^i) are locally Lipschitzian, then so also are the coefficients of (D).

Theorem 39. *Let Z be as in (H1) and let the functions (f_α^i) in (H2) have locally Lipschitz first partial derivatives. Then for almost all ω there exists a function $X(t,\omega,x)$ which is continuously differentiable in the open set $\{x : \zeta(x,\omega) > t\}$, where ζ is the explosion time (cf. Theorem 38). If (f_α^i) are globally Lipschitz then $\zeta = \infty$. Let $D_k(t,\omega,x) \equiv \frac{\partial}{\partial x_k}X(t,\omega,x)$. Then for each x the process $(X(\cdot,\omega,x), D(\cdot,\omega,x))$ is identically càdlàg, and it is the solution of equations (D) on $[0,\zeta(x,\cdot))$.*

Proof. We will give the proof in several steps. In *Step 1* we will reduce the problem to one where the coefficients are globally Lipschitz. We then resolve the first system (for X) of (D), and in *Step 2* we will show that, given X, there exists a "nice" solution D of the second system of equations, which depends continuously on x. In *Step 3* we will show that D_k^i is the partial derivative in x_k of X^i in the distributional sense.[13] Then since it is continuous (in x), we can conclude that it is the true partial derivative.

Step 1: Choose a constant N. Then the open set $\{x : \zeta(x,\omega) > N\}$ is a countable union of closed balls. Therefore it suffices to show that if B is one of these balls, then on the set $\Gamma = \{\omega : \forall x \in B, \zeta(x,\omega) > N\}$ the function $x \to X(t,\omega,x)$ is continuously differentiable on B. However by Theorem 38 we know that for each $\omega \in \Gamma$, the image of X as x runs through B is compact in $\mathcal{D}^n$ with $0 \le t \le N$, hence it is contained in a ball of radius R in $\mathbf{R}^n$, for R sufficiently large.

We fix the radius R and we denote by K the ball of radius R of $\mathbf{R}^n$ centered at 0. Let

$$\Lambda = \{\omega : \text{ for } x \in B \text{ and } 0 \le t \le N, X(t,\omega,x) \in K\}.$$

We then condition on Λ. That is, we replace P by P_Λ, where $P_\Lambda(A) \equiv P(A|\Lambda) = \frac{P(A\cap\Lambda)}{P(\Lambda)}$. Then $P_\Lambda \ll P$, so Z is still a semimartingale with respect to P_Λ (Theorem 2 of Chap. II). This allows us to make, without loss of

[13] These derivatives are also known as *derivatives in the generalized function sense.*

generality, the following simplifying assumption: *if $x \in B$ then $\zeta(x,\omega) > N$, and $X(t,\omega,x) \in K$, $0 \le t \le N$.*

Next let $h : \mathbf{R}^n \to \mathbf{R}$ be $\mathcal{C}^\infty$ with compact support and such that $h(x) = 1$ if $x \in K$ and replace f with fh. Let Z be implicitly stopped at the (constant stopping) time N. (That is, Z^N replaces Z.) With these assumptions and letting P_Λ replace P, *we can therefore assume – without loss of generality – that the coefficients in the first equation in* (D) *are globally Lipschitz and bounded.*

Step 2: In this step we assume that the simplifying assumptions of Step 1 hold. We may also assume by Theorem 5 that $Z \in \mathcal{S}(\beta)$ for a β which will be specified later. If we were to proceed to calculate formally the derivative with respect to x_k of X^i, we would get:

$$\begin{aligned}\frac{\partial X_t^i}{\partial x_k} &= 1 + \sum_{\alpha=1}^m \frac{\partial}{\partial x_k}\int_0^t f_\alpha^i(X_{s-})dZ_s^\alpha \\ &= 1 + \sum_{\alpha=1}^m \int_0^t \sum_{j=1}^n \frac{\partial f_\alpha^i}{\partial x_j}(X_{s-})\frac{\partial}{\partial x_k}(X_{s-}^j)dZ_s^\alpha.\end{aligned}$$

Therefore our best candidate for the partial derivative with respect to x_k is the solution of the system:

$$(*4) \qquad D_{kt}^i = \delta_k^i + \sum_{\alpha=1}^m \int_0^t \sum_{j=1}^n \frac{\partial f_\alpha^i}{\partial x_j}(X_{s-})D_{ks-}^j dZ_s^\alpha,$$

and let D be the matrix (D_k^i). Naturally $X_s = X(s,\omega,x)$, and we can make explicit this dependence on x by rewriting the above equation as:

$$D_{kt}^{ix} = \delta_k^i + \int_0^t \frac{\partial f_\alpha^i}{\partial x_j}(X_{s-}^x)D_{ks-}^{jx}dZ_s^\alpha,$$

where the summations over α and j are implicit (Einstein convention). We now show that $D_k = (D_k^1, \dots, D_k^n)$ is continuous in x. Fix $x, y \in \mathbf{R}^n$ and let $V_s(\omega) = D_k(s,\omega,x) - D_k(s,\omega,y)$. Then:

$$\begin{aligned}V_t^i &= \int_0^t \left\{\frac{\partial f_\alpha^i}{\partial x_j}(X_{s-}^x)D_{ks-}^{jx} - \frac{\partial f_\alpha^i}{\partial x_j}(X_{s-}^y)D_{ks-}^{jy}\right\}dZ_s^\alpha \\ &= \int_0^t \left\{\frac{\partial f_\alpha^i}{\partial x_j}(X_{s-}^x) - \frac{\partial f_\alpha^i}{\partial x_j}(X_{s-}^y)\right\}D_{ks-}^{jx}dZ_s^\alpha \\ &\quad + \int_0^t \frac{\partial f_\alpha^i}{\partial x_j}(X_{s-}^y)\left\{D_{ks-}^{jx} - D_{ks-}^{jy}\right\}dZ_s^\alpha \\ &= H_t^i(x,y,k) + \int_0^t \frac{\partial f_\alpha^i}{\partial x_j}(X_{s-}^y)V_{s-}^j dZ_s^\alpha \\ &= H_t^i + \int_0^t V_{s-}^j dY_{js}^{iy},\end{aligned}$$

where $Y^{iy}_{js} = \sum_{\alpha=1}^m \int_0^s \frac{\partial f^i_\alpha}{\partial x_j}(X^y_{u-})dZ^\alpha_u$. Note that by Step 1 we know that $\frac{\partial f^i_\alpha}{\partial x_j}(X^y_{u-})$ is bounded; therefore since $Z^\alpha \in \mathcal{S}(\beta)$, then the Y^i_j are in $\mathcal{S}(c\beta)$ for a constant c. If β is small enough, by Lemma 2 (preceding Theorem 7) we have that for each $p \geq 2$ there exists a constant $C_p(Z)$ such that

$$\|V\|_{\underline{\underline{S}}^p} \leq C_p(Z)\|H\|_{\underline{\underline{S}}^p}.$$

However we can also estimate $\|H\|_{\underline{\underline{S}}^p}$. If we let

$$J^i_{\alpha s} = \sum_{j=1}^n \left\{ \frac{\partial f^i_\alpha}{\partial x_j}(X^x_{s-}) - \frac{\partial f^i_\alpha}{\partial x_j}(X^y_{s-}) \right\} D^{jx}_{ks-},$$

then $H^i_t = \sum_{\alpha=1}^m \int_0^t J^i_{\alpha s} dZ^\alpha_s$, and therefore by Emery's inequality (Theorem 3) we have that

$$\|H\|_{\underline{\underline{S}}^p} \leq c_p \|J\|_{\underline{\underline{S}}^p} \|Z\|_{\underline{\underline{H}}^\infty},$$

which in turn implies

$$\|V\|_{\underline{\underline{S}}^p} \leq \hat{C}_p(Z)\|J\|_{\underline{\underline{S}}^p}.$$

We turn our attention to estimating $\|J\|_{\underline{\underline{S}}^p}$. Consider first the terms

$$\frac{\partial f^i_\alpha}{\partial x_j}(X^x_{s-}) - \frac{\partial f^i_\alpha}{\partial x_j}(X^y_{s-}).$$

By the simplifying assumptions of Step 1, the functions $\frac{\partial f^i_\alpha}{\partial x_j}$ are Lipschitz in K, and X^x takes its values in K. Therefore Theorem 37 applies, and as we saw in its proof (inequality (***)) we have that:

$$\left\| \frac{\partial f^i_\alpha}{\partial x_j}(X^x_{s-}) - \frac{\partial f^i_\alpha}{\partial x_j}(X^y_{s-}) \right\|_{\underline{\underline{S}}^{2p}} \leq K\|X^x - X^y\|_{\underline{\underline{S}}^{2p}} \leq \widetilde{K}(p, Z)\|x - y\|.$$

Next consider the terms D^{jx}_{ks}. We have seen that these terms are solutions of the system of equations:

$$D^{jx}_{kt} = \delta^j_k + \int_0^t D^{\ell x}_{ks-} \frac{\partial f^j_\alpha}{\partial x_\ell}(X^x_{s-})dZ^\alpha_s$$

and therefore they can be written as solutions of the exponential system:

$$D^{jx}_{kt} = \delta^j_k + \int_0^t D^{\ell x}_{ks-} dY^{jx}_{\ell s},$$

with $Y^{jx}_{\ell s} = \int_0^s \frac{\partial f^j_\alpha}{\partial x_\ell}(X^x_{u-})dZ^\alpha_u$. As before, by Lemma 2

$$\|D^{jx}_k\|_{\underline{\underline{S}}^{2p}} \leq C_{2p}(Z).$$

Recalling the definition of J and using the Cauchy-Schwarz inequality:

$$\begin{aligned}\|J\|_{\underline{\underline{S}}^p} &\leq \|\frac{\partial f_\alpha^i}{\partial x_j}(X_{s-}^x) - \frac{\partial f_\alpha^i}{\partial x_j}(X_{s-}^y)\|_{\underline{\underline{S}}^{2p}} \|D_k^{jx}\|_{\underline{\underline{S}}^{2p}} \\ &\leq \hat{C}_{2p}(Z)\|x-y\|,\end{aligned}$$

which in turn combined with previous estimates yields:

$$\|V\|_{\underline{\underline{S}}^p} \leq C(p,Z)\|x-y\|.$$

Since V was defined to be $V_s(\omega) = D_k(s,\omega,x) - D_k(s,\omega,y)$, we have shown that (with $p > n$):

$$E\left\{\sup_{s \leq N} \|D_k(s,\omega,x) - D_k(s,\omega,y)\|^p\right\} \leq C(p,Z)^p \|x-y\|^p,$$

and therefore by Kolmogorov's lemma (Theorem 53 of Chap. IV) we have the continuity in x of $D_k(t,\omega,x)$.

Step 3: In this step we first show that $D_k(t,\omega,x)$, the solution of equations (*4) (and hence also the solution of the n^2 equations of the second line of (D)) is the partial derivative of X in the variable x_k in the sense of distributions (i.e., generalized functions). Since in Step 2 we established the continuity of D_k in x, we can conclude that D_k is the true partial derivative.

Let us first note that with the continuity established in Step 2, by increasing the compact ball K, we can assume further that $D_k(s,\omega,x) \in K$ also, for $s \leq N$ and all $x \in B$.

We now make use of Cauchy's method of approximating solutions of differential equations, established for stochastic differential equations in Theorem 16 and its Corollary. Note that by our simplifying assumptions, $Y = (X, D)$ takes its values in a compact set, and therefore the coefficients are (globally) Lipschitz. The process Y is the solution of a stochastic differential equation, which we write in vector and matrix form:

$$Y_t = y + \int_0^t f(Y_{s-})dZ_s.$$

Let σ_r be a sequence of partitions tending to the identity, and with the notation of Theorem 16 let $Y(\sigma) = (X(\sigma), D(\sigma))$ denote the solution of the equation of the form:

$$Y_t = y + \int_0^t f(Y^{\sigma+})_s^\sigma dZ_s.$$

For each (σ_r) the equations (D) become difference equations, and thus trivially:

$$\frac{\partial X(\sigma_r)}{\partial x_k} = D_k(\sigma_r).$$

The proof of the theorem will be finished if we can find a subsequence r_q such that $\lim_{q\to\infty} X(\sigma_{r_q}) = X$ and $\lim_{q\to\infty} D_k(\sigma_{r_q}) = D_k$, in the sense of distributions, considered as functions of x.

We now enlarge our space Ω exactly as in Sect. 6 (immediately preceding Theorem 31):

$$\overline{\Omega} = \mathbf{R}^{2n} \times \Omega$$

$$\overline{\mathcal{F}}_t^0 = \mathcal{B}^{2n} \otimes \mathcal{F}_t,\ \overline{\mathcal{F}}_t = \bigcap_{u>t} \overline{\mathcal{F}}_t^0,$$

where $\mathcal{B}^{2n}$ denotes the Borel sets of $\mathbf{R}^{2n}$. Let λ be normalized Lebesgue measure of K. Finally define

$$\overline{P} = \lambda \times P.$$

We can as in the proof of Step 2 assume that $Z \in \mathcal{S}(\beta)$ for β small enough and then,

$$\lim_{r\to\infty} (X(\sigma(r)), D(\sigma(r)) = (X, D) \text{ in } \underline{\underline{S}}^2,$$

by Theorem 16. Therefore there exists a subsequence r_q such that

$$M = \sum_{q=1}^{\infty} \sup_t \|(X(r_q), D(r_q)) - (X, D)\| \in L^1(d\overline{P}).$$

The function $M = M(\omega, x)$ is in $L^1(\lambda \times P)$, and therefore for P-almost all ω the function $x \mapsto M(\omega, x) \in L^1(d\lambda)$. For ω not in the exceptional set, and t fixed it follows that

$$\lim_{q\to\infty} (X(r_q), D(r_q)) = (X, D)$$

λ a.e., and furthermore it is bounded by the function $M(\omega,\cdot)+ \|(X(t,\omega,\cdot), D(t,\omega,\cdot))\|$ which is integrable by hypothesis. This gives convergence in the distributional sense, and the proof is complete. □

We state the following Corollary to Theorem 39 as a theorem.

Theorem 40. *Let Z be as in (H1) and let the functions (f_α^i) in (H2) have locally Lipschitz derivatives up to order N, for some N, $0 \le N \le \infty$. Then there exists a solution $X(t, w, x)$ to:*

$$X_t^i = x_i + \sum_{\alpha=1}^{m} \int_0^t f_\alpha^i(X_{s-}) dZ_s^\alpha,\ 1 \le i \le n,$$

which is N times continuously differentiable in the open set $\{x : \zeta(x, w) > t\}$, where ζ is the explosion time of the solution. If the coefficients (f_α^i) are globally Lipschitz, then $\zeta = \infty$.

Proof. If $N = 0$, then Theorem 40 is exactly Theorem 38. If $N = 1$, then Theorem 40 is Theorem 39. If $N > 1$, then the coefficients of equations (D) have locally Lipschitz derivatives of order $N-1$ at least. Induction yields $(X, D) \in \mathcal{C}^{N-1}$, whence $X \in \mathcal{C}^N$. □

Note that the coefficients (f^i_α) in Theorem 40 are locally Lipschitz of order N if, for example, they have $N+1$ continuous partial derivatives; that is, if $f^i_\alpha \in \mathcal{C}^{N+1}(\mathbf{R}^n)$, for each i and α, then (f^i_α) are locally Lipschitz of order N.

8. Flows as Diffeomorphisms: The Continuous Case

In this section we will study a system of differential equations of the form:

$$(*) \qquad X^i_t = x_i + \sum_{\alpha=1}^m \int_0^t F^i_\alpha(X)_{s-} dZ^\alpha_s, \quad 1 \le i \le n,$$

where the *semimartingales* Z^α *are assumed to have continuous paths with* $Z_0 = 0$. The continuity assumption leads to pleasing results. In Sect. 10 we consider the general case where the semimartingale differentials can have jumps. The flow of an equation such as (*) is considered to be an $\mathbf{R}^n$-valued function $\varphi : \mathbf{R}^n \to \mathbf{R}^n$ given by $\varphi(x) = X(t, \omega, x)$, for each (t, w). We first consider the possible injectivity of φ, of which there are two forms.

Definition. The flow φ of equation (*) is said to be **weakly injective** if for each fixed $x, y \in \mathbf{R}^n$, $x \neq y$,

$$P(\{w : \exists t : X(t, \omega, x) = X(t, \omega, y)\}) = 0.$$

Definition. The flow φ of equation (*) is said to be **strongly injective** (or, simply, **injective**) if for almost all ω the function $\varphi : x \to X(t, \omega, x)$ is injective for all t.

For convenience we recall here a definition from Sect. 3 of this chapter.

Definition. An operator F from $\mathbf{D}^n$ into $\mathbf{D}$ is said to be **process Lipschitz** if for any $\mathbf{X}, \mathbf{Y} \in \mathbf{D}^n$ the following two conditions are satisfied:

(i) for any stopping time T, $\mathbf{X}^{T-} = \mathbf{Y}^{T-}$ implies $F(\mathbf{X})^{T-} = F(\mathbf{Y})^{T-}$;
(ii) there exists an adapted process $K \in \mathbf{L}$ such that

$$\|F(\mathbf{X})_t - F(\mathbf{Y})_t\| \le K_t \|\mathbf{X}_t - \mathbf{Y}_t\|.$$

Actually, process Lipschitz is only slightly more general than random Lipschitz. The norm symbols in the above definition denote Euclidean norm, and not sup norm. Note that if F is process Lipschitz then F is also functional Lipschitz and all the theorems we have proven for functional Lipschitz coefficients hold as well for process Lipschitz coefficients. If f is a function which is Lipschitz (as defined at the beginning of Sect. 3) then f induces a process Lipschitz operator. Finally, observe that by Theorem 37 we know that the flow of equation (*) is continuous from $\mathbf{R}^n$ into $\mathbf{R}^n$ or from $\mathbf{R}^n$ into $\mathcal{D}^n$ a.s., where $\mathcal{D}^n$ has the topology of uniform convergence on compacts.

Theorem 41. *Let Z^α be continuous semimartingales, $1 \leq \alpha \leq m$, H a vector of adapted càdlàg processes, and F an $n \times m$ matrix of process Lipschitz operators. Then the flow of the solution of*

$$X_t = x + H_t + \int_0^t F(X)_{s-} dZ_s \quad {}^{14}$$

is weakly injective.

Proof. Let $x, y \in \mathbf{R}^n$, $x \neq y$. Let X^x, X^y denote the solutions of the above equation with initial conditions x, y respectively. We let $u = x - y$ and $U = X^x - X^y$. We must show $P(\{\omega : \exists t : U_t(\omega) = 0\}) = 0$. Set $V = F(X^x)_- - F(X^y)_-$. Then $V \in \mathbf{L}$ and $|V| \leq K|U|$. Further, the processes U and V are related by

$$U_t = u + \int_0^t V_s dZ_s$$

Let $T = \inf\{t > 0 : U_t = 0\}$; the aim is to show $P(\{T = \infty\}) = 1$. Since U is continuous the stopping time T is the limit of a sequence of increasing stopping times S^k strictly less than T. Therefore the process $1_{[0,T)} = \lim_{k\to\infty} 1_{[0,S_k]}$ is predictable.

We use Itô's formula (Theorem 32 of Chap. II) on $[0, T)$ for the function $f(x) = \log \|x\|$. Note that:

$$\frac{\partial f}{\partial x_i} = \frac{1}{\|x\|}\frac{\partial \|x\|}{\partial x_i} = \frac{1}{\|x\|}\frac{x_i}{\|x\|} = \frac{x_i}{\|x\|^2};$$

$$\frac{\partial^2 f}{\partial x_i^2} = \frac{1}{\|x\|^2} - \frac{2x_i}{\|x\|^3}\frac{\partial \|x\|}{\partial x_i} = \frac{1}{\|x\|^2} - \frac{2x_i^2}{\|x\|^4};$$

$$\frac{\partial^2 f}{\partial x_i \partial x_j} = -\frac{2x_i x_j}{\|x\|^4}.$$

Therefore on $[0, T)$:

$$\log\|U_t\| - \log\|u\| = \sum_i \int_0^t \frac{1}{\|U_s\|^2} U_s^i dU_s^i + \frac{1}{2}\sum_i \int_0^t \frac{1}{\|U_s\|^2} d[U^i, U^i]_s$$
$$- \sum_{i,j} \int_0^t \frac{U_s^i U_s^j}{\|U_s\|^4} d[U^i, U^j]_s.$$

[14] We are using the Einstein convention on sums.

Since $dU^i = \sum_\alpha V^{i,\alpha} dZ^\alpha$, the foregoing equals:

$$= \sum_{i,\alpha} \int_0^t \frac{U_s^i V_s^{i,\alpha}}{\|U_s\|^2} dZ_s^\alpha + \frac{1}{2} \sum_{i,\alpha,\beta} \int_0^t \frac{V_s^{i,\alpha} V_s^{i,\beta}}{\|U_s\|^2} d[Z^\alpha, Z^\beta]_s$$

$$- \sum_{i,j,\alpha,\beta} \int_0^t \frac{U_s^i U_s^j V_s^{i,\alpha} V_s^{j,\beta}}{\|U_s\|^4} d[Z^\alpha, Z^\beta]_s.$$

All the integrands on the right side are predictable and since $\|V\| \leq K\|U\|$ they are moreover bounded by K and K^2 in absolute value. However on $\{T < \infty\}$ the left side of the equation, $\log \|U_t\| - \log \|u\|$, tends to $-\infty$ as t increases to T; the right side is a well defined non-exploding semimartingale on all of $[0, \infty)$. Therefore $P(\{T < \infty\}) = 0$, and the proof is complete. □

In the study of strong injectivity the stochastic exponential of a semimartingale (introduced in Theorem 36 of Chap. II) plays an important role. Recall that if Z is a *continuous* semimartingale, then $X_0 \mathcal{E}(Z)$ denotes the (unique) solution of the equation

$$X_t = X_0 + \int_0^t X_s dZ_s,$$

and $\mathcal{E}(Z)_t = \exp(Z_t - \frac{1}{2}[Z, Z]_t)$. In particular, $P(\{\inf_{s \leq t} \mathcal{E}(Z)_s > 0\}) = 1$.

Theorem 42. *For $x \in \mathbf{R}^n$, let H^x be in $\mathbf{D}^k$ such that they are locally bounded uniformly in x. Assume further that there exists a sequence of stopping times $(T_\ell)_{\ell \geq 1}$ increasing to ∞ a.s. such that $\|(H_-^x - H_-^y)^{T_\ell}\|_{\underline{\underline{S}}^r} \leq K\|x - y\|$, each $\ell \geq 1$, for a constant K and for some $r > n$. Let $Z = (Z^1, \ldots, Z^k)$ be k semimartingales. Then the functions*

$$x \mapsto \int_0^t H_s^x dZ_s$$

$$x \mapsto [H^x \cdot Z, H^x \cdot Z]_t$$

have versions which are continuous as functions from $\mathbf{R}^n$ into $\mathcal{D}$, with $\mathcal{D}$ having the topology of uniform convergence on compacts.

Proof. By Theorem 5 there exists an arbitrarily large stopping time T such that $Z^{T-} \in \underline{\underline{H}}^\infty$. Thus without loss of generality we can assume that $Z \in \underline{\underline{H}}^\infty$, and that H^x is bounded by some constant K, uniformly in x. Further we assume $\|H_-^x - H_-^y\|_{\underline{\underline{S}}^r} \leq K\|x - y\|$. Then

$$E\{\sup_t \| \int_0^t H_{s-}^x dZ_s - \int_0^t H_{s-}^y dZ_s \|^r\} \leq C E\{\sup_t \|H_{t-}^x - H_{t-}^y\|^r\} \|Z\|_{\underline{\underline{H}}^\infty}^r$$

$$\leq \widetilde{K} \|x - y\|^r \|Z\|_{\underline{\underline{H}}^\infty}^r,$$

where we have used Emery's inequality (Theorem 3). The result for $\int H^x_{s-}dZ_s$ now follows from Kolmogorov's lemma (Theorem 53 of Chap. IV). For the second result:

$$\begin{aligned}
&\|[H^x_- \cdot Z, H^x_- \cdot Z] - [H^y_- \cdot Z, H^y_- \cdot Z]\|_{\underline{\underline{S}}^r} \\
&\qquad = \| \int \{(H^x_{s-})^2 - (H^y_{s-})^2\} d[Z,Z]_s \|_{\underline{\underline{S}}^r} \\
&\qquad = \| \sum_{i,j=1}^{k} \int (H^{xi}_{s-}) + H^{yi}_{s-}(H^{xj}_{s-} - H^{yj}_{s-}) d[Z^i, Z^j]_s \|_{\underline{\underline{S}}^r}, \\
&\qquad \leq 2K \|Z\|^2_{\underline{\underline{H}}^\infty} \|H^x - H^y\|_{\underline{\underline{S}}^r},
\end{aligned}$$

and the result follows. □

Theorem 43. *Let F be a matrix of process Lipschitz operators and X^x the solution of (*) with initial condition x, for continuous semimartingales Z^α, $1 \leq \alpha \leq m$. Fix $x, y \in \mathbb{R}^n$. For r in $\mathbb{R}$ there exist for every $x, y \in \mathbb{R}^n$ with $x \neq y$ (uniformly) locally bounded predictable processes $H^\alpha(x,y)$, $J^{\alpha,\beta}(x,y)$, which depend on r, such that:*

$$\|X^x_t - X^y_t\|^r = \|x-y\|^r \mathcal{E}(\Lambda_r(x,y))_t$$

where

$$\Lambda_r(x,y)_t = \int_0^t H^\alpha_s(x,y) dZ^\alpha_s + \int_0^t J^{\alpha,\beta}_s(x,y) d[Z^\alpha, Z^\beta]_s.$$

Proof. Fix $x, y \in \mathbb{R}^n$ and let $U = X^x - X^y$, $V = F(X^x)_- - F(X^y)_-$. Itô's formula applies since U is never zero by weak injectivity (Theorem 36); using the Einstein convention:

$$\begin{aligned}
\|U\|^r = \|x-y\|^r &+ \int r\|U_s\|^{r-2} U^i_s dU^i_s \\
&+ \frac{1}{2} \int r\{(r-2)\|U_s\|^{r-4} U^i_s U^j_s + \delta^i_j \|U_s\|^{r-2}\} d[U^i, U^j]_s.
\end{aligned}$$

Let $(\cdot,\cdot)$ denote Euclidean inner product on $\mathbb{R}^n$. It suffices to take:

$$H^\alpha_s(x,y) = r\|U_s\|^{-2}(U_s, V^\alpha_s),$$

(where V^α is the α-column of V); and:

$$J^{\alpha,\beta}_s(x,y) = \frac{1}{2} r\{(r-2)\|U_s\|^{-4}(U_s, V^\alpha_s)(U_s, V^\beta_s) + \|U_s\|^{-2}(V^\alpha_s, V^\beta_s)\}.$$

One checks that these choices work by observing that $dU_t^i = \sum_{\alpha=1}^m V_t^{i,\alpha} dZ_t^\alpha$. Finally the above allows us to conclude that

$$\|U_t\|^r = \|x-y\|^r + \int_0^t \|U_s\|^r d\Lambda_r(x,y)_s,$$

and the result follows. □

Before giving a key corollary to Theorem 43, we need a lemma. Let $\widetilde{H}^\infty$ be the space of *continuous* semimartingales X with $X_0 = 0$ such that X has a (unique) decomposition

$$X = N + A$$

where N is a continuous local martingale, A is a continuous process of finite variation, $N_0 = A_0 = 0$, *and such that* $[N,N]_\infty$ *and* $\int_0^\infty |dA_s|$ *are in* L^∞. Further, let us define

$$\|X\|_{\widetilde{H}^\infty} = \|[N,N]_\infty^{1/2} + \int_0^\infty |dA_s|\|_{L^\infty}.$$

Lemma. *For every $p, a < \infty$, there exists a constant $C(p,a) < \infty$ such that if $\|X\|_{\widetilde{H}^\infty} \le a$, then $\|\mathcal{E}(X)\|_{\underline{\underline{S}}^p} \le C(p,a)$.*

Proof. Let $X = N + A$ be the (unique) decomposition of X. Then

$$\begin{aligned}\|\mathcal{E}(X)\|_{\underline{\underline{S}}^p}^p &= E\{\sup_t e^{p(X_t - \frac{1}{2}[X,X]_t)}\} \\ &\le E\{e^{pX^*}\} \quad {}^{15} \\ &\le E\{e^{pN^* + pa}\} \\ &= e^{pa} E\{e^{pN^*}\},\end{aligned}$$

since $|A_t| \le a$, a.s. We therefore need to prove only an exponential maximal inequality for continuous martingales. By Theorem 41 of Chap. II, since N is a continuous martingale, it is a time change of a Brownian motion: That is, $N_t = B_{[N,N]_t}$, where B is a Brownian motion defined with a different filtration. Therefore since $[N,N]_\infty \le a^2$, we have

$$N^* \le B^*_{[N,N]_\infty} \le B^*_{a^2},$$

and hence $E\{e^{pN^*}\} \le E\{e^{pB^*_{a^2}}\}$. Using the reflection principle (Theorem 33 of Chap. I) we have:

$$E\{e^{pB^*_{a^2}}\} \le 2E\{e^{pB_{a^2}}\} = 2e^{\frac{p^2a^2}{2}}.$$ □

Note that in the course of the proof of the Lemma we obtained $C(p,a) = 2^{1/p} e^{a + pa^2/2}$.

[15] Recall that $X^* = \sup_t |X_t|$.

Corollary. *Let $-\infty < r < \infty$ and $p < \infty$, and let $(\Lambda_r(x,y)_t)_{t\geq 0}$ be as given in Theorem 43. Then $\mathcal{E}(\Lambda_r(x,y))$ is locally in $\underline{\underline{S}}^p$, uniformly in x, y.*

Proof. We need to show that there exists a sequence of stopping times T_ℓ increasing to ∞ a.s., and constants $C_\ell < \infty$, such that $\|\mathcal{E}(\Lambda_r(x,y))^{T_\ell}\|_{\underline{\underline{S}}^p} \leq C_\ell$ for all $x, y \in \mathbb{R}^n$, $x \neq y$.

By stopping, we may assume that Z^α and $[Z^\alpha, Z^\beta]$ are in $\underline{\underline{\widetilde{H}}}^\infty$ and that $|H^\alpha|$ and $|J^{\alpha,\beta}| \leq b$ for all (x,y), $1 \leq \alpha, \beta \leq m$. Therefore $\|\Lambda_r(x,y)\|_{\widetilde{H}^\infty} \leq C$ for a constant C, since if $X \in \underline{\underline{\widetilde{H}}}^\infty$ and if K is bounded, predictable, then $K \cdot X \in \widetilde{H}^\infty$ and $\|K \cdot X\|_{\widetilde{H}^\infty} \leq \|K\|_{\underline{\underline{S}}^\infty} \|X\|_{\widetilde{H}^\infty}$, as can be proved exactly analogously to Emery's inequalities (Theorem 3). The result now follows by the preceding Lemma. □

Comment. A similar result in the right continuous case is proved by a different method in the proof of Theorem 62 in Sect. 10.

Theorem 44. *Let Z^α be continuous semimartingales, $1 \leq \alpha \leq m$, and F an $n \times m$ matrix of process Lipschitz operators. Then the flow of the solution of*

$$(*) \qquad X_t = x + \int_0^t F(X)_{s-} dZ_s$$

is strongly injective on $\mathbb{R}^n$.

Proof. It suffices to show that for any compact set $C \subset \mathbb{R}^n$, for each N, there exists an event of probability zero outside of which for every $x, y \in C$ with $x \neq y$,

$$\inf_{s \leq N} \|X(s,\omega,x) - X(s,\omega,y)\| > 0.$$

Let x, y, s have rational coordinates. By Theorem 43 a.s.

$$\|X(s,\omega,x) - X(s,\omega,y)\|^r = \|x-y\|^r \mathcal{E}(\Lambda_r(x,y))_s.$$

The left side of the equation is continuous (Theorem 37). As for the right side, $\mathcal{E}(\Lambda_r(x,y))$ will be continuous if we can show that the processes $H_s^\alpha(x,y)$ and $J_s^{\alpha,\beta}(x,y)$, given in Theorem 43, verify the hypotheses of Theorem 42. To this end, let B be any relatively compact subset of $\mathbb{R}^n \times \mathbb{R}^n \setminus \{(x,x)\}$ (e.g., $B = B_1 \times B_2$ where B_1, B_2 are open balls in $\mathbb{R}^n$ with disjoint closures). Then $\|x-y\|^r$ is bounded on B for any real number r. Without loss we take $r = 1$ here. Let $U(x,y) = X^x - X^y$ and $V(x,y) = F(X^x)_- - F(X^y)_-$, and $V^\alpha(x,y)$ the α-column of V. Then for (x,y) and (x',y') in B we have:

(**)

$$\begin{aligned}
H^\alpha(x,y) - H^\alpha(x',y') = {} & (\|U(x,y)\|^{-2} - \|U(x',y')\|^{-2})(U(x,y), V^\alpha(x,y)) \\
& + \|U(x',y')\|^{-2}(U(x,y) - U(x',y'), V^\alpha(x,y)) \\
& + \|U(x',y')\|^{-2}(U(x',y'), V^\alpha(x,y) - V^\alpha(x',y')).
\end{aligned}$$

The first term on the right side of (**) above is dominated in absolute value by

$$\frac{|\|U(x,y)\|-\|U(x',y')\||(\|U(x,y)\|+\|U(x',y')\|)}{\|U(x,y)\|^2\|U(x',y')\|^2}\|U(x,y)\|\|V^\alpha(x,y)\|$$
$$\leq K\|U(x,y)-U(x',y')\|(\|U(x,y)\|+\|U(x',y')\|)\|U(x',y')\|^{-2},$$

where we are assuming (by stopping), that F has a Lipschitz constant K. Since $U(x,y)-U(x',y')=U(x,x')-U(y,y')$, the above is less than

$$\begin{aligned}&K(\|U(x,x')\|+\|U(y,y')\|)(\|U(x,y)\|+\|U(x',y')\|)\|U(x',y')\|^{-2}\\&\quad=K(\|x-x'\|\mathcal{E}(\Lambda_1(x,x'))+\|y-y'\|\mathcal{E}(\Lambda_1(y,y')))\cdot\\&\qquad(\|x-y\|\mathcal{E}(\Lambda_1(x,y))+\|x'-y'\|\mathcal{E}(\Lambda_1(x',y')))\|x'-y'\|^{-2}\mathcal{E}(\Lambda_{-2}(x',y'))\\&\quad\leq K_1\|(x,y)-(x',y')\|(\mathcal{E}(\Lambda_1(x,x'))+\mathcal{E}(\Lambda_1(y,y')))(\mathcal{E}(\Lambda_1(x,y))\\&\qquad+\mathcal{E}(\Lambda_1(x',y')))\mathcal{E}(\Lambda_{-2}(x',y')).\end{aligned}$$

By the Lemma following Theorem 43, and by Hölder's and Minkowski's inequalities we may, for any $p<\infty$, find stopping times T_ℓ increasing to ∞ a.s. such that the last term above is dominated in $\underline{\underline{S}}^p$ norm by $K_\ell\|(x,y)-(x',y')\|$ for a constant K_ℓ corresponding to T_ℓ. We get analogous estimates for the second and third terms on the right side of (**) by similar (indeed, slightly simpler) arguments. Therefore H^α satisfies the hypotheses of Theorem 42, for $(x,y)\in B$. The same is true for $J^{\alpha,\beta}$, and therefore Theorem 42 shows that Λ_r and $[\Lambda_r,\Lambda_r]$ are continuous in (x,y) on B. (Actually we are using a local version of Theorem 42 with $(x,y)\in B\subset\mathbb{R}^{2n}$ instead of all of $\mathbb{R}^{2n}$; this is not a problem since Theorem 42 extends to the case $x\in W$ open in $\mathbb{R}^n$, because Kolmogorov's lemma does – recall that continuity is a local property.) Finally since Λ_r and $[\Lambda_r,\Lambda_r]$ are continuous in $(x,y)\in B$ we deduce that $\mathcal{E}(\Lambda_r(x,y))$ is continuous in $\{(x,y)\in\mathbb{R}^{2n}:x\neq y\}$.

We have shown that both sides of

$$\|X(s,w,x)-X(s,w,y)\|^r=\|x-y\|^r\mathcal{E}(\Lambda_r(x,y))_s$$

can be taken jointly continuous. Therefore except for a set of probability zero the equality holds for all (x,y,s) in $\mathbb{R}^n\times\mathbb{R}^n\times\mathbb{R}_+$. The result follows because $\mathcal{E}(\Lambda_r(x,y))_t$ is defined for all t finite and it is never zero. □

Theorem 45. *Let Z^α be continuous semimartingales, $1\leq\alpha\leq m$, and let F be an $n\times m$ matrix of process Lipschitz operators. Let X be the solution of (*). Then for each $N<\infty$ and almost all ω*

$$\lim_{\|x\|\to\infty}\inf_{s\leq N}\|X(s,\omega,x)\|=\infty.$$

Proof. By Theorem 43 the equality

$$\|X_t^x-X_t^y\|^r=\|x-y\|^r\mathcal{E}(\Lambda_r(x,y))_t$$

is valid for all $r\in\mathbb{R}$.

For $x \neq 0$ let $Y^x = \|X^x - X^0\|^{-1}$ (note that Y^x is well defined by Theorem 41). Then:
(***)

$$|Y^x| = \|x\|^{-1}\mathcal{E}(\Lambda_{-1}(x,0))$$
$$|Y^x - Y^y| \leq \|X^x - X^y\|\|X^x - X^0\|^{-1}\|X^y - X^0\|^{-1}$$
$$= \|x-y\|\|x\|^{-1}\|y\|^{-1}\mathcal{E}(\Lambda_1(x,y))\mathcal{E}(\Lambda_{-1}(x,0))\mathcal{E}(\Lambda_{-1}(y,0)).$$

Define $Y^\infty = 0$. The mapping $x \mapsto \frac{x}{\|x\|^2}$ inspires a distance d on $\mathbf{R}^n \setminus \{0\}$ by $d(x,y) = \frac{\|x-y\|}{\|x\|\|y\|}$. Indeed,

$$\left\|\frac{x}{\|x\|^2} - \frac{y}{\|y\|^2}\right\|^2 = \left(\frac{\|x-y\|}{\|x\|\|y\|}\right)^2.$$

By Hölder's inequality we have that

$$\|\mathcal{E}(\Lambda_1(x,y))\mathcal{E}(\Lambda_{-1}(x,0)\mathcal{E}(\Lambda_{-1}(y,0))\|_{\underline{\underline{S}}^r}$$
$$\leq \|\mathcal{E}(\Lambda_1(x,y))\|_{\underline{\underline{S}}^{3r}}\|\mathcal{E}(\Lambda_{-1}(x,0))\|_{\underline{\underline{S}}^{3r}}\|\mathcal{E}(\Lambda_{-1}(y,0))\|_{\underline{\underline{S}}^{3r}}$$

and therefore by the Corollary to Theorem 43 we can find a sequence of stopping times $(T_\ell)_{\ell\geq 1}$ increasing to ∞ a.s. such that there exist constants C_ℓ with (using (***)):

$$\|(Y^x - Y^y)^{T_\ell}\|_{\underline{\underline{S}}^r} \leq d(x,y)C_\ell.$$

Next set:

$$\hat{Y}^x = \begin{cases} Y^{\frac{x}{\|x\|^2}}, & 0 < \|x\| < \infty \\ Y^\infty = 0, & \|x\| = 0. \end{cases}$$

Then $\|(\hat{Y}^x - \hat{Y}^y)^{T_\ell}\|^r_{\underline{\underline{S}}^r} \leq C_\ell^r\|x-y\|^r$ on $\mathbf{R}^n$, and by Kolmogorov's lemma (Theorem 53 of Chap. IV), there exists a jointly continuous version of $(t,x) \mapsto \hat{Y}^x_t$, on $\mathbf{R}^n$. Therefore $\lim_{\|x\|\to 0} \hat{Y}^x$ exists and equals 0. Since $(\hat{Y}^x)^{-1} = \|X^{\frac{x}{\|x\|^2}} - X^0\|$, we have the result. □

Theorem 46. *Let Z^α be continuous semimartingales, $1 \leq \alpha \leq m$, and F be an $n \times m$ matrix of process Lipschitz operators. Let X be the solution of (*). Let $\varphi : \mathbf{R}^n \mapsto \mathbf{R}^n$ be the flow $\varphi(x) = X(t,\omega,x)$. Then for almost all ω one has that for all t the function φ is surjective and moreover it is a homeomorphism from $\mathbf{R}^n$ to $\mathbf{R}^n$.*

Proof. As noted preceding Theorem 41, the flow φ is continuous from $\mathbf{R}^n$ to $\mathcal{D}^n$, topologized by uniform convergence on compacts; hence for a.a. ω it is continuous from $\mathbf{R}^n$ to $\mathbf{R}^n$ for all t.

The flow φ is injective a.s. for all t by Theorem 44.

Next observe that the image of $\mathbf{R}^n$ under φ, denoted $\varphi(\mathbf{R}^n)$, is closed. Indeed, let $\overline{\varphi(\mathbf{R}^n)}$ denote its closure and let $y \in \overline{\varphi(\mathbf{R}^n)}$. Let (x_k) denote a sequence such that $\lim_{k\to\infty} \varphi(x_k) = y$. By Theorem 45 $\limsup_{k\to\infty}\|x_k\| < \infty$, and hence the sequence (x_k) has a limit point $x \in \mathbf{R}^n$. Continuity implies $\varphi(x) = y$, and we conclude that $\varphi(\mathbf{R}^n) = \overline{\varphi(\mathbf{R}^n)}$; that is, $\varphi(\mathbf{R}^n)$ is closed. Then, as we have seen, the set $\{x_k\}$ is bounded. If x_k does not converge to x, there must exist a limit point $z \neq x$. But then $\varphi(z) = y = \varphi(x)$, and this violates the injectivity, already established. Therefore φ^{-1} is continuous.

Since φ is a homeomorphism from $\mathbf{R}^n$ to $\varphi(\mathbf{R}^n)$, the subspace $\varphi(\mathbf{R}^n)$ of $\mathbf{R}^n$ is homeomorphic to a manifold of dimension n in $\mathbf{R}^n$; therefore by the theorem of the invariance of the domain (see, e.g. Greenberg [**1**, p. 82]), the space $\varphi(\mathbf{R}^n)$ is open in $\mathbf{R}^n$. But $\varphi(\mathbf{R}^n)$ is also closed and non-empty. There is only one such set in $\mathbf{R}^n$ that is open and closed and non-empty and it is the entire space $\mathbf{R}^n$. We conclude that $\varphi(\mathbf{R}^n) = \mathbf{R}^n$. □

Comment. The proof of Theorem 46 can be simplified as follows: extend φ to the Alexandrov compactification $\overline{\mathbf{R}}^n = \mathbf{R}^n \cup \{\infty\}$ of $\mathbf{R}^n$ to $\overline{\varphi}$ by

$$\overline{\varphi}(x) = \begin{cases} \varphi(x) & x \in \mathbf{R}^n \\ \infty & x = \infty. \end{cases}$$

Then $\overline{\varphi}$ is continuous on $\overline{\mathbf{R}}^n$ by Theorem 45, and obviously it is still injective. Since $\overline{\mathbf{R}}^n$ is compact, $\overline{\varphi}$ is a homeomorphism of $\overline{\mathbf{R}}^n$ onto $\overline{\varphi}(\overline{\mathbf{R}}^n)$. However $\overline{\mathbf{R}}^n$ is topologically the sphere S^n, and thus it is not homeomorphic to any proper subset (this is a consequence of the Jordan-Brouwer separation theorem (e.g., Greenberg [**1**, p.79])). Hence $\overline{\varphi}(\overline{\mathbf{R}}^n) = \overline{\mathbf{R}}^n$.

We next turn our attention to determining when the flow is a diffeomorphism of $\mathbf{R}^n$. Recall that a **diffeomorphism** of $\mathbf{R}^n$ is a bijection (one to one and onto) which is $\mathcal{C}^\infty$ and which has an inverse that is also $\mathcal{C}^\infty$. Clearly the hypotheses on the coefficients need to be the intersection of those of Sect. 7 and process Lipschitz.

First we introduce a useful concept, that of *right stochastic exponentials*, which arises naturally in this context. For given n, let Z be an $n \times n$ matrix of given semimartingales. If X is a solution of

$$X_t = I + \int_0^t X_{s-} dZ_s,$$

where X is an $n \times n$ matrix of semimartingales and I is the identity matrix, *then* $X = \mathcal{E}(Z)$, *the (matrix-valued) exponential of* Z. Since the space of $n \times n$ matrices is not commutative, it is also possible to consider *right stochastic integrals*, denoted

$$(Z : H)_t = \int_0^t (dZ_s) H_s,$$

where Z is an $n \times n$ matrix of semimartingales and H is an $n \times n$ matrix of (integrable) predictable processes. If $'$ (prime) denotes matrix transpose, then

$$(Z : H) = (H' \cdot Z')',$$

and therefore right stochastic integrals can be defined in terms of stochastic integrals. Elementary results concerning right stochastic integrals are collected in the next theorem. Note that $\int Y_- dZ$ and $[Y, Z]$ denote $n \times n$ matrix valued processes here.

Theorem 47. *Let Y, Z be given $n \times n$ matrices of semimartingales, H an $n \times n$ matrix of locally bounded predictable processes. Then:*

$$Y_t Z_t - Y_0 Z_0 = \int_0^t Y_{s-} dZ_s + \int_0^t (dY_s) Z_{s-} + [Y, Z]_t$$

$$[H \cdot Y, Z] = H \cdot [Y, Z]$$

$$[Y, Z : H] = [Y, Z] : H$$

Moreover if F is an $n \times n$ matrix of functional Lipschitz operators, then there exists a unique $n \times n$ matrix of D*-valued processes which is the solution of:*

$$X_t = I + \int_0^t (dZ_s) F(X)_{s-}.$$

Proof. The first three identities are easily proved by calculating the entries of the matrices and using the results of Chap. II. Similarly the existence and uniqueness result for the stochastic integral equation is a simple consequence of Theorem 7. □

Theorem 47 allows the definition of the right stochastic exponential.

Definition. The **right stochastic exponential** of an $n \times n$ matrix of semimartingales Z, denoted $\mathcal{E}^R(Z)$, is the (unique) matrix-valued solution of the equation

$$X_t = I + \int_0^t (dZ_s) X_{s-}.$$

We illustrate the relation between left and right stochastic exponentials in the continuous case. The general case is considered in Sect. 10 (see Theorem 63). Note that $\mathcal{E}^R(Z) = \mathcal{E}(Z')'$.

Theorem 48. *Let Z be an $n \times n$ matrix of continuous semimartingales with $Z_0 = 0$. Then $\mathcal{E}(Z)$ and $\mathcal{E}^R(-Z + [Z, Z])$ are inverses; that is, $\mathcal{E}(Z)\mathcal{E}^R(-Z + [Z, Z]) = I$.*

Proof. Let $U = \mathcal{E}(Z)$ and $V = \mathcal{E}^R(-Z + [Z, Z])$. Since $U_0 V_0 = I$, it suffices to show that $d(U_t V_t) = 0$, all $t > 0$. Note that

$$dV = (-dZ + d[Z, Z])V$$

$$(dU)V = (U dZ)V$$

$$d[U, V] = U d[Z, V] = -U d[Z, Z] V.$$

Using Theorem 47 and the preceding:

$$\begin{aligned} d(UV) &= UdV + (dU)V + d[U,V] \\ &= U(-dZ + d[Z,Z])V + UdZV - Ud[Z,Z]V \\ &= 0, \end{aligned}$$

and we are done. □

The next theorem is a special case of Theorem 40 (of Sect. 7), but we state it here as a separate theorem for ease of reference.

Theorem 49. *Let $(Z^1,\dots,Z^m)$ be continuous semimartingales with $Z_0^i = 0$, $1 \le i \le m$, and let (f_α^i), $1 \le i \le n$, $1 \le \alpha \le m$, be functions mapping $\mathbb{R}^n$ to $\mathbb{R}$, with locally Lipschitz partial derivatives up to order N, $1 \le N \le \infty$, and bounded first derivatives. Then there exists a solution $X(t,\omega,x)$ to*

$$X_t^i = x_i + \sum_{\alpha=1}^m \int_0^t f_\alpha^i(X_s)dZ_s^\alpha, \quad 1 \le i \le n,$$

such that its flow $\varphi(x) := x \to X(x,t,\omega)$ is N times continuously differentiable on $\mathbb{R}^n$. Moreover the first partial derivatives satisfy the linear equation

$$D_{kt}^i = \delta_k^i + \sum_{\alpha=1}^m \sum_{j=1}^n \int_0^t \frac{\partial f_\alpha^i}{\partial x_j}(X_s)D_{ks}^j dZ_s^\alpha \qquad (1 \le i \le n).$$

where δ_k^j is Kronecker's delta.

Observe that since the first partial derivatives are bounded, the coefficients are globally Lipschitz and it is not necessary to introduce an explosion time. Also, the value $N = \infty$ is included in the statement. The explicit equation for the partial derivatives comes from Theorem 39.

Let D denote the $n \times n$ matrix valued process:

$$(\otimes) \qquad D_t = (D_{kt}^i)_{1\le i\le n, 1\le k\le n}.$$

The process D is the right stochastic exponential $\mathcal{E}^R(Y)$, where Y is defined by:

$$dY_s^{i,j} = \sum_{\alpha=1}^m \frac{\partial f_\alpha^i}{\partial x_j}(X_s)dZ_s^\alpha.$$

Combining Theorems 48 and 49 and the above observation we have the important following result:

Theorem 50. *With the hypotheses and notation of Theorem 49, the matrix D_t is non-singular for all $t > 0$ and $x \in \mathbb{R}^n$, a.s.*

Theorem 51. *Let $(Z^1, \dots, Z^m)$ be continuous semimartingales and let (f^i_α), $1 \le i \le n$, $1 \le \alpha \le m$, be functions mapping $\mathbb{R}^n$ to $\mathbb{R}$, with partial derivatives of all orders, and bounded first partials. Then the flow of the solution of*

$$X^i_t = x_i + \sum_{\alpha=1}^{m} \int_0^t f^i_\alpha(X_s) dZ^\alpha_s, \; 1 \le i \le n,$$

is a diffeomorphism from $\mathbb{R}^n$ to $\mathbb{R}^n$.

Proof. Let φ denote the flow of X. Since $(f^i_\alpha)_{1 \le i \le n, 1 \le \alpha \le m}$ have bounded first partials, they are globally Lipschitz, and hence there are no finite explosions. Moreover since they are $\mathcal{C}^\infty$, the flow is $\mathcal{C}^\infty$ on $\mathbb{R}^n$ by Theorem 49. The coefficients (f^i_α) are trivially process Lipschitz, hence by Theorem 46 the flow φ is a homeomorphism; in particular it is a bijection of $\mathbb{R}^n$. Finally, the matrix D_t (defined in ($\otimes$) preceding Theorem 50) is non-singular by Theorem 50, thus φ^{-1} is $\mathcal{C}^\infty$ by the inverse function theorem. Since φ^{-1} is also $\mathcal{C}^\infty$, we conclude φ is a diffeomorphism of $\mathbb{R}^n$. □

9. General Stochastic Exponentials and Linear Equations

Let Z be a given continuous semimartingale with $Z_0 = 0$ and let $\mathcal{E}(Z)_t$ denote the unique solution of the **stochastic exponential equation**:

$$(*) \qquad X_t = 1 + \int_0^t X_s dZ_s.$$

Then $X_t = \mathcal{E}(Z)_t = e^{Z_t - \frac{1}{2}[Z,Z]_t}$ (cf. Theorem 36 of Chap. II). It is of course unusual to have a closed form solution of a stochastic differential equation, and it is therefore especially nice to be able to give an explicit solution of the stochastic exponential equation *when it also has an exogeneous driving term.* That is, we want to consider equations of the form

$$(**) \qquad X_t = H_t + \int_0^t X_{s-} dZ_s,$$

where $H \in \mathbb{D}$ (càdlàg and adapted), and Z is a continuous semimartingale. A unique solution of (**) exists by Theorem 7. It is written $\mathcal{E}_H(Z)$.

Theorem 52. *Let H be a semimartingale and let Z be a continuous semimartingale with $Z_0 = 0$. Then the solution $\mathcal{E}_H(Z)$ of equation (**) is given by:*

$$\mathcal{E}_H(Z)_t = \mathcal{E}(Z)_t \{H_0 + \int_{0+}^t \mathcal{E}(Z)_s^{-1} d(H_s - [H, Z]_s)\}.$$

Proof. We use the method of "variation of constants". Assume the solution is of the form $X_t = C_t U_t$, where $U_t = \mathcal{E}(Z)_t$, the normal stochastic exponential. The process C is càdlàg while U is continuous. Using integration by parts:

$$\begin{aligned} dX_t &= C_{t-}dU_t + U_t dC_t + d[C,U]_t \\ &= C_{t-}U_t dZ_t + U_t dC_t + U_t d[C,Z]_t \\ &= X_{t-}dZ_t + U_t d\{C_t + [C,Z]_t\}. \end{aligned}$$

If X is the solution of (**), then equating the above with (**) yields

$$dH_t + X_{t-}dZ_t = X_{t-}dZ_t + U_t d\{C_t + [C,Z]_t\}$$

or

$$dH_t = U_t d\{C_t + [C,Z]_t\}.$$

Since U is an exponential it is never zero and $\frac{1}{U}$ is locally bounded. Therefore

$$(***) \qquad \frac{1}{U_t} dH_t = dC_t + d[C,Z]_t.$$

Calculating the quadratic covariation of each side with Z and noting that $[[C,Z],Z] = 0$, we conclude

$$[\frac{1}{U} \cdot H, Z] = [C,Z].$$

Therefore equation (***) becomes

$$\frac{1}{U} dH = dC + \frac{1}{U} d[H,Z],$$

and $C_t = \int_0^t \frac{1}{U_s} d(H_s - [H,Z]_s)$. Recall that $U_t = \mathcal{E}(Z)_t$ and $X_t = C_t U_t$, and the theorem is proved. □

Since $\mathcal{E}(Z)_t^{-1} = \frac{1}{\mathcal{E}(Z)_t}$ appears in the formula for $\mathcal{E}_H(Z)$, it is worthwhile to note that (for Z a continuous semimartingale)

$$d\left(\frac{1}{\mathcal{E}(Z)}\right) = \frac{dZ - d[Z,Z]}{\mathcal{E}(Z)}$$

and also

$$\frac{1}{\mathcal{E}(Z)} = \mathcal{E}(-Z + [Z,Z]).$$

A more complicated formula for $\mathcal{E}_H(Z)$ exists when Z is not continuous (see Yoeurp-Yor [1]). The next theorem generalizes Theorem 52 to the case where H is not necessarily a semimartingale.

Theorem 53. *Let H be càdlàg, adapted (i.e. $H \in \mathbb{D}$), and let Z be a continuous semimartingale with $Z_0 = 0$. Let $X_t = \mathcal{E}_H(Z)_t$ be the solution of*

$$X_t = H_t + \int_0^t X_{s-} dZ_s.$$

Then $X_t = \mathcal{E}_H(Z)_t$ is given by:

$$X_t = H_t + \mathcal{E}(Z)_t \int_0^t \mathcal{E}(Z)_s^{-1}(H_{s-}dZ_s - H_{s-}d[Z,Z]_s).$$

Proof. Let $Y_t = X_t - H_t$. Then Y satisfies

$$\begin{aligned} Y_t &= \int_0^t H_{s-}dZ_s + \int_0^t Y_{s-}dZ_s \\ &= K_t + \int_0^t Y_{s-}dZ_s, \end{aligned}$$

where K is the semimartingale $H_- \cdot Z$. By Theorem 52,

$$Y_t = \mathcal{E}(Z)_t\{K_0 + \int_{0+}^t \mathcal{E}(Z)_s^{-1}d(K_s - [K,Z]_s)\}$$

and since $K_0 = 0$ and $[K,Z]_t = \int_0^t H_{s-}d[Z,Z]_s$:

$$Y_t = \mathcal{E}(Z)_t \int_0^t \mathcal{E}(Z)_s^{-1}(H_{s-}dZ_s - H_{s-}d[Z,Z]_s),$$

from which the result follows. □

Theorem 54 uses the formula of Theorem 52 to give a pretty result on the comparison of solutions of stochastic differential equations.

Lemma. *Let F be functional Lipschitz such that if $X_t(\omega) = 0$ then $F(X)_{t-}(\omega) > 0$ for continuous processes X. Let C be a continuous increasing process and let X be the solution of*

$$X_t = x_0 + \int_{0+}^t F(X)_{s-}dC_s,$$

with $x_0 > 0$. Then $P(\{\exists t > 0 : X_t \leq 0\}) = 0$.

Proof. Let $T = \inf\{t > 0 : X_t = 0\}$. Since $X_0 \geq 0$ and X is continuous, $X_s \geq 0$ for all $s < T$ on $\{T < \infty\}$. The hypotheses then imply that $F(X)_{T-} > 0$ on $\{T < \infty\}$, which is a contradiction. □

Comment. In the previous lemma if one allows $x_0 = 0$, then it is necessary to add the hypothesis that C be strictly increasing at 0. One then obtains the same conclusion.

Theorem 54 (Comparison Theorem). *Let $(Z^\alpha)_{1\leq\alpha\leq m}$ be continuous semimartingales with $Z_0^\alpha = 0$, and let F_α be process Lipschitz. Let A be a continuous, adapted process with increasing paths, strictly increasing at $t = 0$. Let G and H be process Lipschitz functionals such that $G(X)_{t-} > H(X)_{t-}$*

for any continuous semimartingale X. Finally, let X and Y be the unique solutions of

$$X_t = x_0 + \int_{0+}^{t} G(X)_{s-}dA_s + \int_0^t F(X)_{s-}dZ_s$$

$$Y_t = y_0 + \int_{0+}^{t} H(Y)_{s-}dA_s + \int_0^t F(Y)_{s-}dZ_s$$

where $x_0 \geq y_0$ and F and Z are written in vector notation. Then $P(\{\exists t > 0 : X_t \leq Y_t\}) = 0$.

Proof. Let:

$$U_t = X_t - Y_t$$

$$N_t = \int_0^t \{F(X)_{s-} - F(Y)_{s-}\}(X_s - Y_s)^{-1} 1_{\{X_s \neq Y_s\}} dZ_s$$

$$C_t = x_0 - y_0 + \int_{0+}^{t} \{G(X)_{s-} - H(Y)_{s-}\}dA_s.$$

Then $U_t = C_t + \int_0^t U_{s-}dN_s$, and by Theorem 52

$$U_t = \mathcal{E}(N)_t\{(x_0 - y_0) + \int_{0+}^{t} \mathcal{E}(N)_s^{-1} dC_s\}.$$

Next set:

$$V_t = \frac{1}{\mathcal{E}(N)_t} U_t$$

and define the operator K on continuous processes W by

$$K(W) = G(W\mathcal{E}(N) + Y) - H(Y).$$

Note that since G and H are process Lipschitz, if $W_t = 0$ then $G(W\mathcal{E}(N) + Y)_{t-} = G(Y)_{t-}$. Therefore K has the property that $W_t(w) = 0$ implies that $K(W)_{t-} > 0$. Note further that $K(V) = G(U+Y) - H(Y) = G(X) - H(Y)$.

Next observe that:

$$\begin{aligned}
V_t &= \frac{1}{\mathcal{E}(N)_t} U_t \\
&= x_0 - y_0 + \int_{0+}^{t} \mathcal{E}(N)_s^{-1} dC_s \\
&= x_0 - y_0 + \int_{0+}^{t} \mathcal{E}(N)_s^{-1} \{G(X)_{s-} - H(Y)_{s-}\}dA_s \\
&= x_0 - y_0 + \int_{0+}^{t} K(V)_{s-}\mathcal{E}(N)_s^{-1} dA_s \\
&= x_0 - y_0 + \int_0^t K(V)_{s-}dD_s,
\end{aligned}$$

where $D_t = \int_{0+}^t \mathcal{E}(N)_s^{-1} dA_s$. Then D is a continuous, adapted, increasing process which is strictly increasing at zero. Since V satisfies an equation of the type given in the lemma, we conclude that a.s. $V_t > 0$ for all t. Since $\mathcal{E}(N)_t^{-1}$ is strictly positive (and finite) for all $t \geq 0$, we conclude $U_t > 0$ for all $t > 0$, hence $X_t > Y_t$ for all $t > 0$. □

Comment. If $x_0 > y_0$ (i.e., $x_0 = y_0$ is not allowed), then the hypothesis that A is strictly increasing at 0 can be dropped.

The theory of flows can be used to generalize the formula of Theorem 52. In particular, the homeomorphism property is used to prove Theorem 55.

Consider the system of linear equations given by:

$$X_t = H_t + \sum_{j=1}^{m} \int_0^t A_{s-}^j X_{s-} dZ_s^j \tag{*4}$$

where $H = (H^i)$, $1 \leq i \leq n$ is a vector of n semimartingales, X takes values in $\mathbf{R}^n$, and A^j is an $n \times n$ matrix of adapted, càdlàg processes. The processes Z^j, $1 \leq j \leq m$, are given, continuous semimartingales which are zero at zero.

Define the operators F_j on $\mathbf{D}^n$ by

$$F(X)_t = A_t^j X_t$$

where A^j is the $n \times n$ matrix specified above. The operators F_j are essentially process Lipschitz. (The Lipschitz processes can be taken to be $\|A_t^j\|$ which is càdlàg, not càglàd, but this is unimportant since one takes $F(X)_{t-}$ in the equation.)

Before examining equation (*4), consider the simpler system ($1 \leq i, k \leq n$):

$$U_t^{i,k} = \delta_k^i + \sum_{j=1}^{m} \int_0^t \sum_{\ell=1}^{n} (A_{i,\ell}^j)_{s-} U_{s-}^{\ell,k} dZ_s^j$$

where

$$\delta_k^i = \begin{cases} 1 & i = k \\ 0 & i \neq k. \end{cases}$$

Letting I denote the $n \times n$ identity matrix and writing the preceding in matrix notation yields:

$$U_t = I + \sum_{j=1}^{m} \int_0^t A_{s-}^j U_{s-} dZ_s^j, \tag{*5}$$

where U takes its values in the space of $n \times n$ matrices of adapted processes in $\mathbf{D}$.

Theorem 55. *Let A^j, $1 \leq j \leq m$, be $n \times n$ matrices of càdlàg, adapted processes, and let U be the solution of (*5). Let X_t^x be the solution of (*4)* *where $H_t = x (x \in \mathbb{R}^n)$. Then $X_t^x = U_t x$ and for almost all ω, for all t and x the matrix $U_t(\omega)$ is invertible.*

Proof. Note that U_t is an $n \times n$ matrix for each (t, ω) and $x \in \mathbb{R}^n$, so that $U_t x$ is in $\mathbb{R}^n$. If $X_t^x = U_t x$, then since the coefficients are process Lipschitz we can apply Theorem 46 (which says that the flow is a homeomorphism of $\mathbb{R}^n$) to obtain the invertibility of $U_t(\omega)$.

Note that U is also a right stochastic exponential: Indeed, $U = \mathcal{E}^R(V)$, where $V_t = \int_0^t \sum_{j=1}^m A_{s-}^j dZ_s^j$, and therefore the invertibility also follows from Theorem 48.

Thus we need to show only that $X_t^x = U_t x$. Since $U_t x$ solves (*4) with $H_t = x$, we have $U_t x = X_t^x$ a.s. for each x. Note that a.s. the function $x \to U(\omega)x$ is continuous from $\mathbb{R}^n$ into the subspace of $\mathcal{D}^n$ consisting of continuous functions; in particular $(t, x) \to U_t(\omega)x$ is continuous. Also as shown in the proof of Theorem 46, $(x, t) \to X_t^x$ is continuous in x and right continuous in t. Since $U_t x = X_t^x$ a.s. for each fixed x and t, the continuity permits the removal of the dependence of the exceptional set on x and t. □

Let U^{-1} denote the $n \times n$ matrix-valued process with continuous trajectories a.s. defined by $(U^{-1})_t(\omega) = (U_t(\omega))^{-1}$.

Recall equation (*4):

$$X_t = H_t + \sum_{j=1}^m \int_0^t A_{s-}^j X_{s-} dZ_s^j, \tag{*4}$$

where H is a column vector of n semimartingales and $Z_0^j = 0$. Let $[H, Z^j]$ denote the column vector of n components, the ith one of which is $[H^i, Z^j]$.

Theorem 56. *Let H be a column vector of n semimartingales, Z^j $(1 \le j \le m)$ be continuous semimartingales with $Z_0^j = 0$, and let A^j, $1 \le j \le m$ be $n \times n$ matrices of processes in $\mathbb{D}$. Let U be the solution of equation (*5). Then the solution X^H of (*4) is given by*

$$X_t^H = U_t H_0 + U_t \int_{0+}^t U_s^{-1}(dH_s - \sum_{j=1}^m A_{s-}^j d[H, Z^j]_s)$$

Proof. Write X^H as the matrix product UY. Recall that U^{-1} exists by Theorem 48, hence $Y = U^{-1} X^H$ is a semimartingale, that we need to find explicitly. Using matrix notation throughout, we have:

$$d(UY) = dH + \sum_{j=1}^m A_-^j X_- dZ^j.$$

Integration by parts yields (recall that U is continuous)

$$(dU)Y_- + U(dY) + d[U, Y] = dH + \sum_{j=1}^m A_-^j U_- Y_- dZ^j,$$

by replacing X with UY on the right side above. However U satisfies (*5) and therefore

$$(dU)Y_{-} = \sum_{j=1}^{m} A_{-}^{j} U_{-} Y_{-} dZ^{j},$$

and combining this with the preceding gives

$$U(dY) + d[U,Y] = dH,$$

or equivalently

$$dY = U^{-1}dH - U^{-1}d[U,Y]. \tag{*6}$$

Taking the quadratic covariation of the preceding equation with Z, we have:

$$d[Y,Z^{j}] = U^{-1}d[H,Z^{j}],$$

since $[U^{-1}d[U,Y],Z^{j}] = 0$, $1 \leq j \leq m$. However since U satisfies (*5),

$$\begin{aligned} d[U,Y] &= \sum_{j=1}^{m} A_{-}^{j} U_{-} d[Y,Z^{j}], \\ &= \sum_{j=1}^{m} A_{-}^{j} U U^{-1} d[H,Z^{j}] \\ &= \sum_{j=1}^{m} A_{-}^{j} d[H,Z^{j}], \end{aligned}$$

since U equals U_{-}. Substitute the above expression for $d[U,Y]$ into (*6) and we obtain

$$dY = U^{-1}(dH - \sum_{j=1}^{m} A_{-}^{j} d[H,Z^{j}]),$$

and since $X^{H} = UY$, the theorem is proved. □

10. Flows as Diffeomorphisms: The General Case

In this section we study the same equations as in Sect. 8:

$$X_t^i = x_i + \sum_{\alpha=1}^{m} \int_0^t F_\alpha^i(X)_{s-} dZ_s^\alpha, \quad 1 \leq i \leq n, \tag{*}$$

except that the semimartingales $(Z^\alpha)_{1 \leq \alpha \leq m}$ *are no longer assumed to be continuous*. For simplicity we still assume that $\mathbf{Z}_0 = 0$. In the general case it is not always true that the flows of solutions are diffeomorphisms of $\mathbf{R}^n$, as the following example shows.

Example. Consider the exponential equation in $\mathbf{R}^1$.

$$X_t = x + \int_0^t X_{s-} dZ_s.$$

Let Z be a semimartingale, $Z_0 = 0$, such that Z has a jump of size -1 at a stopping time T, $T > 0$ a.s. Then all trajectories, starting at any initial value x, become zero at T and stay there after T, as is trivially seen by the closed form of the solution with initial condition x:

$$X_t = x \exp(Z_t - \frac{1}{2}[Z,Z]_t^c) \prod_{0<s\le t} (1 + \Delta Z_s) e^{-\Delta Z_s}.$$

Therefore injectivity of the flow fails, and the flow cannot be a diffeomorphism of $\mathbf{R}^1$. □

We examine both the injectivity of the flow and when it is a diffeomorphism of $\mathbf{R}^n$. Recall the hypotheses of Sect. 7, to which we add one, denoted (H3).

Hypotheses.

(H1) Z^α *are given semimartingales with* $Z_0^\alpha = 0$, $1 \le \alpha \le m$;
(H2) $f_\alpha^i : \mathbf{R}^n \to \mathbf{R}$ *are given functions,* $1 \le i \le n$, $1 \le \alpha \le m$, *and* $f(x)$ *denotes the* $n \times m$ *matrix* $(f_\alpha^i(x))$.

The system of equations:

$$(*) \qquad X_t^i = x^i + \sum_{\alpha=1}^m \int_0^t f_\alpha^i(X_{s-}) dZ_s^\alpha, \quad 1 \le i \le n$$

may also be written

$$(*) \qquad X_t = x + \int_0^t f(X_{s-}) dZ_s$$

where X_t and x are column vectors in $\mathbf{R}^n$, $f(X_{s-})$ is an $n \times m$ matrix, and Z is a column vector of m semimartingales.

Hypothesis (H3). *f is C^∞ and has bounded derivatives of all orders.*

Note that by Theorem 40 of Sect. 7, under (H3) the flow is C^∞. The key to studying the injectivity (and diffeomorphism properties) is an analysis of the jumps of the semimartingale driving terms.

Choose an $\epsilon > 0$, the actual size of which is yet to be determined. For $(Z^\alpha)_{1\le\alpha\le m}$ we can find stopping times $0 = T_0 < T_1 < T_2 < \cdots$ tending a.s. to ∞ such that

$$Z^{\alpha,j} = (Z^\alpha)^{T_j-} - (Z^\alpha)^{T_{j-1}}$$

have an $\underline{\underline{H}}^\infty$ norm less than ϵ (cf. Theorem 5). Note that by Theorem 1, $[Z^{\alpha,j}, Z^{\alpha,j}]_\infty^{1/2} < \epsilon$ as well, hence the jumps of each $Z^{\alpha,j}$ are smaller than ϵ. Therefore all of the "large" jumps of $Z^{\alpha,j}$ occur only at the times (T_j), $j \ge 1$.

Let $X_t^j(x)$ denote the solution of (*) driven by the semimartingales $Z^{\alpha,j}$. Outside of the interval (T_{j-1}, T_j) the solution is:

$$X_t^j(x) = \begin{cases} x & \text{for } t \le T_{j-1} \\ X_{T_j-}^j(x) & \text{for } t \ge T_j. \end{cases}$$

Next define the *linkage operators*:

$$H^j(x) = x + f(x)\Delta Z_{T_j},$$

using vector and matrix notation. We have the following obvious result.

Theorem 57. *The solution X of (*) is equal to, for $T_n \le t < T_{n+1}$:*

$$X_t(x) = X_t^{n+1}(x_{n+}),$$

where

$$\begin{aligned} x_{0+} &= x \\ x_{1-} &= X_{T_1-}^1(x), \; x_{1+} = H^1(x_{1-}) \\ x_{2-} &= X_{T_2-}^2(x_{1+}), \; x_{2+} = H^2(x_{2-}) \\ \vdots &= \qquad\qquad \vdots \\ x_{n-} &= X_{T_n-}^n(x_{(n-1)+}), \; x_{n+} = H^n(x_{n-}). \end{aligned}$$

Theorem 58. *The flow $\varphi : x \to X_t(x,\omega)$ of the solution X of (*) is a diffeomorphism if the collections of functions*

$$x \to X_t^j(x,\omega) \qquad x \to H^j(x,\omega)$$

are diffeomorphisms.

Proof. By Theorem 57, the solution (*) can be constructed by composition of functions of the types given in the theorem. Since the composition of diffeomorphisms is a diffeomorphism, the theorem is proved. □

We begin by studying the functions $x \to X_{T_j}^j(x,\omega)$ and $x \to X_t^{n+1}(x,\omega)$. Note that by our construction and choice of the times T_j, *we need only to consider the case where $Z = Z^j$ has a norm in $\underline{\underline{H}}^\infty$ smaller than ϵ.*

The following classical result, due to Hadamard, underlies our analysis.

Theorem 59. *Let $g : \mathbf{R}^n \to \mathbf{R}^n$ be C^∞. Suppose*

(i) $\lim_{\|x\|\to\infty} \|g(x)\| = \infty$, *and*
(ii) the Jacobian matrix $g'(x)$ is an isomorphism of $\mathbf{R}^n$ for all x.

Then g is a diffeomorphism of $\mathbf{R}^n$.

Proof. By the inverse function theorem the function g is a local diffeomorphism, and hence it suffices to show it is a bijection of $\mathbf{R}^n$.

To show that g is onto (i.e., a surjection), first note that $g(\mathbf{R}^n)$ is open and non-empty. It thus suffices to show that $g(\mathbf{R}^n)$ is a closed subset of $\mathbf{R}^n$, since $\mathbf{R}^n$ itself is the only nonempty subset of $\mathbf{R}^n$ that is open and closed. Let $(x_i)_{i\geq 1}$ be a sequence of points in $\mathbf{R}^n$ such that $\lim_{i\to\infty} g(x_i) = y \in \mathbf{R}^n$. We will show that $y \in g(\mathbf{R}^n)$. Let $x_i = t_i v_i$, where $t_i > 0$ and $\|v_i\| = 1$. By choosing a subsequence if necessary we may assume that v_i converges to $v \in S^n$, the unit sphere, as i tends to ∞. Next observe that the sequence $(t_i)_{i\geq 1}$ must be bounded by hypothesis (i): for if not, then $t_i = \|x_i\|$ tends to ∞ along a subsequence and then $\|g(x_{i_k})\|$ tends to ∞ by (i), which contradicts that $\lim_{i\to\infty} g(x_i) = y$. Since $(t_i)_{i\geq 1}$ is bounded we may assume $\lim_{i\to\infty} t_i = t_0 \in \mathbf{R}^n$ again by taking a subsequence if necessary. Then $\lim_{i\to\infty} x_i = t_0 v$, and by the continuity of g we have $y = \lim_{i\to\infty} g(x_i) = g(t_0 v)$.

To show g is injective (i.e. one-to-one), we first note that g is a local homeomorphism, and moreover g is finite to one: Indeed, if there exists an infinite sequence $(x_n)_{n\geq 1}$ such that $g(x_n) = y_0$, all n, for some y_0, then by hypothesis (i) the sequence must be bounded in norm and therefore have a cluster point. By taking a subsequence if necessary we can assume that x_n tends to $\hat{x}$ (the cluster point), where $g(x_n) = y_0$, all n. By the continuity of g we have $g(\hat{x}) = y_0$ as well. This then violates the condition that g is a local homeomorphism, and we conclude that g is finite to one.

Since g is a finite-to-one surjective homeomorphism, it is a covering map.[16] However since $\mathbf{R}^n$ is simply connected the only covering space of $\mathbf{R}^n$ is $\mathbf{R}^n$ (the fundamental group of $\mathbf{R}^n$ is trivial). Therefore the fibers $g^{-1}(x)$ for $x \in \mathbf{R}^n$ each consist of one point, and g is injective. □

The next step is to show that the functions $x \to X^j_{T_j}(x,\omega)$ and $x \to X^{n+1}_t(x,\omega)$ of Theorem 58 satisfy the two conditions of Theorem 59 and are thus diffeomorphisms. This is done in Theorems 62 and 64. First we give a result on weak injectivity which is closely related to Theorem 41.

[16] For the algebraic topology used here, the reader can consult, for example, Munkries [1, Chap. 8].

Theorem 60. *Let Z^α be semimartingales, $1 \leq \alpha \leq m$ with $Z_0^\alpha = 0$, and let F be an $n \times m$ matrix of process Lipschitz operators with non-random Lipschitz constant K. Let $H^i \in \mathbb{D}$, $1 \leq i \leq n$ (càdlàg, adapted). If $\sum_{\alpha=1}^m \|Z^\alpha\|_{\underline{\underline{H}}^\infty} < \epsilon$, for $\epsilon > 0$ sufficiently small, then the flow of the solution of*

$$X_t = x + H_t + \int_0^t F(X)_{s-} dZ_s$$

is weakly injective.[17]

Proof. Let $x, y \in \mathbb{R}^n$, and let X^x, X^y denote the solutions of the above equation with initial conditions x, y respectively. Let $u = x - y$, $U = X^x - X^y$, and $V = F(X^x)_- - F(X^y)_-$. Then $V \in \mathbb{L}$ and $|V| \leq K|U_-|$. Also,

$$U_t = u + \int_0^t V_s dZ_s.$$

Therefore $\Delta U_s = \sum_\alpha V_s^\alpha \Delta Z_s^\alpha$ and moreover (using the Einstein convention to leave the summations implicit)

$$\begin{aligned} \|\Delta U_s\| &\leq \|V_s^\alpha\| \|\Delta Z_s^\alpha\| \\ &\leq C\|U_{s-}\|\epsilon \\ &< \frac{1}{2}\|U_{s-}\| \end{aligned}$$

if ϵ is small enough. Consequently $\|U_s\| \geq \frac{1}{2}\|U_{s-}\|$. Define $T = \inf\{t > 0 : U_{t-} = 0\}$. Then $U_{t-} \neq 0$ on $[0, T)$ and the above implies $U_t \neq 0$ on $[0, T)$ as well. Using Itô's formula for $f(x) = \log \|x\|$, as in the proof of Theorem 41 we have:

$$\begin{aligned} \log \|U_t\| - \log \|u\| &= \int_0^t \frac{U_{s-}^i V_s^{i,\alpha}}{\|U_{s-}\|^2} dZ_s^\alpha \\ &+ \frac{1}{2} \int_0^t \frac{V_s^{i,\alpha} V_s^{i,\beta}}{\|U_{s-}\|^2} d[Z^\alpha, Z^\beta]_s^c \\ (**) \qquad &- \int_0^t \frac{U_{s-}^i U_{s-}^j V_s^{i,\alpha} V_s^{j,\beta}}{\|U_{s-}\|^4} d[Z^\alpha, Z^\beta]_s^c \\ &+ \sum_{0<s\leq t} \{\log \|U_s\| - \log \|U_{s-}\| - \frac{U_{s-}^i}{\|U_{s-}\|^2} \Delta U_s^i\}. \end{aligned}$$

For s fixed, let

$$J_s = \{\log \|U_s\| - \log \|U_{s-}\| - \frac{U_{s-}^i}{\|U_{s-}\|^2} \Delta U_s^i\},$$

[17] We are using vector and matrix notation, and the Einstein convention on sums. The Einstein convention is used throughout this section.

so that the last sum on the right side of equation (**) can be written $\sum_{0<s\leq t} J_s$. By Taylor's theorem

$$|J_s| \leq C\|\Delta U_s\|^2 \sup_{\substack{i,j \\ x\in I_s}} |\frac{\partial^2 f}{\partial x_i \partial x_j}(x)|$$

where $f(x) = \log\|x\|$, and I_s denotes the segment with extremities U_s and U_{s-}. Since

$$\sup_{\substack{i,j \\ x\in I_s}} |\frac{\partial^2 f}{\partial x_i \partial x_j}(x)| \leq \sup_{x\in I_s} \left(\frac{1}{\|x\|^2}\right),$$

and since $\|\Delta U_s\| \leq \frac{1}{2}\|U_{s-}\|$, we deduce

$$\sup_{x\in I_s} \left(\frac{1}{\|x\|^2}\right) \leq \frac{4}{\|U_{s-}\|^2}$$

which in turn implies

$$|J_s| \leq C\|U_{s-}\|^2 \|\Delta Z_s\|^2 \frac{4}{\|U_{s-}\|^2}.$$

Therefore

$$\sum_{0<s\leq t} |J_s| \leq C \sum_{0<s\leq t} \|\Delta Z_s\|^2 \leq C \sum_{\alpha} [Z^\alpha, Z^\alpha]_t < \infty$$

on $[0,T)$. Returning to (**), as t increases to T, the left side tends to $-\infty$ on $\{T < \infty\}$ and the right side remains finite. Therefore $P(\{T < \infty\}) = 0$, and U and U_- never vanish, which proves the theorem. □

Theorem 61. *Let $(Z^\alpha)_{1\leq\alpha\leq m}$ be semimartingales, $Z_0^\alpha = 0$, F an $n\times m$ matrix of process Lipschitz operators with a non-random Lipschitz constant, and $H^i \in \mathbb{D}$, $1\leq i\leq n$. If $\sum_{\alpha=1}^m \|Z^\alpha\|_{\underline{H}^\infty} < \epsilon$ for $\epsilon > 0$ sufficiently small, then for $r \in \mathbb{R}$ there exist uniformly locally bounded predictable processes $H^\alpha(x,y)$ and $K^{\alpha,\beta}(x,y)$, which depend on r, such that*

$$\|X^x - X^y\|^r = \|x-y\|^r \mathcal{E}(\Lambda_r(x,y))$$

where X^x is the solution of

$$X_t = x + H_t + \int_0^t F(X)_{s-} dZ_s.$$

The semimartingale Λ_r is given by:

$$\Lambda_r(x,y)_t = \int_0^t H_s^\alpha(x,y) dZ_s^\alpha + \int_0^t K_s^{\alpha,\beta}(x,y) d[Z^\alpha, Z^\beta]_s^c + J_t$$

where $J_t = \sum_{0<s\leq t} A_s$, and where A_s is an adapted process such that $|A_s| \leq C_r(\Delta Z_s^\alpha)^2$.

Proof. Fix $x, y \in \mathbf{R}^n$, $x \neq y$, and let $U = X^x - X^y$, $V = F(X^x)_- - F(X^y)_-$. By Theorem 60 U is never zero. As in the proof of Theorem 43, by Itô's formula:

$$\begin{aligned}\|U_y\|^r = \|x-y\|^r &+ \int_0^t r\|U_{s-}\|^{r-2} U^i_{s-} dU^i_s \\ &+ \frac{1}{2}\int_0^t r\{(r-2)\|U_{s-}\|^{r-4} U^i_{s-} U^j_{s-} + \delta^i_j \|U_{s-}\|^{r-2}\} d[U^i, U^j]^c_s \\ &+ \sum_{0<s\leq t} \{\|U_s\|^r - \|U_{s-}\|^r - r\|U_{s-}\|^{r-2} U^i_{s-} \Delta U^i_s\}.\end{aligned}$$

Let L_s denote the summands of the last sum on the right side of the above equation. If $g(x) = \|x\|^r$, then

$$|L_s| \leq C\|\Delta U_s\|^2 \sup_{\substack{i,j \\ x\in I_s}} |\frac{\partial^2 g}{\partial x_i \partial x_j}|,$$

where I_s is the segment with boundary U_s and U_{s-}. However

$$|\frac{\partial^2 g}{\partial x_i \partial x_j}| = r|(r-2)\|x\|^{r-4} x_i x_j + \delta^i_j \|x\|^{r-2}|$$

which is less than $C_r\|x\|^{r-2}$. However as in the proof of Theorem 60, we have $\|\Delta U_s\| \leq \frac{1}{2}\|U_{s-}\|$, which implies (for all r positive or negative) that $\|x\|^r \leq C_r \|U_{s-}\|^r$ for all x between U_{s-} and U_s. Hence (the constant C changes):

$$\begin{aligned}|L_s| &\leq C\|\Delta U_s\|^2 \|U_{s-}\|^{r-2} \\ &\leq C\|U_{s-}\|^r \|\Delta Z_s\|^2.\end{aligned}$$

Therefore let $A_s = \frac{1}{\|U_{s-}\|^r} L_s$, and we have:

$$\sum_{0<s\leq t} |A_s| \leq \sum_{0<s\leq t} C\|\Delta Z_s\|^2 \leq C\sum_\alpha [Z^\alpha, Z^\alpha]_t,$$

an absolutely convergent series with a bound independent of (x, y). To complete the proof it suffices to take

$$H^\alpha_s(x,y) = r\|U_{s-}\|^{-2}(U_{s-}, V^\alpha_s) 1_{\{U_{s-}\neq 0\}}$$

and

$$\begin{aligned}K^{\alpha,\beta}_s(x,y) = \frac{1}{2} r\{&(r-2)\|U_{s-}\|^4 (U_{s-}, V^\alpha_s)(U_{s-}, V^\beta_s) \\ &+ \|U_{s-}\|^{-2}(V^\alpha_s, V^\beta_s)\} 1_{\{U_{s-}\neq 0\}},\end{aligned}$$

as in the proof of Theorem 43. Note that $1_{\{U_{s-}\neq 0\}}$ is indistinguishable from the zero process by weak injectivity (Theorem 60). These choices for H^α and

$K^{\alpha,\beta}$ are easily seen to work by observing that $U_t^i = \sum_{\alpha=1}^m \int_0^t V_s^{i,\alpha} dZ_s^\alpha$, and the preceding allows us to conclude that

$$\|U_t\|^r = \|x-y\|^r + \int_0^t \|U_{s-}\|^r d\Lambda_r(x,y)_s,$$

and the result follows. □

Theorem 62. *Let $(Z^\alpha)_{1\le\alpha\le m}$ be semimartingales, $Z_0^\alpha = 0$, F an $n\times m$ matrix of process Lipschitz operators with a non-random Lipschitz constant, and $H^i \in \mathbf{D}$, $1\le i\le n$. Let $X = X(t,\omega,x)$ be the solution of:*

$$(*) \qquad X_t = x + \int_0^t F(X)_{s-} dZ_s.$$

If $\sum_{\alpha=1}^m \|Z^\alpha\|_{\underline{\underline{H}}^\infty} < \epsilon$ for $\epsilon > 0$ sufficiently small, then for each $N<\infty$ and almost all ω

$$\lim_{\|x\|\to\infty} \inf_{s\le N} \|X(s,\omega,x)\| = \infty.$$

Proof. The proof is essentially the same as that of Theorem 45, so we only sketch it. For $x\neq 0$ let $Y^x = \|X^x - X^0\|^{-1}$, which is well defined by weak injectivity (Theorem 60). Then:

$$|Y^x| = \|x\|^{-1}\mathcal{E}(\Lambda_{-1}(x,0))$$
$$\begin{aligned}|Y^x - Y^y| &\le \|X^x - X^y\|\|X^x - X^0\|^{-1}\|X^y - X^0\|^{-1} \\ &= \|x-y\|\|x\|^{-1}\|y\|^{-1}\mathcal{E}(\Lambda_1(x,y))\mathcal{E}(\Lambda_{-1}(x,0))\mathcal{E}(\Lambda_{-1}(y,0))\end{aligned}$$

by Theorem 61, where $\Lambda_r(x,y)$ is as defined in Theorem 61. Set $Y^\infty = 0$.

Since $\|Z\|_{\underline{\underline{H}}^\infty} < \epsilon$ each Z^α has jumps bounded by ϵ, and the process J_t defined in Theorem 61 also has jumps bounded by $C_r\epsilon^2$. Therefore we can stop the processes $\Lambda_r(x,y)$ at an appropriately chosen sequence of stopping times $(T_\ell)_{\ell\ge 1}$ increasing to ∞ a.s. such that each $\Lambda_r(x,y)\in\mathcal{S}(\epsilon)$ for a given ϵ, and for each ℓ, uniformly in (x,y). However if Z is a semimartingale in $\mathcal{S}(\epsilon)$, then since $\mathcal{E}(\Lambda_r(x,y))$ satisfies the equation $U_t = 1 + \int_{0+}^t U_{s-} d\Lambda_r(x,y)_s$, by Lemma 2 of Sect. 3 of this chapter we have

$$\|\mathcal{E}(\Lambda_r(x,y))\|_{\underline{\underline{S}}^p} \le C(p,z) < \infty,$$

where $C(p,z)$ is a constant depending on p and $z = \|\Lambda_r(x,y)\|_{\underline{\underline{H}}^\infty} \le k_\ell$, the bound for ℓ, provided of course that ϵ is sufficiently small. We conclude that for these T_ℓ there exist constants C_ℓ such that:

$$\|(Y^x - Y^y)^{T_\ell}\|_{\underline{\underline{S}}^p} \le C_\ell d(x,y),$$

where $p>n$, and where d is the distance on $\mathbf{R}^n\setminus\{0\}$ given by $d(x,y) = \frac{\|x-y\|}{\|x\|\|y\|}$.

Set:

$$\hat{Y}^x = \begin{cases} Y^{\frac{x}{\|x\|^2}}, & 0 < \|x\| < \infty \\ Y^\infty = 0, & \|x\| = 0. \end{cases}$$

Then $\|(\hat{Y}^x - \hat{Y}^y)^{T_\ell}\|^p_{\underline{\underline{S}}^p} \leq C^p_\ell \|x-y\|^p$ on $\mathbf{R}^n$, and by Kolmogorov's lemma (Theorem 53 of Chap. IV) we conclude that $\lim_{\|x\|\to 0} \hat{Y}^x$ exists and it is zero. Since $(\hat{Y}^x)^{-1} = \|X^{\frac{x}{\|x\|^2}} - X^0\|$, the result follows. □

If φ is the flow of the solution of (*), Theorem 62 shows that $\lim_{\|x\|\to\infty}\|\varphi(x)\| = +\infty$, and the first condition in Hadamard's theorem (Theorem 59) is satisfied. Theorem 63 allows us to determine when the second condition in Hadamard's theorem is also satisfied (see Theorem 64), but it has an independent interest. First, however, some preliminaries are needed.

For given n, let Z be an $n \times n$ matrix of given semimartingales. Recall that $X = \mathcal{E}(Z)$ denotes *the (matrix-valued) exponential of* Z, and that $\mathcal{E}^R(Z)$ denotes *the (matrix-valued) right stochastic exponential of* Z, which was defined in Sect. 8, following Theorem 47.

Recall that in Theorem 48 we showed that if Z is an $n \times n$ matrix of *continuous* semimartingales with $Z_0 = 0$, then $\mathcal{E}(Z)\mathcal{E}^R(-Z+[Z,Z]) = I$, or equivalently $\mathcal{E}(-Z+[Z,Z])\mathcal{E}^R(Z) = I$. The general case is more delicate.

Theorem 63. *Let Z be an $n \times n$ matrix of semimartingales with $Z_0 = 0$. Suppose that $W_t = -Z_t + [Z,Z]^c_t + \sum_{0<s\leq t}(I+\Delta Z_s)^{-1}(\Delta Z_s)^2$ is a well defined semimartingale. Then*

$$\mathcal{E}(W)_t \mathcal{E}^R(Z)_t = I$$

for all $t \geq 0$.

Proof. Let $U = \mathcal{E}(W)$ and $V = \mathcal{E}^R(Z)$, and let $J_t = \sum_{0<s\leq t}(I + \Delta Z_s)^{-1}(\Delta Z_s)^2$. Then:

$$\begin{aligned} dU &= U_-(-dZ + d[Z,Z]^c + dJ) \\ dV &= (dZ)V_-, \end{aligned}$$

and therefore

$$\begin{aligned} (dU)V_- &= U_-(-dZ + d[Z,Z]^c + dJ)V_- \\ U_- dV &= U_-(dZ)V_- \\ d[U,V] &= U_- d[W,Z]V_- \\ &= -U_- d[Z,Z]V_- + U_- d[J,Z]V_-, \end{aligned}$$

since $d[Z,[Z,Z]^c] = 0$. By Theorem 47

$$d(UV) = U_- dV + (dU)V_- + d[U,V];$$

using the above calculations several terms cancel, yielding

$$d(UV) = U_- d[Z,Z]^c V_- + U_-(dJ)V_- - U_- d[Z,Z]V_- + U_- d[J,Z]V_-.$$

Since $[Z,Z]_t = [Z,Z]^c_t + \sum_{0<s\le t}(\Delta Z_s)^2$, and

$$[J,Z]_t = \sum_{0<s\le t} \Delta J_s \Delta Z_s = \sum_{0<s\le t} (\Delta Z_s)^3 (I+\Delta Z_s)^{-1},$$

the preceding becomes:

$$\begin{aligned} d(UV) = &-U_- d(\sum (\Delta Z_s)^2)V_- + U_- d(\sum (\Delta Z_s)^2 (I+\Delta Z_s)^{-1})V_- \\ &+ U_- d(\sum (\Delta Z_s)^3 (I+\Delta Z_s)^{-1})V_-. \end{aligned}$$

Since

$$(\Delta Z_s)^2 = ((\Delta Z_s)^2 + (\Delta Z_s)^3)(I+\Delta Z_s)^{-1},$$

the above equation implies $d(UV)_t = 0$, all $t \ge 0$. However $U_0 V_0 = I$, and therefore $U_t V_t = I$, all $t \ge 0$. □

Corollary. *Let Z be a square matrix of semimartingales. If $\|Z\|_{\underline{\underline{H}}^\infty} < \epsilon$ for $\epsilon > 0$ sufficiently small, then $\mathcal{E}^R(Z)_t$ is invertible for all $t \ge 0$.*

Proof. If $\|Z\|_{\underline{\underline{H}}^\infty} < \epsilon$ then the jumps of Z are bounded by $\epsilon > 0$ and therefore the process W of Theorem 63 is a well defined semimartingale. □

Theorem 64. *Let $(Z^\alpha)_{1\le\alpha\le m}$ be semimartingales with $Z_0 = 0$, and let f be a matrix of coefficients satisfying Hypotheses (H2) and (H3). Let X be the unique solution of*

$$(*) \qquad X_t = x + \int_0^t f(X_{s-})dZ_s.$$

The Jacobian matrix $D^i_k(t,\omega,x) = \frac{\partial}{\partial x_k} X^i(t,\omega,x)$ is invertible for each $t \ge 0$ provided $\|Z\|_{\underline{\underline{H}}^\infty} < \epsilon$ for sufficiently small $\epsilon > 0$.

Proof. By Theorem 39 (in Sect. 7) the Jacobian matrix D satisfies the right stochastic exponential equation

$$D_t = I + \int_0^t (\frac{\partial f^i_\alpha}{\partial x_k}(X_{s-})dZ^\alpha_s)D_{s-},$$

and the matrix semimartingale differential $\frac{\partial f^i_\alpha}{\partial x_k}(X_{s-})dZ^\alpha_s$ satisfies the hypotheses of the Corollary of Theorem 63, whence the result. □

Before stating the principal result of this section, we need to define two subsets of $\mathbf{R}^m$; recall that under Hypotheses (H1), (H2), and (H3), that $Z =$

$(Z^\alpha)_{1\le\alpha\le m}$ is a given m-tuple of semimartingales and that $f(x) = (f^i_\alpha(x))$ is an $n \times m$ matrix of $\mathcal{C}^\infty$ functions. Let:

$$\mathcal{D} = \{z \in \mathbb{R}^m : H(x) = x + f(x)z \text{ is a diffeomorphism of } \mathbb{R}^n\}$$

$$\mathcal{I} = \{z \in \mathbb{R}^m : H(x) = x + f(x)z \text{ is injective in } \mathbb{R}^n\}.$$

Clearly $\mathcal{D} \subset \mathcal{I}$.

Theorem 65. *Let Z and f be as given in Hypotheses (H1), (H2), and (H3), and let X be the solution of*

$$(*) \qquad X_t = x + \int_0^t f(X_{s-})dZ_s.$$

The flow of X is a.s. a diffeomorphism of $\mathbb{R}^n$ (respectively: trajectories of X from different initial points a.s. never meet) for all t if and only if all the jumps of Z belong to $\mathcal{D}$ (respectively: all the jumps of Z belong to $\mathcal{I}$).

Proof. Recall the processes $X^j_{T_j}(x,\omega)$ and $X^{n+1}_t(x,\omega)$ defined in Theorem 57, and the linkage operator $H^j(x) = x + f(x)\Delta Z_{T_j}$, defined immediately preceding Theorem 57. By hypothesis the linkage operators $H^j(x)$ are clearly diffeomorphisms of $\mathbb{R}^n$ (resp: injective), and by Theorems 62 and 64, Hadamard's conditions are satisfied (Theorem 59), and therefore the functions $x \to X^j_{T_j}(x,\omega)$ and $x \to X^{n+1}_t(x,\omega)$ are diffeomorphisms of $\mathbb{R}^n$ if $\epsilon > 0$ is taken small enough in the definition of the stopping times $(T_j)_{j\ge1}$, which it is always possible to do. Therefore by Theorem 58 the flow $\varphi : x \to X_t(x,\omega)$ is a.s. a diffeomorphism of $\mathbb{R}^n$ for each $t > 0$.

The necessity is perhaps the more surprising part of the theorem. First observe that by Hadamard's theorem (Theorem 59) *the set $\mathcal{D}$ contains a neighborhood of the origin*: Indeed, if z is small enough and x is large enough then $\|f(x)z\| \le \frac{\|x\|}{2}$ since f is Lipschitz, which implies that $\|H(x)\| \ge \frac{\|x\|}{2}$ and thus condition (i) of Hadamard's theorem is satisfied; on the other hand $H'(x) = I + f'(x)Z$ is invertible for all x for $\|z\|$ small enough because $f'(x)$ is bounded; therefore condition (ii) of Hadamard's theorem is satisfied. Since $f(x)$ is $\mathcal{C}^\infty$ (Hypothesis (H3)), we conclude that $\mathcal{D}$ contains a neighborhood of the origin.

To prove necessity, set

$$\Gamma_1 = \{\omega : \text{ there exists } s > 0 \text{ with } \Delta Z_s(\omega) \in \mathcal{D}^c\};$$

$$\Gamma_2 = \{\omega : \text{ there exists } s > 0 \text{ with } \Delta Z_s(\omega) \in \mathcal{I}^c\}.$$

Suppose $P(\Gamma_1) > 0$. Since $\mathcal{D}$ contains a neighborhood of the origin, there exists an $\epsilon > 0$ such that all the jumps of Z less than ϵ are in $\mathcal{D}$. We also take ϵ so small that all the functions $x \to X^x_{T_i}(x)$ are diffeomorphisms as soon as the linkage operators H^k are, all $k \le i$.

Since the jumps of Z smaller than ϵ are in $\mathcal{D}$, the jumps of Z that are in $\mathcal{D}^c$ must take place at the times T_i. Let

$$\Lambda^j = \{\omega : \Delta Z_{T_i} \in \mathcal{D}, \text{ all } i < j, \text{ and } \Delta Z_{T_j} \in \mathcal{D}^c\}.$$

Since $P(\Gamma_1) > 0$, there must exist a j such that $P(\Lambda^j) > 0$. Then for $\omega \in \Lambda^j$, $x \to X_{T_j-}(x,\omega)$ is a diffeomorphism, but $H^j(x,\omega)$ is not a diffeomorphism. Let $\omega_0 \in \Lambda_j$ and $t_0 = T_j(\omega_0)$. Then $x \to X_{t_0}(x,\omega_0)$ is not a diffeomorphism, and therefore

$$P(\{\omega : \exists t \text{ such that } x \to X_t(x,\omega) \text{ is not a diffeomorphism}\}) > 0,$$

and we are done. The proof of the necessity of the jumps belonging to $\mathcal{I}$ to have injectivity is analogous. □

Corollary. *Let Z and f be as given in Hypotheses (H1), (H2), and (H3), and let X be the solution of*

$$X_t = x + \int_0^t f(X_{s-})dZ_s.$$

Then different trajectories of X can meet only at the jumps of Z.

Proof. We saw in the proof of Theorem 65 that two trajectories can intersect only at the times T_j that slice the semimartingales Z^α into pieces of $\underline{\underline{H}}^\infty$ norm less than ϵ. If the Z^α do not jump at T_{j_0} for some j_0, however, and paths of X intersect there, one need only slightly alter the construction of T_{j_0} (cf. the proof of Theorem 5, where the times T_j were constructed), so that T_{j_0} is not included in another sequence that ϵ slices $(Z^\alpha)_{1\leq\alpha\leq m}$, to achieve a contradiction. (Note that if, however, $(Z^\alpha)_{1\leq\alpha\leq m}$ has a large jump at T_{j_0}, then it cannot be altered.) □

Bibliographic Notes

The extension of the $\underline{\underline{H}}^p$ norm from martingales to semimartingales was implicit in Protter [**1**] and first formally proposed by Emery [**1**]. A comprehensive account of this important norm for semimartingales can be found in Dellacherie-Meyer [**2**]. Emery's inequalities (Theorem 3) were first established in Emery [**1**], and later extended by Meyer [**12**].

Existence and uniqueness of solutions of stochastic differential equations driven by general semimartingales was first established by Doléans-Dade [**4**] and Protter [**2**], building on the result for continuous semimartingales in Protter [**1**]. Before this Kazamaki [**1**] published a preliminary result, and of course the literature on stochastic differential equations driven by Brownian motion and Lebesgue measure, as well as Poisson processes, was extensive. See, for example, the book of Gihman-Skorohod [**1**] in this regard. These results were improved and simplified by Doléans-Dade-Meyer [**2**] and Emery [**3**]; our approach is inspired by Emery [**3**]. Métiver-Pellaumail [**2**] have an alternative approach. See also Métivier [**1**]. Other treatments can be found in Doléans-Dade [**5**] and Jacod [**1**].

The stability theory is due to Protter [**4**], Emery [**3**], and also to Métivier-Pellaumail [**3**]. The semimartingale topology is due to Emery [**2**] and Métivier-Pellaumail [**3**]. A pedagogic treatment is in Dellacherie-Meyer [**2**].

The generalization of Fisk-Stratonovich integrals to semimartingales is due to Meyer [**8**]. The treatment here of Fisk-Stratonovich differential equations is new. The idea of quadratic variation is due to Wiener [**3**]. Theorem 18, which is a random Itô's formula, appears in this form for the first time. It has an antecedent in Doss-Lenglart [**1**], and for a very general version (containing some quite interesting consequences), see Sznitman [**1**]. Theorem 19 generalizes a result of Meyer [**8**], and Theorem 22 extends a result of Doss-Lenglart [**1**]. Theorem 24 and its Corollary is from Itô [**7**]. Theorem 25 is inspired by the work of Doss [**1**] (see also Ikeda-Watanabe [**1**] and Sussman [**1**]). The treatment of approximations of the Fisk-Stratonovich integrals was inspired by Yor [**2**]. For an interesting application see Rootzen [**1**].

The results of Sect. 6 are taken from Protter [**3**] and Çinlar-Jacod-Protter-Sharpe [**1**]. A comprehensive pedagogic treatment when the Markov solutions are diffusions can be found in Stroock-Varadhan [**1**] or Williams [**1**] and Rogers-Williams [**1**].

Work on flows of stochastic differential equations goes back to 1961 and the work of Blagovescenskii-Freidlin [**1**] who considered the Brownian case. For recent work on flows of stochastic differential equations, see Kunita [**1–3**], Ikeda-Watanabe [**1**] and the references therein. There are also flows results for the Brownian case in Gihman-Skorohod [**1**], but they are L^2 rather than almost sure results. Much of our treatment is inspired by the work of Meyer [**14**] and that of Uppman [**1, 2**] for the continuous case, however results are taken from other articles as well. For example, the example following Theorem 38 is due to Leandre [**1**], while the proof of Theorem 41, the non-confluence of solutions in the continuous case, is due to Emery [**5**]; an alternative proof is in Uppman [**2**]. For the general (right continuous) case, we follow the work of Leandre [**2**]. A similar result was obtained by Fujiwara-Kunita [**1**].

References

L. Arnold

1. Stochastic differential equations: theory and applications. Wiley, New York, 1974

J. Azema

1. Sur les fermés aléatoires. Séminaire Proba. XIX. Lecture Notes in Mathematics, vol. 1123, pp. 397–495. Springer, Berlin Heidelberg New York, 1985

J. Azema and M. Yor

1. Etude d'une martingale remarquable. Séminaire Proba. XXIII. Lecture Notes in Mathematics, vol. 1372, pp. 88–130. Springer, Berlin Heidelberg New York, 1989

M. T. Barlow, S. D. Jacka and M. Yor

1. Inequalities for a pair of processes stopped at a random time. Proc. London Math. Society **52** (1986), 142–172

K. Bichteler

1. Stochastic integrators. Bull. American Math. Soc. **1** (1979), 761–765
2. Stochastic integration and L^p-theory of semimartingales. Annals of Probability **9** (1981), 49–89

P. Billingsley

1. Convergence of probability measures. Wiley, New York, 1968

Yu. N. Blagovescenskii and M. I. Freidlin

1. Certain properties of diffusion processes depending on a parameter. Sov. Math. (Translation of Doklady) **2** (1961), 633–636

R. M. Blumenthal and R. K. Getoor

1. Markov processes and potential theory. Academic Press, New York, 1968

N. Bouleau

1. Formules de changement de variables. Ann. Inst. Henri Poincaré **20** (1984), 133–145

N. Bouleau and M. Yor

1. Sur la variation quadratique des temps locaux de certaines semimartingales. C. R. Acad. Sc. Paris **292** (1981), 491–494

L. Breiman

1. Probability. Addison-Wesley, Reading Mass., 1968

P. Brémaud

1. Point processes and queues: martingale dynamics. Springer, New York, 1981

J. L. Bretagnolle

1. Processus à accroissements indépendants. Ecole d'Eté de Probabilités. Lecture Notes in Mathematics, vol. 307, pp. 1–26. Springer, Berlin Heidelberg New York, 1973

C. S. Chou, P. A. Meyer and C. Stricker

1. Sur les intégrales stochastiques de processus prévisibles non bornées. Séminaire Prob. XIV. Lecture Notes in Mathematics, vol. 784, pp. 128–139. Springer, Berlin Heidelberg New York, 1980

K. L. Chung and R. J. Williams

1. Introduction to stochastic integration. Birkhäuser, Boston Basel Stuttgart, 1983

E. Çinlar

1. Introduction to stochastic processes. Prentice Hall, Englewood Cliffs, 1975

E. Çinlar, J. Jacod, P. Protter and M. Sharpe

1. Semimartingales and Markov processes. Z. für Wahrscheinlichkeitstheorie **54** (1980), 161–220

P. Courrège

1. Intégrale stochastique par rapport à une martingale de carré intégrable. Séminaire Brelot-Choquet-Dény, 7 année (1962–63). Institut Henri Poincaré, Paris

K. E. Dambis

1. On the decomposition of continuous submartingales. Theory Proba. Appl. **10** (1965), 401–410

C. Dellacherie

1. Capacités et processus stochastiques. Springer, Berlin Heidelberg New York, 1972
2. Un survol de la théorie de l'intégrale stochastique. Stochastic Processes and Their Applications **10** (1980), 115–144
3. Mesurabilité des débuts et théorème de section. Séminaire de Proba. XV. Lecture Notes in Mathematics, vol. 850, pp. 351–360. Springer, Berlin Heidelberg New York, 1981

C. Dellacherie and P. A. Meyer

1. Probabilities and potential. North Holland, Amsterdam New York, 1978
2. Probabilities and potential B. North Holland, Amsterdam New York, 1982

C. Doléans-Dade

1. Intégrales stochastiques dépendant d'un paramètre. Publ. Inst. Stat. Univ. Paris **16** (1967), 23–34
2. Quelques applications de la formule de changement de variables pour les semimartingales. Z. für Wahrscheinlichkeitstheorie **16** (1970), 181–194
3. Une martingale uniformément intégrable mais non localement de carré intégrable. Séminaire Proba. V. Lecture Notes in Mathematics, vol. 191, pp. 138–140. Springer, Berlin Heidelberg New York, 1971
4. On the existence and unicity of solutions of stochastic differential equations. Z. für Wahrscheinlichkeitstheorie **36** (1976), 93–101
5. Stochastic processes and stochastic differential equations. Stochastic Differential Equations, 5–75. Centro Internazionale Matematico Estivo (Cortona), Liguori Editore, Naples, 1981

C. Doléans-Dade and P. A. Meyer

1. Intégrales stochastiques par rapport aux martingales locales. Séminaire Proba. IV. Lecture Notes in Mathematics, vol. 124, pp. 77–107. Springer, Berlin Heidelberg New York, 1970

2. Equations différentielles stochastiques. Séminaire Proba. XI. Lecture Notes in Mathematics, vol. 581, pp. 376–382. Springer, Berlin Heidelberg New York, 1977

J. L. Doob
1. Stochastic processes. Wiley, New York, 1953
2. Classical potential theory and its probabilistic counterpart. Springer, Berlin Heidelberg New York, 1984

H. Doss
1. Liens entre équations différentielles stochastiques et ordinaires. Ann. Inst. H. Poincaré **13** (1977), 99–125

H. Doss and E. Lenglart
1. Sur l'existence, l'unicité et le comportement asymptotique des solutions d'équations différentielles stochastiques. Ann. Institut Henri Poincaré **14** (1978), 189–214

L. Dubins and G. Schwarz
1. On continuous martingales. Proceedings National Acad. Sci. USA **53** (1965), 913–916

D. Duffie and C. Huang
1. Multiperiod security markets with differential information. J. Math. Econ. **15** (1986), 283–303

R. J. Elliott
1. Stochastic calculus and applications. Springer, Berlin Heidelberg New York, 1982

M. Emery
1. Stabilité des solutions des équations différentielles stochastiques; applications aux intégrales multiplicatives stochastiques. Z. Wahrscheinlichkeitstheorie **41** (1978), 241–262
2. Une topologie sur l'espace des semimartingales. Séminaire Proba. XIII. Lecture Notes in Mathematics, vol. 721, pp. 260–280. Springer, Berlin Heidelberg New York, 1979
3. Equations différentielles stochastiques lipschitziennes: étude de la stabilité. Séminaire Proba. XIII. Lecture Notes in Mathematics **721**, 281–293. Springer, Berlin Heidelberg New York, 1979
4. Compensation de processus à variation finie non localement intégrables. Séminaire Proba. XIV. Lecture Notes in Mathematics, vol. 784, pp. 152–160. Springer, Berlin Heidelberg New York, 1980
5. Non confluence des solutions d'une équation stochastique lipschitzienne. Séminaire Proba. XV. Lecture Notes in Mathematics, vol. 850, pp. 587–589. Springer, Berlin Heidelberg New York, 1981
6. On the Azéma martingales. Séminaire Proba. XXIII. Lecture Notes in Mathematics, vol. 1372, pp. 66–87. Springer, Berlin Heidelberg New York, 1989

S. Ethier and T. G. Kurtz
1. Markov processes: characterization and convergence. John Wiley, New York, 1986

D. L. Fisk
1. Quasimartingales. Transactions of the American Math. Soc. **120** (1965,)369–389

H. Föllmer
1. Calcul d'Itô sans probabilités. Séminaire Proba. XV. Lecture Notes in Mathematics, vol. 850, pp. 144–150. Springer, Berlin Heidelberg New York, 1981

T. Fujiwara and H. Kunita
1. Stochastic differential equations of jump type and Lévy processes in diffeomorphisms group. J. Math. Kyoto Univ. **25** (1985), 71–106

R. K. Getoor and M. Sharpe

1. Conformal martingales. Invent. Math. **16** (1972), 271–308

I. I. Gihman and A. V. Skorohod

1. Stochastic differential equations. Springer, Berlin Heidelberg New York, 1972

R. D. Gill and S. Johansen

1. Product-integrals and counting processes. To appear

I. V. Girsanov

1. On transforming a certain class of stochastic processes by absolutely continuous substitutions of measures. Theory Probab. Appl. **5** (1960), 285–301

M. Greenberg

1. Lectures on algebraic topology. Benjamin, New York, 1967

A. Gut

1. An introduction to the theory of asymptotic martingales. Amarts and Set Function Processes. Lecture Notes in Mathematics, vol. 1042, pp. 1–49. Springer, Berlin Heidelberg New York, 1983

T. Hida

1. Brownian motion. Springer, Berlin Heidelberg New York, 1980

N. Ikeda and S. Watanabe

1. Stochastic differential equations and diffusion processes. North-Holland, Amsterdam, 1981

K. Itô

1. Stochastic integral. Proc. Imp. Acad. Tokyo **20** (1944), 519–524
2. On stochastic integral equations. Proc. Japan Acad. **22** (1946), 32–35
3. Stochastic differential equations in a differentiable manifold. Nagoya Math. J. **1** (1950), 35–47
4. On a formula concerning stochastic differentials. Nagoya Math. J. **3** (1951), 55–65
5. Stochastic differential equations. Memoirs of the American Math. Soc. **4** (1951)
6. Multiple Wiener integral. J. Math. Soc. Japan **3** (1951), 157–169
7. Stochastic differentials. Applied Mathematics and Optimization **1** (1974), 374–381
8. Extension of stochastic integrals. Proceedings of International Symposium on Stochastic Differential Equations, Kyoto (1976), 95–109

K. Itô and M. Nisio

1. On stationary solutions of a stochastic differential equation. J. Math. Kyoto Univ. **4** (1964), 1–75

K. Itô and S. Watanabe

1. Transformation of Markov processes by multiplicative functionals. Ann. Inst. Fourier **15** (1965), 15–30

J. Jacod

1. Calcul stochastique et problème de martingales. Lecture Notes in Mathematics, vol. 714. Springer, Berlin Heidelberg New York, 1979
2. Grossissement initial, hypothèse (H'), et théorème de Girsanov. Grossissement de filtrations: exemples et applications. Lecture Notes in Mathematics, vol. 1118, pp. 15–35. Springer, Berlin Heidelberg New York, 1985

J. Jacod and P. Protter

1. Time reversal on Lévy processes. Ann. Probability **16** (1988), 620–641

J. Jacod and A. N. Shiryaev

1. Limit theorems for stochastic processes. Springer, Berlin Heidelberg New York, 1987

S. Janson and M. J. Wichura

1. Invariance principles for stochastic area and related stochastic integrals. Stochastic Processes and their Appl. **16** (1983), 71–84

T. Jeulin

1. Semimartingales et grossissement d'une filtration. Lecture Notes in Mathematics, vol. 833. Springer, Berlin Heidelberg New York, 1980

T. Kailath, A. Segall and M. Zakai

1. Fubini-type theorems for stochastic integrals. Sankhya (Series A) **40** (1978), 138–143

G. Kallianpur and C. Striebel

1. Stochastic differential equations occuring in the estimation of continuous parameter stochastic processes. Theory of Probability and its Appl. **14** (1969), 567–594

I. Karatzas and S. Shreve

1. Brownian motion and stochastic calculus. Springer, Berlin Heidelberg New York, 1988

N. Kazamaki

1. Note on a stochastic integral equation. Séminaire Proba. VI. Lecture Notes in Mathematics, vol. 258, pp. 105–108. Springer, Berlin Heidelberg New York, 1972

J. F. C. Kingman

1. An intrinsic description of local time. J. London Math. Soc. **6** (1973), 725–731

J. F. C. Kingman and S. J. Taylor

1. Introduction to measure and probability. Cambridge, Cambridge, 1966

P. E. Kopp

1. Martingales and stochastic integrals. Cambridge, Cambridge, 1984

H. Kunita

1. On the decomposition of solutions of stochastic differential equations. Stochastic Integrals. Lecture Notes in Mathematics, vol. 851, pp. 213–255. Springer, Berlin Heidelberg New York, 1981
2. Stochastic differential equations and stochastic flow of diffeomorphisms. Ecole d'Eté de Probabilités de Saint Flour XII. Lecture Notes in Mathematics, vol. 1097, pp. 143–303. Springer, Berlin Heidelberg New York, 1984
3. Lectures on stochastic flows and applications. Tata Institute, Bombay; Springer, Berlin Heidelberg New York, 1986

H. Kunita and S. Watanabe

1. On square integrable martingales. Nagoya Math. J. **30** (1967), 209–245

T. G. Kurtz and P. Protter

1. Weak limit theorems for stochastic integrals and stochastic differential equations. To appear

A. U. Kussmaul

1. Stochastic integration and generalized martingales. Pitman, London, 1977

R. Leandre

1. Un exemple en théorie des flots stochastiques. Séminaire Proba. XVII. Lecture Notes in Mathematics, vol. 986, pp. 158–161. Springer, Berlin Heidelberg New York, 1983
2. Flot d'une équation différentielle stochastique avec semimartingale directrice discontinue. Séminaire Proba. XIX. Lecture Notes in Mathematics, vol. 1123, pp. 271–274. Springer, Berlin Heidelberg New York, 1985

E. Lenglart

1. Transformation des martingales locales par changement absolument continu de probabilités. Z. für Wahrscheinlichkeitstheorie **39** (1977), 65–70
2. Une caractérisation des processus prévisibles. Séminaire Proba. XI. Lecture Notes in Mathematics, vol. 581, pp. 415–417. Springer, Berlin Heidelberg New York, 1977
3. Semimartingales et intégrales stochastiques en temps continu. Revue du CETHEDEC-Ondes et Signal **75** (1983), 91–160
4. Appendice à l'exposé précedent: inegalités de semimartingales. Sém. de Proba. XIV, Lecture Notes in Mathematics, vol. 784, pp. 49–52. Springer, Berlin Heidelberg New York, 1980

G. Letta

1. Martingales et Intégration Stochastique. Scuola Normale Superiore, Pisa, 1984

P. Lévy

1. Processus stochastiques et mouvement Brownien. Gauthier-Villars, Paris, 1948; second edition 1965
2. Wiener's random function, and other Laplacian random functions. Proc. of the Second Berkeley Symp. on Math. Stat. and Proba. (1951), 171–187

B. Maisonneuve

1. Une mise au point sur les martingales continues définies sur un intervalle stochastique. Séminaire Proba. XI. Lecture Notes in Mathematics, vol. 581, pp. 435–445. Springer, Berlin Heidelberg New York, 1977

H. P. McKean

1. Stochastic Integrals. Academic Press, New York, 1969

E. J. McShane

1. Stochastic calculus and stochastic models. Academic Press, New York, 1974
2. Stochastic differential equations. J. Multivariate Analysis **5** (1975), 121–177

G. Maruyama

1. On the transition probability functions of the Markov process. Nat. Sci. Rep. Ochanomizu Univ. **5** (1954), 10–20

M. Métivier

1. Semimartingales: a course on stochastic processes. de Gruyter, Berlin New York, 1982

M. Métivier and J. Pellaumail

1. On Doléans-Föllmer's measure for quasimartingales. Illinois J. Math. **19** (1975), 491–504
2. On a stopped Doob's inequality and general stochastic equations. Annals of Probability **8** (1980), 96–114
3. Stochastic integration. Academic Press, New York, 1980

P. A. Meyer

1. A decomposition theorem for supermartingales. Illinois J. Math. **6** (1962), 193–205
2. Decomposition of supermartingales: the uniqueness theorem. Illinois J. Math. **7** (1963), 1–17
3. Probability and potentials. Blaisdell, Waltham, 1966
4. Intégrales stochastiques I. Séminaire Proba. I. Lecture Notes in Mathematics, vol. 39, pp. 72–94. Springer, Berlin Heidelberg New York, 1967
5. Intégrales stochastiques II. Séminaire Proba. I. Lecture Notes in Mathematics, vol. 39, pp. 95–117. Springer, Berlin Heidelberg New York, 1967
6. Intégrales stochastiques III. Séminaire Proba. I. Lecture Notes in Mathematics, vol. 39, pp. 118–141. Springer, Berlin Heidelberg New York, 1967

7. Intégrales stochastiques IV. Séminaire Proba. I. Lecture Notes in Mathematics, vol. 39, pp. 142–162. Springer, Berlin Heidelberg New York, 1967
8. Un cours sur les intégrales stochastiques. Séminaire Proba. X. Lecture Notes in Mathematics, vol. 511, pp. 246–400. Springer, Berlin Heidelberg New York, 1976
9. Le théorème fondamental sur les martingales locales. Séminaire Proba. XI. Lecture Notes in Mathematics, vol. 581, pp. 463–464. Springer, Berlin Heidelberg New York, 1977
10. Sur un théorème de C. Stricker. Séminaire Proba. XI. Lecture Notes in Mathematics, vol. 581, pp. 482–489. Springer, Berlin Heidelberg New York, 1977
11. Sur un théorème de J. Jacod. Séminaire Proba. XII. Lecture Notes in Mathematics, vol. 649, pp. 57–60. Springer, Berlin Heidelberg New York, 1978
12. Inégalités de normes pour les intégrales stochastiques. Séminaire Proba. XII. Lecture Notes in Mathematics, vol. 649, pp. 757–762. Springer, Berlin Heidelberg New York, 1978
13. Caracterisation des semimartingales, d'après Dellacherie. Séminaire Proba. XIII. Lecture Notes in Mathematics, vol. 721, pp. 620–623. Springer, Berlin Heidelberg New York, 1979
14. Flot d'une équation différentielle stochastique. Séminaire Proba. XV. Lecture Notes in Mathematics, vol. 850, pp. 103–117. Springer, Berlin Heidelberg New York, 1981
15. Géométrie différentielle stochastique. Colloque en l'Honneur de Laurent Schwartz. Asterisque **131** (1985), 107–114
16. Construction de solutions d'équation de structure. Séminaire Proba. XXIII. Lecture Notes in Mathematics, vol. 1372, pp. 142–145. Springer, Berlin Heidelberg New York, 1989

P. W. Millar
1. Stochastic integrals and processes with stationary independent increments. Proc. of the Sixth Berkeley Symp. on Math. Stat. and Proba. (1972), 307–331

J. R. Munkries
1. Topology: a first course. Prentice-Hall, Englewood Cliffs, 1975

S. Orey
1. F-processes. Proc. of the Fifth Berkeley Symp on Math. Stat. and Proba. II **1** (1965), 301–313

Y. Ouknine
1. Généralisation d'un lemme de S. Nakao et applications. Stochastics **23** (1988), 149–157

Z. R. Pop-Stojanovic
1. On McShane's belated stochastic integral. SIAM J. App. Math. **22** (1972), 87–92

P. Protter
1. On the existence, uniqueness, convergence, and explosions of solutions of systems of stochastic integral equations. Annals of Probability **5** (1977), 243–261
2. Right-continuous solutions of systems of stochastic integral equations. J. Multivariate Analysis **7** (1977), 204–214
3. Markov solutions of stochastic differential equations. Z. für Wahrscheinlichkeitstheorie **41** (1977), 39–58
4. H^p stability of solutions of stochastic differential equations. Z. für Wahrscheinlichkeitstheorie **44** (1978), 337–352
5. A comparison of stochastic integrals. Annals of Probability **7** (1979), 176–189
6. Semimartingales and stochastic differential equations. Statistics Department, Purdue University Technical Report #85-25, 1985
7. Stochastic integration without tears. Stochastics **16** (1986), 295–325

M. H. Protter and C. B. Morrey
1. A first course in real analysis. Springer, Berlin Heidelberg New York, 1977

K. M. Rao
1. On decomposition theorems of Meyer. Math. Scand. **24** (1969), 66–78
2. Quasimartingales. Math. Scand. **24** (1969), 79–92

L. C. G. Rogers and D. Williams
1. Diffusions, Markov processes, and martingales, volume 2. Itô Calculus, Wiley, New York, 1987

H. Rootzen
1. Limit distributions for the error in approximations of stochastic integrals. Annals of Probability **8** (1980), 241–251

L. Schwartz
1. Semimartingales and their stochastic calculus on manifolds (edited by I. Iscoe). Les Presses de l'Université de Montréal, 1984

M. Sharpe
1. General theory of Markov processes. Academic Press, New York, 1988

R. L. Stratonovich
1. A new representation for stochastic integrals. SIAM J. Control **4** (1966), 362–371. (Translation of Vestn. Mosc. Univ., ser. I. Mat. Mech. (1964) 3–12)

C. Stricker
1. Quasimartingales, martingales locales, semimartingales, et filtrations naturelles. Z. für Wahrscheinlichkeitstheorie **39** (1977), 55–64

C. Stricker and M. Yor
1. Calcul stochastique dépendant d'un paramètre. Z. für Wahrscheinlichkeitstheorie **45** (1978), 109–134

D. W. Stroock and S. R. S. Varadhan
1. Multidimensional diffusion processes. Springer, Berlin Heidelberg New York, 1979

H. Sussman
1. On the gap between deterministic and stochastic ordinary differential equations. Annals of Probability **6** (1978), 19–41

A. S. Sznitman
1. Martingales dépendant d'un paramètre: une formule d'Itô. Z. für Wahrscheinlichkeitstheorie **60** (1982), 41–70

F. Treves
1. Topological vector spaces, distributions and kernels. Academic Press, New York, 1967

A. Uppman
1. Deux applications des semimartingales exponentielles aux équations différentielles stochastiques. C. R. Acad. Sci. Paris **290** (1980), 661–664
2. Sur le flot d'une équation différentielle stochastique. Séminaire Proba. XVI. Lecture Notes in Mathematics, vol. 920, pp. 268–284. Springer, Berlin Heidelberg New York, 1982

J. Walsh
1. A diffusion with a discontinuous local time. Asterisque **52-53** (1978), 37–45

D. Williams
1. Diffusions, Markov processes, and martingales, volume 1: Foundations. Wiley, New York, 1979

N. Wiener

1. Differential-space. J. of Mathematics and Physics **2** (1923), 131–174
2. The Dirichlet problem. J. of Mathematics and Physics **3** (1924), 127–146
3. The quadratic variation of a function and its Fourier coefficients. J. Mathematics and Physics **3** (1924), 72–94

E. Wong

1. Stochastic processes in information and dynamical systems. McGraw Hill, New York, 1971

J.-A. Yan

1. Caractérisation d'une classe d'ensembles convexes de L^1 ou H^1. Séminaire Proba. XIV. Lecture Notes in Mathematics, vol. 784, pp. 220–222. Springer, Berlin Heidelberg New York, 1980

C. Yoeurp and M. Yor

1. Espace orthogonal à une semimartingale; applications. Unpublished (1977)

M. Yor

1. Sur les intégrales stochastiques optionnelles et une suite remarquable de formules exponentielles. Séminaire Proba. X. Lecture Notes in Mathematics, vol. 511, pp. 481–500. Springer, Berlin Heidelberg New York, 1976
2. Sur quelques approximations d'intégrales stochastiques. Séminaire Proba. XI. Lecture Notes in Mathematics, vol. 581, pp. 518–528. Springer, Berlin Heidelberg New York, 1977
3. Sur la continuité des temps locaux associés à certaines semimartingales. Temps Locaux. Asterisque **52-53** (1978), 23-36
4. Un exemple de processus qui n'est pas une semimartingale. Temps Locaux. Asterisque **52-52** (1978), 219–222
5. Remarques sur une formule de Paul Lévy. Séminaire Proba. XIV. Lecture Notes in Mathematics, vol. 784, pp. 343–346. Springer, Berlin Heidelberg New York, 1980
6. Rappels et préliminaires généraux. Asterisque **52-53** (1978), 17–22
7. Inégalités entre processus minces et applications. C. R. Acad. Sc. Paris **286** (1978), 799–801

W. Zheng

1. Une remarque sur une même intégrale stochastique calculée dans deux filtrations. Séminaire Proba. XVIII. Lecture Notes in Mathematics, vol. 1059, pp. 172–178. Springer, Berlin Heidelberg New York, 1984

Symbol Index

Subject Index

www.ingramcontent.com/pod-product-compliance
Ingram Content Group UK Ltd.
Pitfield, Milton Keynes, MK11 3LW, UK
UKHW021833190726
13853UKWH00003B/1293